Optics Demystified

Demystified Series

Accounting Demystified
Advanced Calculus Demystified
Advanced Physics Demystified
Advanced Statistics Demystified
Algebra Demystified
Alternative Energy Demystified
Anatomy Demystified
Astronomy Demystified
Audio Demystified
Biochemistry Demystified
Biology Demystified
Biotechnology Demystified
Business Calculus Demystified
Business Math Demystified
Business Statistics Demystified
C++ Demystified
Calculus Demystified
Chemistry Demystified
Circuit Analysis Demystified
College Algebra Demystified
Complex Variables Demystified
Corporate Finance Demystified
Databases Demystified
Diabetes Demystified
Differential Equations Demystified
Digital Electronics Demystified
Discrete Mathematics Demystified
Dosage Calculations Demystified
Earth Science Demystified
Electricity Demystified
Electronics Demystified
Engineering Statistics Demystified
Environmental Science Demystified
Everyday Math Demystified
Fertility Demystified
Financial Planning Demystified
Fluid Mechanics Demystified
Forensics Demystified
French Demystified
Genetics Demystified
Geometry Demystified
German Demystified
Global Warming and Climate Change Demystified
Hedge Funds Demystified
Investing Demystified
Italian Demystified
Java Demystified
JavaScript Demystified
Lean Six Sigma Demystified
Linear Algebra Demystified
Macroeconomics Demystified
Management Accounting Demystified
Math Proofs Demystified
Math Word Problems Demystified
MATLAB® Demystified
Medical Billing and Coding Demystified
Medical Charting Demystified
Medical-Surgical Nursing Demystified
Medical Terminology Demystified
Meteorology Demystified
Microbiology Demystified
Microeconomics Demystified
Nanotechnology Demystified
Nurse Management Demystified
OOP Demystified
Optics Demystified
Options Demystified
Organic Chemistry Demystified
Pharmacology Demystified
Physics Demystified
Physiology Demystified
Pre-Algebra Demystified
Precalculus Demystified
Probability Demystified
Project Management Demystified
Psychology Demystified
Quantum Field Theory Demystified
Quantum Mechanics Demystified
Real Estate Math Demystified
Relativity Demystified
Robotics Demystified
Sales Management Demystified
Signals and Systems Demystified
Six Sigma Demystified
Spanish Demystified
Statistics Demystified
String Theory Demystified
Technical Analysis Demystified
Technical Math Demystified
Thermodynamics Demystified
Trigonometry Demystified
Vitamins and Minerals Demystified

Optics Demystified

Stan Gibilisco

New York Chicago San Francisco Lisbon London
Madrid Mexico City Milan New Delhi San Juan
Seoul Singapore Sydney Toronto

The McGraw-Hill Companies

Library of Congress Cataloging-in-Publication Data

Gibilisco, Stan.
Optics demystified / Stan Gibilisco.
p. cm.
Includes index.
ISBN 978-0-07-149449-6 (alk. paper)
1. Optics—Popular works. I. Title.
QC358.5.G53 2009
535—dc22 2009019104

McGraw-Hill books are available at special quantity discounts to use as premiums and sales promotions, or for use in corporate training programs. To contact a representative please e-mail us at bulksales@mcgraw-hill.com.

Optics Demystified

1 2 3 4 5 6 7 8 9 0 DOC/DOC 0 1 4 3 2 1 0 9

ISBN 978- 0-07-149449-6
MHID 0-07-149449-9

Sponsoring Editor
Judy Bass

Acquisitions Coordinator
Michael Mulcahy

Editorial Supervisor
David E. Fogarty

Project Manager
Vasundhara Sawhney

Copy Editor
Priyanka Sinha

Proofreader
Bhavna Gupta

Production Supervisor
Pamela A. Pelton

Composition
International Typesetting
and Composition

Art Director, Cover
Jeff Weeks

To Samuel, Tim, and Tony

ABOUT THE AUTHOR

Stan Gibilisco is an electronics engineer, researcher, and mathematician who has authored numerous titles for the McGraw-Hill *Demystified* series, along with more than 30 other books and dozens of magazine articles. His work has been published in several languages.

CONTENTS

PREFACE

This book is for people who want to learn about optics or refresh their knowledge of the field. The course can be used for self-teaching or as a supplement in a classroom, tutored, or home-schooling environment.

Each chapter ends with a multiple-choice quiz. You may (and should) refer back to the text when taking these quizzes. Because the quizzes are "open-book," some of the questions are rather difficult, but one choice is always best. When you're done with the quiz at the end of a chapter, give your list of answers to a friend. Have the friend tell you your score, but not which questions you got wrong. The answers are in the back of the book. Stick with a chapter until you get all of the quiz answers correct.

The book concludes with a multiple-choice, "closed-book" final exam. Don't look back into the chapters when taking this test. A satisfactory score is at least 75 answers correct, but I suggest you shoot for 90. With the exam, as with the quizzes, have a friend tell you your score without letting you know which questions you missed. That way, you won't subconsciously memorize the answers. The questions are similar in format to those you'll encounter in standardized tests.

Suggestions for future editions are welcome.

Stan Gibilisco

Optics Demystified

CHAPTER 1

The Nature of Light

The behavior of light has baffled scientists for centuries. Light rays can pass through glass, but are blocked by cardboard. Light rays can be reflected, bent, and split into colors. Light rays act like particle streams in some experiments, and like wave trains in other experiments.

The Speed of Light

Casual observations suggest that the speed of light is infinite. We can aim a flashlight at a mirror, switch on the light, and instantly see the reflection. Even if we put a large mirror on a distant hill and shine a powerful lantern at it, we'll see the reflected beam as soon as we activate the lantern. Nevertheless, light rays travel at finite speed. A ray of light takes a little more than 1 second (1 s) to get from the earth to the moon.

EARLY EXPERIMENTS

The first serious attempt to measure the speed of light was conducted by two people with kerosene lamps and shutters, standing on hilltops several kilometers apart on a direct line of sight. The experimenters agreed that at a certain time, one of them would open the lamp shutter in view of the other. The instant the other person saw the light from the distant hill, he would open the shutter of his lantern. Allowances were made for human reaction time. When the first experimenter opened the shutter, he saw the light from his companion's lantern immediately. Light traversed the distance between the hills with a round-trip time too short to measure.

In the late 1600s, a Danish astronomer named *Ole Roemer* noted discrepancies in the orbital period of one of Jupiter's moons. When Jupiter was on the far side of the sun, the timing of the orbit was delayed by several minutes compared to when Jupiter was exactly opposite the sun. This discrepancy was repeated year after year. Roemer attributed this delay to the fact that the light from Jupiter travels farther to reach the earth when Jupiter and the earth are on opposite sides of the sun, as compared to when Jupiter and the earth are on the same side of the sun. He realized that the difference in distances was exactly twice the radius of the earth's orbit. Unfortunately, Roemer didn't know the earth's orbital radius. If he had, he would have come up with a figure of around 227,000 kilometers per second (km/s) for the speed of light through space.

THE FIZEAU WHEEL

As generations passed, scientists developed increasingly sophisticated schemes for measuring the speed of light. One ingenious machine, devised and built by *Armand Fizeau* in 1849, had a rotating, serrated wheel through which a focused light beam passed. The rotating wheel broke the light beam up into pulses. The beam was aimed at a mirror several kilometers away, which reflected the rays back through the serrated wheel to the observer's eye.

Before the *Fizeau wheel* could be operated, the hardware had to be carefully set up and aligned (Fig. 1-1). The distance between the lamp and the mirror had to be considerable. Fizeau correctly reasoned that as he increased this distance, the accuracy of his measurements would improve. Once the device was aligned and working, Fizeau set the wheel so the light beam would pass through one of the notches, out to the mirror, and back again through the same notch. Then the wheel was slowly turned. At first, the rate of rotation was so slow that the light beam returned through the same notch. As the rate of rotation increased, the wheel would rotate enough so the opaque part of the wheel blocked the returning ray. The rotational speed was increased gradually until the light beam reappeared. Then Fizeau knew that he was looking at the reflected beam through the next serration on the wheel.

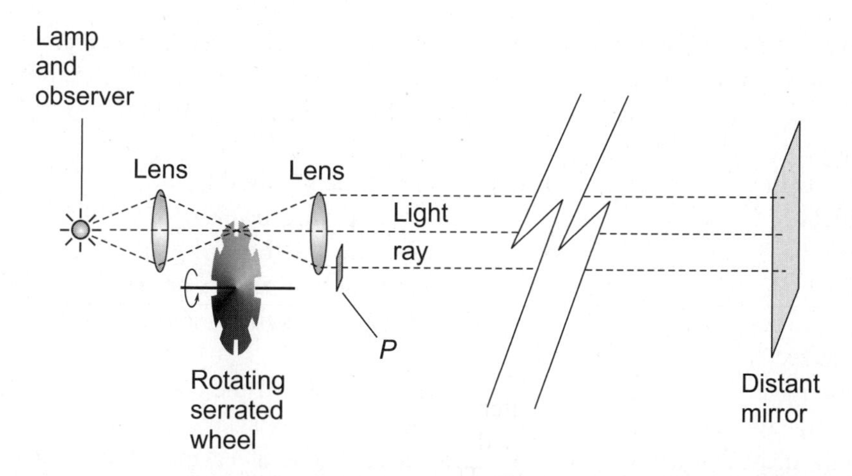

Figure 1-1 Simplified illustration of the Fizeau wheel. As the light ray travels to and from the distant mirror, the wheel rotates from one notch to the next. The significance of point *P* is discussed in Solution 1-1.

Fizeau measured the angular speed of the wheel in rotations per second. He took the reciprocal of that figure, getting the number of seconds per rotation. He divided that result by the number of serrations around the wheel's circumference, obtaining the number of seconds per serration. He deduced that the resultant figure was equal to the time, in seconds, required for the light beam to make the round trip to and from the mirror. If he called this time interval t, and if he set the distance between the wheel and the mirror to d meters, then the speed of light c, in meters per second, would come out as

$$c = 2d/t$$

When Fizeau conducted his experiment in this way and made his calculations, he obtained a figure of $c = 3.13 \times 10^8$ meters per second (m/s).

Another French physicist, *Jean Foucault*, built a more sophisticated system that used rotating mirrors rather than a serrated wheel. His results produced a figure of $c = 2.98 \times 10^8$ m/s—within 1 percent of the value we accept today.

TIME AND DISTANCE VERSUS THE SPEED OF LIGHT

As published on the Web site of the National Institute of Standards and Technology (NIST) in 2007, the speed of light in a vacuum is

$$c = 2.99792458 \times 10^8 \text{ m/s}$$

Today, this is taken as an exact value, because c is regarded as a *universal constant*, inherent in the nature of time and space. It has this value as seen from any nonaccelerating reference frame, regardless of the speed of the observer relative to the source.

According to the NIST, 1 s is the length of time required for 9.192631770×10^9 oscillations in the transition between the two hyperfine levels of the ground state of a cesium-133 atom. That's a lot of jargon, but we can think of it as 9.192631770×10^9 "vibrations" in a certain radioactive material that always behaves in the same way. The cesium atom is considered as an "absolute natural metronome" that beats at a perfectly constant rate at all times.

The meter is defined on the basis of the second and the speed of light, taking advantage of the fact that for anything in motion at a constant speed along a straight line, the displacement is equal to the speed multiplied by the elapsed time. One meter (1 m) is thus defined as the distance that a ray of light travels in $1/(2.99792458 \times 10^8)$ of a second through a vacuum.

We have established the values of three important universal constants: the speed of light, the second, and the meter. The relationship among them can be expressed in three ways:

$$t = d/c$$

$$d = ct$$

$$c = d/t$$

where t is the elapsed time in seconds, d is the distance in meters that a beam of light travels, and c is the speed of light in meters per second.

LARGE DISTANCE UNITS

In astronomy, large units of distance are based on the speed of light and the second. One *minute* (1 min) is precisely 60 s, and one *year* (1 y) is approximately 3.155693 $\times 10^7$ s. On that basis, we can calculate distance units called the *light-second* (lt · s), the *light-minute* (lt · min), and the *light-year* (lt · y). These are as follows, accurate to three significant figures in meters (m), kilometers (km), and statute miles (mi):

$$\begin{aligned} 1 \text{ lt} \cdot \text{s} &= 3.00 \times 10^8 \text{ m} \\ &= 3.00 \times 10^5 \text{ km} \\ &= 1.86 \times 10^5 \text{ mi} \end{aligned}$$

$$\begin{aligned} 1 \text{ lt} \cdot \text{min} &= 1.80 \times 10^{10} \text{ m} \\ &= 1.80 \times 10^7 \text{ km} \\ &= 1.12 \times 10^7 \text{ mi} \end{aligned}$$

$$1 \text{ lt} \cdot \text{y} = 9.46 \times 10^{15} \text{ m}$$
$$= 9.46 \times 10^{12} \text{ km}$$
$$= 5.88 \times 10^{12} \text{ mi}$$

Of these units, the light-year is the most familiar to lay people. In popular science literature, it's often quoted in approximate terms such as "10 trillion kilometers" or "6 trillion miles," where a "trillion" means 1,000,000,000,000 or 10^{12}.

PROBLEM 1-1

Imagine that we set up a rotating-wheel apparatus to reenact the experiment of Armand Fizeau. We place a mirror 1.0000×10^4 m (or 10.000 km) from the lamp-and-wheel system. The wheel has 100 evenly spaced notches around its circumference. We switch on a motor, and the wheel begins to rotate. As we gradually increase the angular speed of the wheel, the reflected beam disappears. The reflected beam first reappears when the wheel reaches an angular speed of ρ_1 rotations per second (r/s). We increase the angular speed a little more, and the reflected beam remains visible until the wheel reaches the slightly higher angular speed of ρ_2 r/s. Beyond that speed, the beam disappears again. We take the average of these two rotation rates (call it ρ) and use it for the purpose of calculating c, the speed of light. Suppose our experiment works out well. Consider $c = 3.00 \times 10^8$ m/s. What is the value of ρ, the average angular speed of the wheel at which the reflected beam first reappears, in wheel rotations per second (r/s)? Are there other angular speeds at which we should also expect the beam to be visible? If so, what are those angular speeds?

SOLUTION 1-1

Let's find the time interval t between the emission of a light-beam pulse from the wheel and its return from the mirror, just before the beam passes through the wheel again. We use the middle of each pulse as the basis for our timing. Imagine that we place a photovoltaic cell at point P as shown in Fig. 1-1 and then connect an oscilloscope to the cell so we can view the returning light pulses on the oscilloscope display (Fig. 1-2). The distance between P and our eye (the whole length of the wheel apparatus) is negligible compared with the distance between P and the mirror. The time interval t, as we define it, is from the middle of any given pulse to the middle of the next pulse.

The distance d between the wheel apparatus and the mirror is 1.0000×10^4 m. We know that $c = 3.00 \times 10^8$ m/s. We can rearrange the above equation relating distance, time, and the measured speed of light to obtain

$$t = 2d/c$$

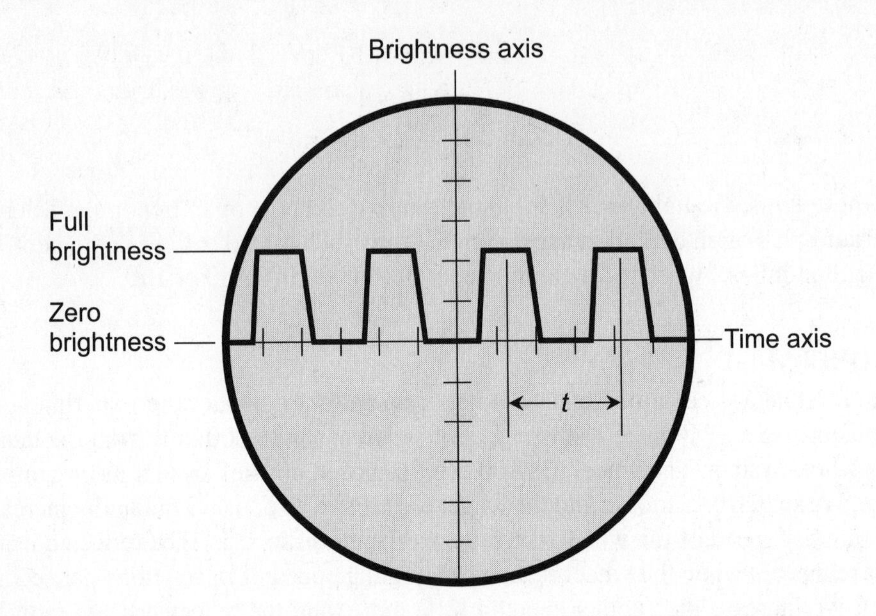

Figure 1-2 Oscilloscope display of returned light pulses in a latter-day Fizeau-wheel experiment before passing through the wheel the second time. The horizontal axis portrays time. The vertical axis shows the relative signal strength.

When we plug in the numbers, we get

$$\begin{aligned} t &= (2 \times 1.0000 \times 10^4)/(3.00 \times 10^8) \\ &= (2.0000 \times 10^4)/(3.00 \times 10^8) \\ &= 6.67 \times 10^{-5}\ \text{s} \end{aligned}$$

There are 100 notches in the wheel. That means the wheel must complete one rotation in a time interval equal to $100t$, or 6.67×10^{-3} s, for the returning beam to go precisely through the notch immediately adjacent to the notch it passed through on its way out. The rate of rotation, ρ, in rotations per second, is therefore

$$\begin{aligned} \rho &= 1/(100t) \\ &= 1/(100 \times 6.67 \times 10^{-5}) \\ &= 1/(6.67 \times 10^{-3}) \\ &= 150\ \text{r/s} \end{aligned}$$

This is the slowest angular speed at which the beam will pass cleanly through the wheel on its return (except for when the wheel is stationary). The beam will also be visible at any angular speed equal to a whole-number multiple of ρ. If the angular

speed is 2ρ, the beam will return after the wheel has rotated through two notches. If the angular speed is $n\rho$ (where n is a whole number), then the beam will return after the wheel has turned through n notches. We will therefore see clean beam returns at angular speeds of 150 r/s, 300 r/s, 450 r/s, 600 r/s, and so on.

Newton's Particle Theory

The fact that light travels through a vacuum in straight lines supports the notion that light consists of fast-moving particles or *corpuscles*. One of the first known scientists to seriously propose and explore the *particle theory* of light, also called the *corpuscular theory*, was *Pierre Gassendi*, an Italian, in the early 1600s. Most of the credit for the particle theory, however, is given to the natural scientist and mathematician *Isaac Newton*.

NEWTON'S CONCEPT

In 1704, Newton published his particle theory in a work called *Opticks*. The book was widely accepted, and Newton revised it several times. His concept dominated optics for more than a century thereafter.

The basic notion is simple: visible light consists of a barrage of tiny, high-speed particles, too small to be seen. Even if they weren't so small, we couldn't see them directly, because the very image of a light particle would be composed of other particles of the same type. To detect light particles, we must employ some form of nonoptical device, or else deduce their existence on the basis of experimental data.

Despite their submicroscopic size and high speed, Newton believed that light particles must have nonzero *mass*, and therefore a certain amount of *weight* in a gravitational field. He believed that visible light rays would be bent by gravitation as they travel through space, although the effect would be too small to notice under ordinary circumstances. Observers in the 1900s discovered that light rays are indeed bent by the intense gravitation near the sun. The effect can be observed and measured.

If such an experiment had been done in Newton's time, it would have lent support to his theory. But when the experiments were conducted, Albert Einstein had developed his general theory of relativity, and Newton's particle theory no longer dominated. Einstein explained the bending of light rays by suggesting that space itself is "curved" under the influence of gravitation. That means there can be no perfectly straight lines in space near the sun, or near any other source of gravitation!

PARTICLE REFLECTION

Newton's particle theory, even though it is technically obsolete, can explain one of the fundamental rules of optics. When a light ray strikes a smooth, flat, reflective surface, the *angle of incidence* is always equal to the *angle of reflection*, as shown in Fig. 1-3. It's as if the light particles "bounce" off of a reflective surface in the same way that hard rubber balls bounce off of concrete when hurled at high speed.

Light particles come off of a reflective surface at the same speed as their incident speed, as well as leaving at an angle equal to the incident angle. This observation implies that the collisions between light particles and reflective surfaces are perfectly elastic. Otherwise, the return speed would be less than the incident speed, because some of the particles' *kinetic energy* would be lost in the collisions.

When we talk about light rays as they encounter or pass through surfaces or boundaries, the angle of incidence and the angle of reflection are expressed or measured with respect to a line perpendicular to the surface at the point where the reflection occurs, not with respect to the surface itself. That perpendicular line is called the *normal*.

PARTICLE REFRACTION

When light rays pass from air into water or vice versa, the rays are bent. You've noticed this effect if you've observed the lines on the bottom of a swimming pool

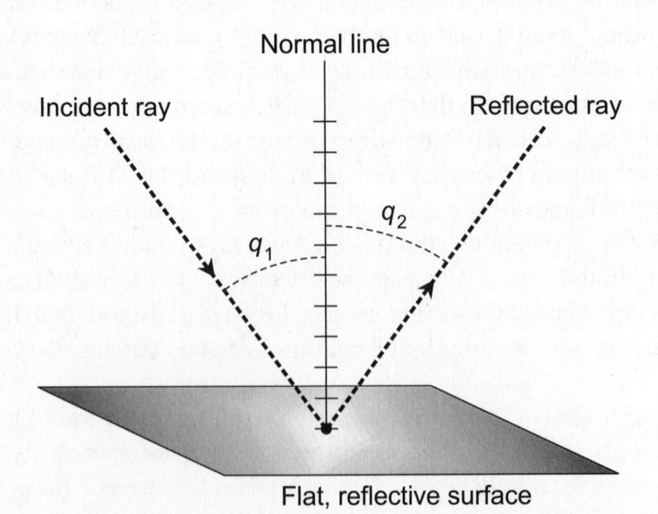

Figure 1-3 Newton's particle theory explains the reflection angle for light. The angle of incidence (q_1) always equals the angle of reflection (q_2) relative to a line normal (perpendicular) to the surface at the point where the light strikes it.

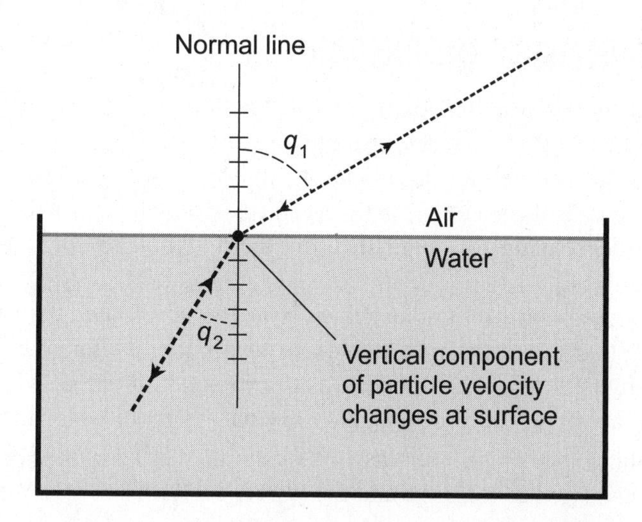

Figure 1-4 Newton explained the refraction of light between air and water (the difference in the angles q_1 and q_2) in terms of a change in the vertical component of light velocity near the surface. He reasoned that this acceleration was caused by the gravitational pull of the water on the light particles.

when the water is calm. If you've done any spear fishing, you know that you must adjust your aim to take this optical effect into account. The phenomenon is more vivid if you dive underwater and look up through a calm surface at the world above.

When a ray of light passes from air into water at an angle different from the normal to the surface, the path of the ray is bent at the surface, even though it's straight in the air and straight in the water (Fig. 1-4). Newton used his particle theory to explain this bending, known as *refraction*, in terms of gravitational effects between light particles and the water.

Because water is denser than air, Newton reasoned that water would impose a gravitational force on light particles as they approached the surface from above. This force would momentarily increase the vertical component of the light-particle velocity, while having no effect on the horizontal component. The result would be a momentary increase in the speed of light near the surface, along with a change in their direction.

When a light beam passed from the water into the air, the opposite effect would occur. As the beam approached the surface, the gravitational force of the water would pull back on the particles for a short while, slowing the vertical component of their velocity while having no effect on the horizontal component. The result would be a brief decrease in the speed of light near the point where the particles left the water, along with a change in the direction of the ray.

NEWTON'S THEORY DISPROVED

Newton's theory of refraction explained the bending effect as well as the tendency for the bending to be more pronounced at larger angles of incidence than at smaller angles (and zero if the angle of incidence is zero). However, the predicted changes in light speed near the surface—increasing for beams entering the water and decreasing for beams leaving the water—did not stand up to later observations.

In reality, the speed of light in water is *uniformly lower* than the speed of light in air. When a ray of light passes from air into water, the ray does not speed up. It slows down, and it continues to propagate more slowly in the water than it did in the air. Conversely, when a ray passes from water into air, the ray speeds up, and it continues to propagate at a *uniformly higher* speed in the air than it did in the water. This fact was conclusively demonstrated by Jean Foucault in 1850.

PROBLEM 1-2

Nowadays, scientists believe that visible light (as well as all other forms of radiation) is, in fact, composed of high-speed particles. Why, then, did Newton's predictions fail to correctly explain the refractive behavior of light between different media such as air and water?

SOLUTION 1-2

Newton's theory was not entirely wrong. It was merely an oversimplification. Newton did not notice, or did not consider, another important aspect of light: its wavelike properties. Newton also made a conceptual error when he ascribed *mass* or *weight* to light particles in his attempt to explain refraction at a boundary as a result of gravitational effects. Later in this chapter, we'll see that it is difficult or impossible to define the mass of a light corpuscle.

The Wave Theory

The Dutch physicist *Christian Huygens* was among the first scientists to suggest that visible light is a wave disturbance, like the ripples on a pond or the vibrations of a violin string. Huygens showed that light waves interact with each other in the same way as waves on water, and in the same way as sound waves from musical instruments. This interaction explained the ripples or concentric rings seen around images in high-powered optical instruments, another phenomenon that the particle theory could not explain.

FUNDAMENTAL PROPERTIES OF WAVES

All waves, no matter what the medium or mode, have certain basic interdependent properties. The *wavelength* is the distance between identical points on two adjacent waves. The *frequency* is the number of wave cycles that occur, or that pass a given point, per unit time. The *propagation speed* is the rate at which the disturbance travels through the medium carrying it.

Sometimes it's easier to talk about a wave's *period* rather than its frequency. The period T (in seconds) of a wave is the reciprocal of the frequency f (in oscillations per second, or *hertz*). Mathematically, the following formulas hold:

$$f = 1/T$$

$$T = 1/f$$

The period of a wave is related to the wavelength λ (in meters) and the propagation speed c (in meters per second) as follows:

$$\lambda = cT$$

This gives rise to other formulas:

$$\lambda = c/f$$

$$c = f\lambda$$

$$c = \lambda/T$$

In addition to the frequency, period, wavelength, and propagation speed, waves have another property: *amplitude*. This is the "strength" of the wave. In a water wave, for example, the amplitude is the height as measured between adjacent *crests* and *troughs*.

REFRACTION REVISITED

The refraction of light rays at the surface of a lake or pool can be explained by the wave theory. When a light beam strikes the surface from either direction, the beam is bent as shown in Fig. 1-5. The extent of the bending depends on the angle at which the wavefronts strike the surface. Wavefronts parallel to the boundary, representing rays perpendicular to that boundary (an angle of incidence equal to zero), are not bent. As the angle of incidence increases, so does the extent of the bending. Wavefronts striking the water surface at a large enough angle from within the water do not pass through the surface boundary, but are reflected instead.

The wave theory offers an elegant explanation of the phenomenon of color, and particularly the way in which a glass prism breaks light into a "rainbow." According to the wave theory, so-called *white light* covers a broad range of wavelengths. The

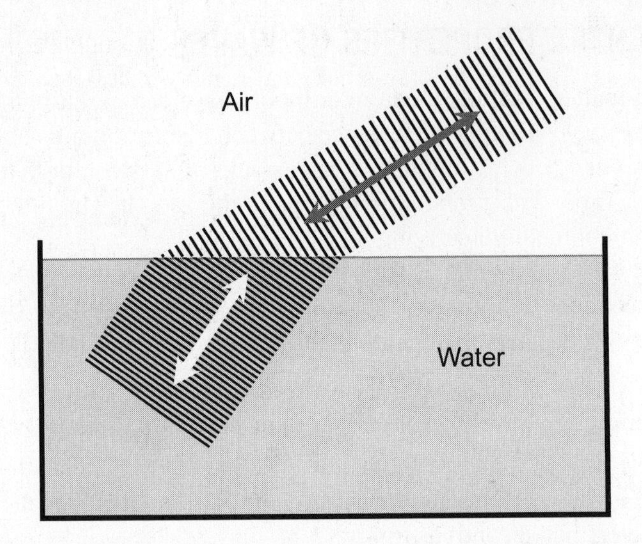

Figure 1-5 The wave theory provides an explanation for refraction that accurately explains both the change in speed and the change in direction of a light beam at a boundary between air and water.

longest waves appear red, the shortest waves appear violet, and intermediate colors, in order of wavelength from longest to shortest, appear orange, yellow, green, blue, and indigo. All visible-light wavelengths are a tiny fraction of a meter (Table 1-1). The longest waves are bent to the smallest extent when they strike the surface of a prism at a sharp angle, while the shortest waves are bent to the greatest extent. This is

Table 1-1 Approximate wavelength ranges in nanometers (nm) for visible light of familiar rainbow colors. A displacement of 1 nm is equal to 10^{-9} m.

Wavelength, nm	Color
750–670	Red
670–600	Orange
600–560	Yellow
560–500	Green
500–470	Blue
470–440	Indigo
440–400	Violet

what happens when white light, such as sunlight, passes through a glass prism and is projected onto a screen. Red light is bent the least, and violet light is bent the most.

Although the speed of light in water, glass, or other dense transparent medium is lower than the speed of light in air, the frequency and period are not affected. As a result, the wavelength is directly proportional the speed of propagation. For example, if the speed of a ray of colored light in a certain medium is 80 percent of the speed of light in the air, then the wavelength of that ray as it travels through that medium is 80 percent of its wavelength in air.

A similar effect can be observed when sunlight or moonlight passes through ice crystals in certain types of clouds, or through water droplets in a rain shower. When you see a rainbow against a dark cloud with the sun at your back, or when you see a multicolored halo around the sun or moon when it shines through thin, high clouds, you are looking at white light that has been broken down into its constituent wavelengths as a result of refraction by water droplets or ice crystals.

THE DOUBLE-SLIT EXPERIMENT

Newton's version of the particle theory suffered from another shortcoming, more serious than its failure to explain refraction. In its pure form, the particle theory can't explain the *interference* effects that are observed with visible light. Rays having a well-defined, specific wavelength and therefore a well-defined color (called *monochromatic*) don't add up as we would expect them to if they were composed of simple, high-speed particles like a barrage of bullets.

An English physicist named *Thomas Young* devised an experiment to demonstrate the wavelike nature of light. He directed a beam of light, having a certain color and originating at a nearly perfect point source, at a barrier with two narrow slits cut in it. At some distance beyond the barrier, Young placed a flat sheet of photographic film. The experiment was conducted in an otherwise dark environment so that the film would be exposed only to the light from the experimental source. A series of bright and dark parallel bands, called an *interference pattern*, appeared on the film (Fig. 1-6). This demonstration, which can be repeated in any well-equipped high-school physics lab, became known as the *double-slit experiment*.

The pattern of bands seemed to conclusively disprove the particle theory, which suggested that only two bright lines would appear on the film. The pattern showed that the light beam was *diffracted* as it passed through the slits. The crests and troughs (maxima and minima) from the two diffracted wavefronts appeared to alternately add and cancel as they arrived at various points on the film. This effect could take place with a wave disturbance, but not with any sort of particles known or theorized at the time.

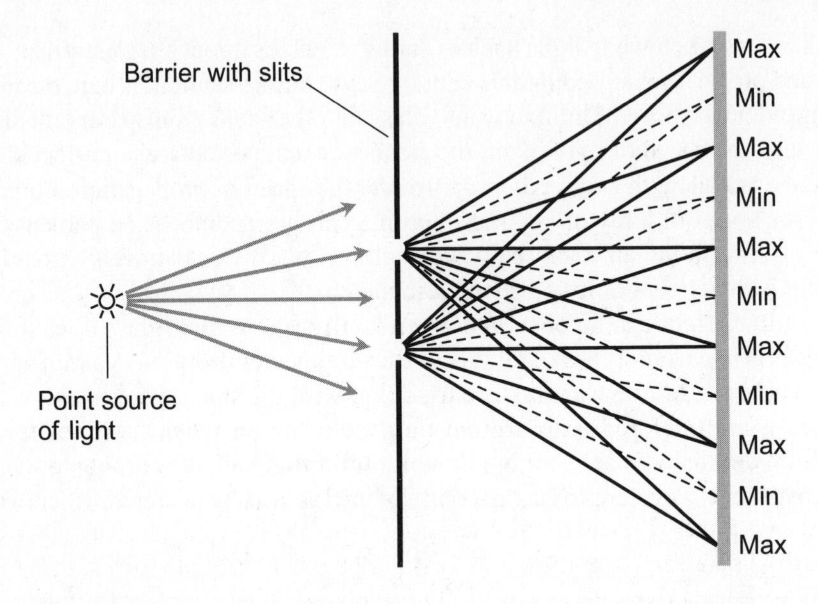

Figure 1-6 The double-slit experiment provides support for the wave theory of light by showing that interference patterns, characteristic of all wave disturbances, are produced in a projected image of two monochromatic light beams.

THE ETHER CONUNDRUM

By the late 1800s, physicists had firmly established that light has wavelike properties, and in some ways resembles sound. Of course, light travels much faster than sound. Also, light can travel through a vacuum, while sound cannot. Sound waves, as well as water waves or vibrations in a string, require a material medium to propagate. Most scientists thought that light waves must also require some sort of material medium to carry them. What could exist everywhere, even in interstellar space? No one knew what such a substance would look like or feel like, and no one knew exactly how it could ever be isolated. But they were certain it had to exist. This hypothetical stuff, this "cosmic jelly," was given the name of *luminiferous ether*, or simply *ether*.

If the ether existed, how could it pass through solid matter to get inside an evacuated chamber? Was it evenly distributed throughout the universe? How could it be detected? Maybe scientists could do an experiment to see if the ether "blew" against the earth as our planet orbited the sun. This suspected phenomenon became known as the *ether wind*. If there was an observable, measurable "current" in the medium through which light traveled, then the speed of a light beam through space would depend on its direction. This effect would occur for the same reason that a passenger on a fast-moving train would measure the speed of sound waves coming from the front as higher than the speed of sound waves coming from behind.

In 1887, an experiment was done by *Albert Michelson* and *Edward Morley* in an attempt to find out how fast, and from what direction, the ether wind "blew." To their surprise, they found that the speed of light was exactly the same in all directions. If the ether existed, then according to the results obtained by Michelson and Morley, it was not moving relative to the earth. But the ether was supposed to be stationary with respect to the universe as a whole! The notion that the earth was not moving with respect to the "cosmic jelly" seemed preposterous, a throwback to the old *geocentric theory* in which the earth was imagined to be standing still at the center of the universe. Could the earth be stationary with respect to the ether because all of the aspects of its complex spatial motion happened to cancel out? Perhaps, but that would be an incredible coincidence.

All sorts of convoluted explanations were offered to explain the results of the Michelson-Morley experiment. One of the more interesting theories suggested that the earth drags a certain amount of ether along with itself. As a result, the ether would be "stationary" near the earth even though "flowing" at great distances from the earth. This seemed plausible to some, but others regarded it as a desperate attempt to keep a dying theory alive. The explanation also ran contrary to known principles of fluid dynamics. A solid object passing through a liquid or gas doesn't pull the liquid or gas along near itself in a uniform mass. Instead, turbulence occurs. This turbulence should have shown up as slight fluctuations in the speed of light, even on the surface of the earth, but the measured speed of light was the same at all times and in all directions.

EINSTEIN'S CONCEPT

Albert Einstein took the results of the Michelson-Morley experiment literally, and decided that the constancy of the speed of light is an inherent property of the universe. Einstein rejected the notion of luminiferous ether because it was too complicated and cumbersome. Einstein hypothesized that in a vacuum, the speed of light is a physical constant, having the same value as measured from any nonaccelerating point of view. It wouldn't matter, then, whether the observer was stationary or moving relative to a light source; the observed speed of the light from that source would always be the same.

By the time Einstein was creating his theories, the speed of light was known with almost as much accuracy as we know it today. Using sophisticated test equipment, scientists could measure the speed at which light arrives at a point on the earth's surface from stars in all directions. The speed of the light from a distant star was found to be the same when the earth moved toward that star in its orbit around the sun, or away from that same star 6 months later. The speed of the light from every star in the sky was the same as the speed from every other star. This was true even though the speed of the earth relative to various distant stars could vary to a much larger extent than the experimental error in the measurement of light speed.

THE PHOTON

Certain materials that conduct electricity to a limited extent in total darkness become better conductors when exposed to visible light. This phenomenon is called the *photoelectric effect*. It occurs because of *ionization* of the atoms in the material. In the course of his work on relativity, Einstein concluded that the wave theory of light, taken in its pure form, could not adequately account for the photoelectric effect. The particle theory, however, offered an explanation. Suppose that visible-light corpuscles could knock electrons loose from atoms, making it easier for *electric current* to flow through a substance?

According to the wave theory, the photoelectric effect should occur in any electrical conductor or semiconductor if the incident light is intense enough. Lab experiments revealed that things don't happen that way. Instead, the occurrence (or absence) of the photoelectric effect correlates more closely with the frequency or wavelength of the radiant energy than with its intensity. A metal that shows pronounced photoelectric behavior for visible light at moderate intensity might not exhibit the effect when bombarded with waves having a lower frequency (longer wavelength) than visible light, no matter how powerful that radiation. Conversely, materials that do not ionize under bright visible light might readily ionize when exposed to low-intensity waves of higher frequency and shorter wavelength.

To fully explain the photoelectric effect, Einstein theorized that visible light rays, *infrared* (IR) rays, *ultraviolet* (UV) rays, *X rays*, *gamma rays*, and even *radio waves* consist of discrete, energy-containing packets of wavelets. Einstein didn't give these particles any particular name, but other scientists coined the term *photon*. The photon theory offered a plausible explanation of the anomalies of the photoelectric effect, and also allowed for all the observed wavelike characteristics of radiant energy. Isaac Newton was at least partially vindicated. He hadn't been entirely wrong, after all.

PHOTON "MASS" AND ENERGY

If we revive Newton's old idea and accept the notion that light rays can behave as particle streams, it is tempting to think that photons have mass. But what, exactly, is meant by the term *mass* in this context? The idea of *rest mass* is meaningless for photons, because they are never at rest. These days, physicists generally agree that photons are *massless*. That doesn't mean that their mass is mathematically zero, but it does suggest that we might as well forget about trying to ascribe mass to them in the conventional sense.

Einstein showed that when a material particle with rest mass m moves at an extremely high speed v, then its mass increases by a *relativistic factor*, becoming a larger mass m_{rel} according to this formula:

$$m_{rel} = m/(1 - v^2/c^2)^{1/2}$$

When $v = c$, the equation "blows up." We have

$$\begin{aligned} m_{rel} &= m/(1 - v^2/c^2)^{1/2} \\ &= m/(1 - c^2/c^2)^{1/2} \\ &= m/(1 - 1)^{1/2} \\ &= m/0 \end{aligned}$$

If we suppose that the rest mass m_{ph} of a photon is equal to zero, then according to relativity theory, its mass $m_{ph\text{-}rel}$ at the speed of light (which is always the speed of a photon in free space) is

$$\begin{aligned} m_{ph\text{-}rel} &= m_{ph}/0 \\ &= 0/0 \end{aligned}$$

That's mathematical nonsense! But if we let m_{ph} be anything larger than zero, then $m_{ph\text{-}rel}$ becomes the quotient of some positive real number divided by zero. Again, nonsense.

While we can't define the rest mass of a photon, we can determine the *kinetic energy* E_{ph} of a photon if we know the frequency f of the wave associated with it:

$$E_{ph} = hf$$

where E_{ph} is in joules (J), f is in hertz (Hz), and h is a tiny number known as *Planck's constant*, given in joule-seconds (J · s). Expressed to six significant figures according to NIST:

$$h = 6.62607 \times 10^{-34}\ \mathrm{J \cdot s}$$

If we know the wavelength λ associated with a photon, we can substitute in the above equation, obtaining:

$$\begin{aligned} E_{ph} &= h\,(3.00 \times 10^8/\lambda) \\ &= 3.00 \times 10^8\ h/\lambda \end{aligned}$$

where E_{ph} is in joules and λ is in meters.

PHOTON MOMENTUM

Although it is difficult to define the mass of a photon, its *momentum* can be determined by experiment, and it is inversely proportional to the wavelength. The momentum p_{ph} of a photon is equal to Planck's constant h divided by the wavelength λ, as follows:

$$p_{ph} = h/\lambda$$

where p_{ph} is in kilogram meters per second (kg · m/s), h is in joule-seconds, and λ is in meters. The momentum of a photon, like its kinetic energy, is therefore directly proportional to its frequency. If we know the wave frequency f, then:

$$p_{ph} = h/(3.00 \times 10^8/f)$$
$$= hf/(3.00 \times 10^8)$$

where p_{ph} is in kilogram meters per second, h is in joule-seconds, and f is in hertz.

PROBLEM 1-3

Visible light occurs at free-space wavelengths ranging from approximately 750 nanometers (nm) down to 400 nm, where 1 nm = 10^{-9} m. What is this wavelength range in terms of frequency and period? Express the answers in hertz and in seconds, to three significant figures. Consider the speed of light in free space to be 3.00×10^8 m/s.

SOLUTION 1-3

First, we must convert the wavelengths to meters. Let λ_1 be the longest visible wavelength and let λ_2 be the shortest. This gives us $\lambda_1 = 7.50 \times 10^{-7}$ m and $\lambda_2 = 4.00 \times 10^{-7}$ m. Expressing these values as a range of frequencies where f_1 is lowest and f_2 is highest:

$$f_1 = c/\lambda_1$$
$$= (3.00 \times 10^8)/(7.50 \times 10^{-7})$$
$$= 4.00 \times 10^{14} \text{ Hz}$$

and

$$f_2 = c/\lambda_2$$
$$= (3.00 \times 10^8)/(4.00 \times 10^{-7})$$
$$= 7.50 \times 10^{14} \text{ Hz}$$

To determine the range of periods where T_1 is the longest and T_2 is the shortest, we can use the formulas for period in terms of frequency:

$$T_1 = 1/f_1$$
$$= 1.00/(4.00 \times 10^{14})$$
$$= 2.50 \times 10^{-15} \text{ s}$$

and

$$T_2 = 1/f_2$$
$$= 1/(7.50 \times 10^{14})$$
$$= 1.33 \times 10^{-15} \text{ s}$$

PROBLEM 1-4

What is the range of photon kinetic-energy values for visible light? Use the frequencies f_1 and f_2 as derived in the solution to Problem 1-3. Express the answers in joules to three significant figures. Consider Planck's constant to be 6.63×10^{-34} J · s.

SOLUTION 1-4

Let E_{ph1} be the energy contained in a photon having frequency f_1, and let E_{ph2} be the energy contained in a photon having frequency f_2. Then we have

$$\begin{aligned} E_{ph1} &= hf_1 \\ &= 6.63 \times 10^{-34} \times 4.00 \times 10^{14} \\ &= 2.65 \times 10^{-19} \text{ J} \end{aligned}$$

and

$$\begin{aligned} E_{ph2} &= hf_2 \\ &= 6.63 \times 10^{-34} \times 7.5 \times 10^{14} \\ &= 4.97 \times 10^{-19} \text{ J} \end{aligned}$$

PROBLEM 1-5

What is the range of photon momentum values for visible light? Use the wavelengths $\lambda_1 = 7.50 \times 10^{-7}$ m and $\lambda_2 = 4.00 \times 10^{-7}$ m. Express the answers in kilogram meters per second to three significant figures. Consider Planck's constant as 6.63×10^{-34} J · s.

SOLUTION 1-5

Let p_{ph1} be the momentum of a photon having wavelength λ_1, and let p_{ph2} be the momentum of a photon having wavelength λ_2. Then:

$$\begin{aligned} p_{ph1} &= h/\lambda_1 \\ &= (6.63 \times 10^{-34})/(7.50 \times 10^{-7}) \\ &= 8.84 \times 10^{-28} \text{ kg} \cdot \text{m/s} \end{aligned}$$

and

$$\begin{aligned} p_{ph2} &= h/\lambda_2 \\ &= (6.63 \times 10^{-34})/(4.00 \times 10^{-7}) \\ &= 1.66 \times 10^{-27} \text{ kg} \cdot \text{m/s} \end{aligned}$$

In passing, we should note that λ_1, f_1, E_{ph1}, and p_{ph1} in the preceding three problems represent parameters for visible red light, while λ_2, f_2, E_{ph2}, and p_{ph2} represent parameters for visible violet light. In between these extremes as the wavelength decreases, we see the colors orange, yellow, green, blue, and indigo.

The Electromagnetic Spectrum

Visible light, and many other forms of radiant energy, are composed of *electromagnetic (EM) waves* that range in frequency from a fraction of 1 Hz to quadrillions of hertz. All EM waves can travel through a vacuum, and sometimes they exert effects over vast distances.

ELECTRIC FIELDS

An electrically charged object is always surrounded by an *electric field.* This field has effects on other charged objects nearby. The electric field around a charged object can be depicted by *lines of flux*. The strength of an electric field is measured in *volts per meter*. Two opposite charges with a potential difference of 1 volt, separated by 1 meter, produce a field of 1 volt per meter.

The greater the electric charge on an object, the stronger the electric field surrounding the object, and the greater the force exerted on other charged objects in the vicinity. Lines of electric flux are not visible or material. They are theoretical, with each line representing a certain amount of electric charge. Electric flux density is measured in *coulombs per square meter*, where 1 coulomb (1 C) is the quantity of electric charge contained in approximately 6.24×10^{18} electrons.

A single charged object produces lines of flux that emanate directly outward from, or inward toward, the charge center. The direction of the field is considered to be outward from the *positive electric pole*, or inward toward the *negative electric pole*. When two charged objects are brought near each other, their electric fields interact. If the charges are both positive or both negative, the two objects repel each other. If the charges are opposite, the two objects attract each other. This force is often large enough to be measured, and in some cases it can be considerable.

Electric-field strength, also called *flux density*, diminishes with increasing distance from a charged object according to the *inverse square law*. If the distance from a charge center doubles, the electric flux density decreases to 1/4 of its previous value. If the distance increases by a factor of 10, the electric flux density becomes 1/100 as great.

MAGNETIC FIELDS

Whenever electric charge carriers move, an electric current flows. This current produces a *magnetic field*. A magnetic field can also occur when the atoms of certain substances are aligned in a certain way. Under such conditions, the "orbiting" electrons produce magnetic fields that effectively reinforce each other. This phenomenon, this ability to be *magnetized*, is called *ferromagnetism*. A material that can be magnetized is called *ferromagnetic*.

In a straight wire that carries electric current, circular *magnetic lines of flux* surround the wire, with the wire at the center of every circle. When a current-carrying wire is coiled up, two well-defined *magnetic poles* develop. You might hear of a certain number of flux lines per unit cross-sectional area, such as 100 lines per square centimeter. This is a relative way of describing the strength of the magnetic field. By convention, the lines of flux are said to emerge from a *magnetic north pole*, and to converge toward a *magnetic south pole*.

ELECTROMAGNETIC FIELDS

Electromagnetic (EM) fields were discovered in the 1800s by physicists who noticed that *alternating current (AC)* can exert effects at a distance through space, even when there is no medium of conduction. The earliest practical use of EM fields was in wireless communication and broadcasting, which later became known as *radio* and *television*. All EM fields contain electric lines of flux and magnetic lines of flux. The electric lines of flux are perpendicular to the magnetic lines of flux at every point in space. The direction in which the EM field propagates is perpendicular to both sets of flux lines.

The controlled acceleration and deceleration of electrons in an electrical conductor can generate *radio waves* at frequencies from about 3000 Hz (3×10^3 Hz) to 3000 GHz (3×10^{12} Hz). Infrared, visible light, ultraviolet, X rays, and gamma rays are forms of EM energy with frequencies higher (and wavelengths shorter) than those of radio waves. The entire range of EM frequencies or wavelengths is known as the *electromagnetic (EM) spectrum*. Theoretically, there is no limit to how low or high the frequency of an EM wave can be, nor is there any limit to how long or short the wavelength can be. Scientists use a logarithmic scale to depict the EM spectrum over several *orders of magnitude* (powers of 10). Figure 1-7 is a simplified rendition of the EM spectrum, with wavelengths shown in meters. The *visible spectrum* is but a thin slice of the EM realm.

PROPAGATION IN THE ATMOSPHERE

At wavelengths shorter than those of the radio waves, EM fields propagate through the earth's lower atmosphere to a greater or lesser extent, depending on various conditions including

- Wavelength
- Relative concentrations of gases
- Presence or absence of rain, snow, fog, and dust
- Extent of human-made or natural air pollution
- Humidity

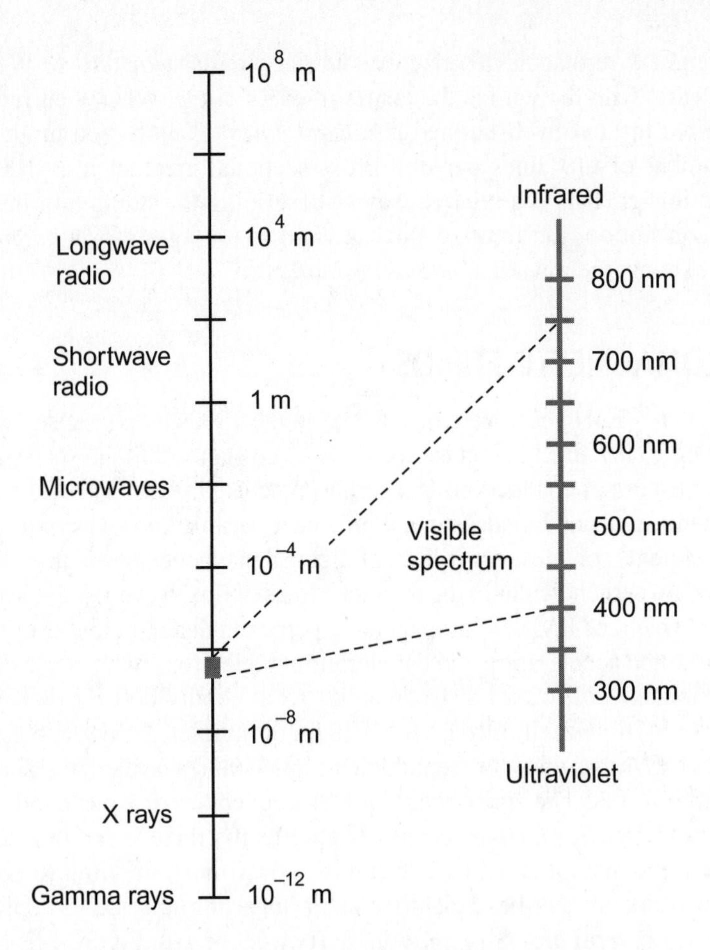

Figure 1-7 The EM spectrum from wavelengths of 10^8 m down to 10^{-12} m, and an exploded view of the visible spectrum within.

Water vapor causes considerable attenuation of IR between approximately 4500 nm and 8000 nm. Carbon dioxide (CO_2) gas interferes with the transmission of IR at wavelengths ranging from about 14,000 nm to 16,000 nm. Rain, snow, fog, and dust also interfere with the propagation of IR. In clear air, however, emissions in the near IR (wavelengths from the visible red up to about 4500 nm) propagate quite well over long distances through the atmosphere.

Visible light waves propagate fairly well through the atmosphere at all wavelengths. Scattering increases toward the short-wavelength end of the spectrum (blue and violet). This scattering explains why a clear daytime sky appears blue as viewed from the earth's surface. For free-space optical communications, red is the preferred visible-light color for this reason. Rain, snow, fog, and dust interfere with the transmission of visible light through the air.

Ultraviolet at the longer wavelengths can penetrate the air fairly well, although some scattering takes place. Ozone gas (O_3) in the upper atmosphere protects life on the surface from extreme solar UV radiation. At shorter wavelengths, atmospheric attenuation of UV increases. Rain, snow, fog, and dust interfere with UV propagation in the same way as they interfere with visible light.

X rays and gamma rays do not propagate well for long distances through the earth's lower atmosphere because the sheer mass of the air, over paths of any appreciable distance, is sufficient to block most of this radiation. Electromagnetic energy at these wavelengths is dangerous, because it ionizes living tissue. Practically none of the X rays and gamma rays from the sun reach the earth's surface. However, in the unlikely event of an extreme *gamma-ray burst* from an exploding star, considerable X-ray and gamma-ray energy could penetrate the atmosphere. Some scientists suspect that one or more events of this type may have affected the evolution of life in eons past.

Light and Relativity

There are two aspects to relativity theory: the *special theory* and the *general theory*. The special theory involves relative motion, and the general theory involves acceleration and gravitation. Both are based on the hypothesis that the speed of light through a vacuum is a physical constant.

Simultaneity

One of the first results of Einstein's speed-of-light axiom is that there can be no absolute time standard. It is impossible to synchronize two clocks so they agree precisely, unless both clocks or all the observers occupy the same point in space.

Imagine eight clocks arranged at the vertices of a gigantic cube in space that measures exactly 1 lt · min or 1.80×10^7 km along each edge, as shown in Fig. 1-8. Suppose that we want to synchronize the clocks. We position ourselves at the center of the cube, so we are equidistant from all eight clocks. We set them (using remote-control wireless equipment) so that they all appear to agree from our point of view.

Now suppose that we move to some point outside the cube and look back at the clocks. They no longer appear synchronized, because their readings depend on how far their visible images must travel to reach us. At the center of the cube, the images from all eight clocks arrive from the same distance, but this is not the case for any other point in space. We can readjust the clocks so they appear to agree as seen from our new location, but they will appear out of sync again if we move to yet another location.

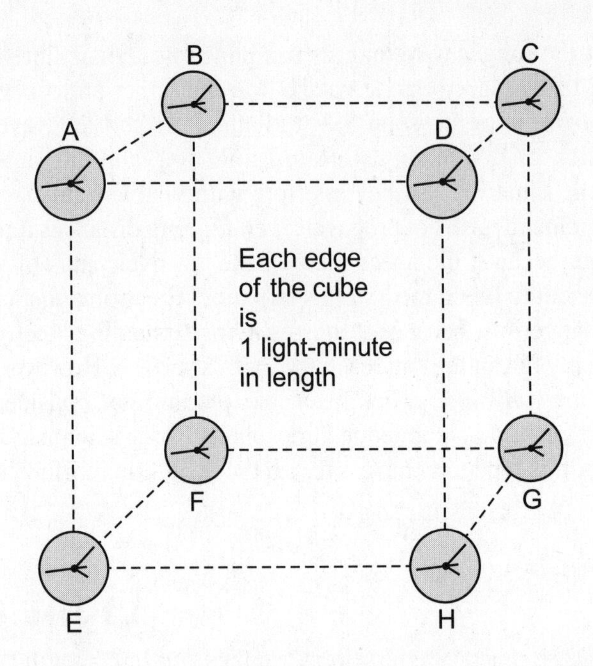

Figure 1-8 A hypothetical set of eight clocks, arranged at the vertices of a cube that measures 1.00 lt · min on each edge. How will we synchronize these clocks?

Any attempt to establish an absolute time standard over a vast region, with which all observers in that region agree, must take each observer's position into account. To some extent, this phenomenon occurs even over small distances, but the discrepancy is too small for us to see, so it doesn't affect our everyday lives. It will become important when and if humans become regular travelers in interplanetary space, however. The effect is a consequence of the constancy and finiteness of the speed of light.

TIME DILATION

Isaac Newton believed that time flows smoothly and uniformly in an absolute sense, and that time constitutes a fundamental constant in the universe. Newtonian physics therefore allowed for the possibility that the speed of light could depend on the motion of the observer. Einstein theorized that it is the speed of light, not time, that is constant. This paradigm shift gave rise to a whole new theoretical dynamic of time, space, and motion when relative speeds are high.

Imagine a bullet-shaped space ship equipped with a laser and sensor on one wall and a mirror on the opposite wall, as shown in Fig. 1-9. Further suppose that the

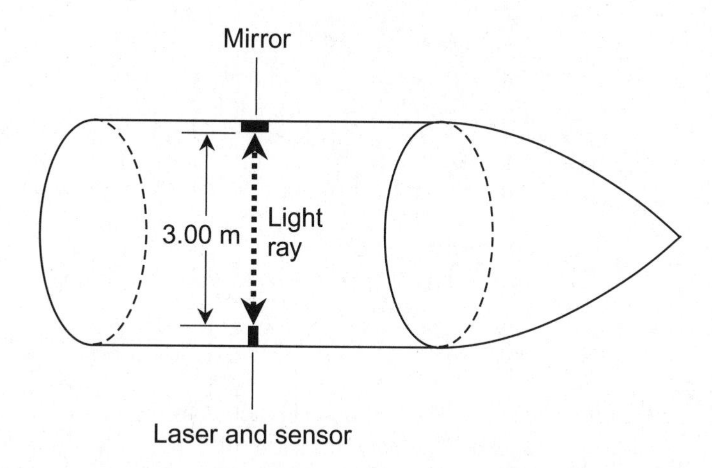

Figure 1-9 A space ship equipped with a laser clock, as seen by an observer in the ship.

laser/sensor and the mirror are positioned so the light ray from the laser must travel perpendicular to the axis of the ship, perpendicular to its walls, and perpendicular to its direction of motion. The laser and mirror are adjusted so they are separated by 3.00 m. Because the speed of light in air is approximately 3.00×10^8 m/s, it takes 1.00×10^{-8} s, or 10.0 nanoseconds (10.0 ns), for the light ray to get across the ship from the laser to the mirror, and another 10.0 ns for the ray to return to the sensor. The ray therefore requires 20.0 ns to make one round trip from the laser/sensor to the mirror and back again.

Suppose that we get the ship moving at a sizable fraction of the speed of light. We measure the time it takes for the laser beam to travel across the ship and back again. We are moving along with the laser, the mirror, the sensor, and all the rest of the equipment on board the spacecraft. We find that the time lag is still 20.0 ns. If we increase our speed so the ship is going 60 percent, 70 percent, or even 99 percent of the speed of light, the time lag is always 20.0 ns as measured from a reference frame inside the ship.

Imagine that a group of scientists on the earth have a telescope that allows them to see inside the ship as it whizzes by at nearly the speed of light. Their view is shown in Fig. 1-10. The laser beam appears to travel in straight lines, and it appears to travel at 3.00×10^8 m/s. But from this reference frame, with respect to which the ship is moving at high speed, the rays must travel farther than 3.00 m to get across the ship. The ship is going so fast that, by the time the ray has reached the mirror from the laser, the ship has moved a significant distance forward. According to these observers, it takes more than 20.0 ns for the laser beam to go across the ship and back.

The laser apparatus in this "mind experiment" measures time based on the speed of light, an absolute constant. According to this standard, time appears to slow

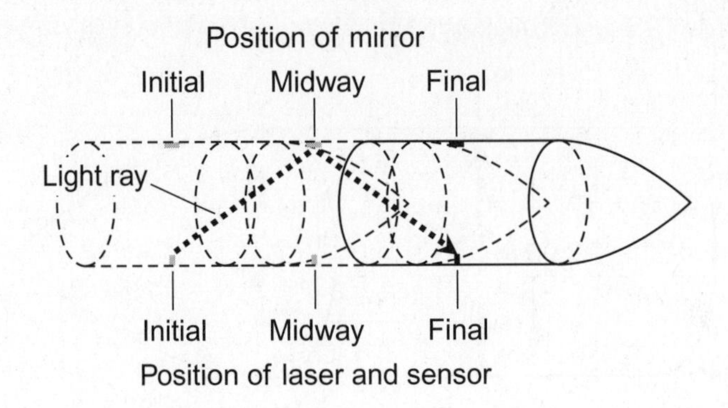

Figure 1-10 This is what an external observer sees as the space ship passes by at a large fraction of the speed of light.

down inside the fast-moving space ship as seen from a nonaccelerating point of view that we call "stationary." The faster the ship goes, the greater the discrepancy becomes. This effect is called *relativistic time dilation*.

There is a mathematical relationship between the speed of the space ship and the extent to which time dilates. Let t_{ship} be the number of seconds that appear to elapse on the moving ship, as precisely 1 s elapses as measured by a clock on an earth-based observatory. Let u be the speed of the ship, as a fraction of the speed of light. Then

$$t_{ship} = (1 - u^2)^{1/2}$$

The *time dilation factor* (call it k) is the reciprocal of this, so we have

$$k = (1 - u^2)^{-1/2}$$

In these formulas, the number 1 is a mathematically exact value.

PROBLEM 1-6

What is the time dilation factor for an object moving at exactly half the speed of light relative to a stationary observer? Express the answer to three significant figures.

SOLUTION 1-6

We can use the formula above for the time dilation factor k. If the object is traveling at half the speed of light, then $u = 0.500$, and

$$\begin{aligned} k &= (1.00 - 0.500^2)^{-1/2} \\ &= (1.00 - 0.250)^{-1/2} \\ &= 0.750^{-1/2} \\ &= 1.15 \end{aligned}$$

EXPERIMENTAL CONFIRMATION OF TIME DILATION

The effect of relativistic time dilation at speeds encountered in everyday life cannot be measured unless atomic clocks, accurate to a minuscule fraction of 1 s, are used to measure the time in both reference frames. Nevertheless, relativistic time dilation has been measured under quite ordinary conditions, and the results concur with Einstein's formulas. In one notable experiment, an atomic clock was placed on an aircraft, and the aircraft was sent up in flight to cruise around for awhile at several hundred kilometers per hour. Another atomic clock was kept at the place where the aircraft took off and landed. Although the aircraft's speed was only a tiny fraction of the speed of light, and the resulting time dilation was exceedingly small, it was large enough to measure. When the aircraft finished the trip, the clocks, which had been synchronized (when placed next to each other) before the trip began, were compared. The clock that had been on the aircraft registered a time slightly earlier than the clock that had stayed behind.

ACCELERATING REFERENCE FRAMES

Einstein noticed something special about accelerating reference frames, compared with those that are not accelerating. This difference is apparent if we consider the situation of an observer who is enclosed in a chamber that is completely sealed and opaque.

Suppose you are in a space ship in which the windows are covered up, and the radar and navigational equipment have been temporarily disabled. You cannot examine the surrounding environment and determine where you are, how fast you are moving, or in what direction you are traveling. But you can tell whether or not the ship is accelerating, because acceleration always produces a force on objects inside the ship.

When the ship's engines are fired and the vessel gains speed in the forward direction, all the objects in the ship, including your body, are subjected to a force directed backward. If the ship's retro rockets are fired so the ship slows down (decelerates), everything in the ship is subjected to a force directed forward. If rockets on the side of the ship are fired so that the ship changes direction without changing speed, then this is a form of acceleration, and it causes lateral forces on everything inside the ship.

As the acceleration to which the space ship is subjected increases, the force on every object inside the ship also increases. If m is the mass of an object in the ship (in kilograms) and a is the acceleration of the ship (in meters per second squared), then the force F (in newtons) is

$$F = ma$$

Now imagine that our space ship, instead of accelerating in deep space, is set down on the surface of a planet. If the windows are kept covered, the radar is shut off, and the navigational aids are placed on standby, then according to Einstein's

hypothesis, the passengers in the vessel will not be able to tell whether the force is caused by gravitation or by acceleration. This so-called *equivalence principle* is the basis of the theory of general relativity.

SPATIAL CURVATURE

Imagine that you are onboard a vessel traveling through deep space. The ship's rockets are fired, and the vessel accelerates at an extreme rate. Suppose a laser is set up to shine a light beam across the ship, but instead of a mirror on the wall opposite the laser, there is a screen. Before the acceleration begins, you align the laser so its beam hits the forward edge of the screen, as shown in Fig. 1-11A. Then you accelerate the vessel. If you accelerate fast enough, the ship "pulls away" from the laser beam as the beam travels across the ship. You, looking at the situation from inside the ship, see the light beam follow a curved path (Fig. 1-11B) and strike the rear edge of the screen. A stationary observer on the outside sees the light beam follow a straight path through space, but the vessel "pulls out" ahead of the beam, as shown in Fig. 1-12.

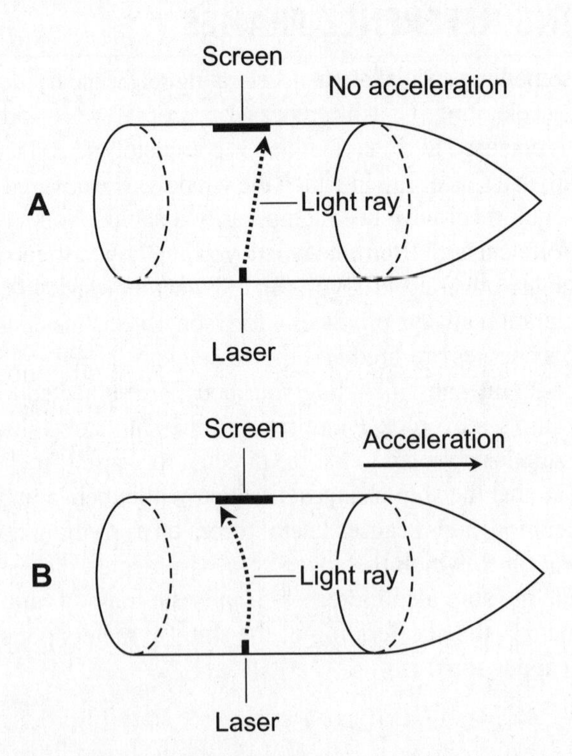

Figure 1-11 At A, a light ray travels in a straight line across a nonaccelerating vessel as seen by an onboard observer. At B, a light ray takes a curved path across a rapidly accelerating vessel as seen by an onboard observer.

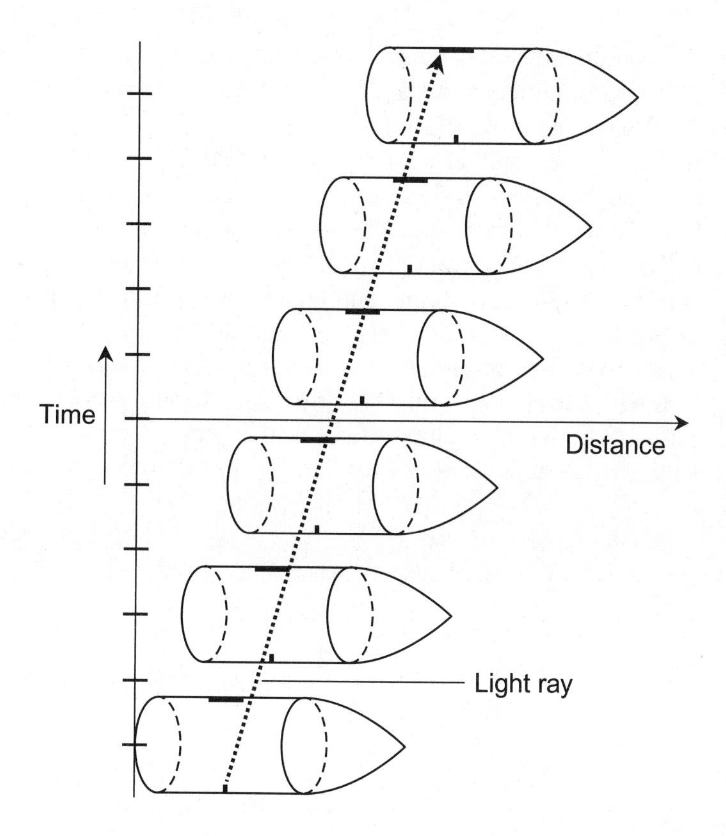

Figure 1-12 When viewed from a nonaccelerating reference frame, an accelerating vessel "pulls away" from the straight-line path of a lateral light ray.

Regardless of the reference frame—stationary, moving but not accelerating, or accelerating—a ray of light always follows the shortest possible path between two points in space. When there is no acceleration, these paths are straight lines, indicating that space is "flat," or *Euclidean*. But when observed from an accelerating point of view, light rays appear curved. The shortest distance between the two points at opposite ends of the laser beam in Fig. 1-11B is not a straight line! This phenomenon gives rise to the idea that space is "warped" in a powerful acceleration field. A mathematician would say that such space is *non-Euclidean*. Because of the principle of equivalence, powerful gravitation causes the same *spatial curvature* as does extreme acceleration.

In real life, the acceleration or gravitation necessary to produce curvature of a laser beam to the extent shown in Fig. 1-11B would be so great that an observer onboard the vessel would be crushed against the floor. No experiment can be conducted to verify these results by direct human observation. However, the effect takes place to a small extent even when a vessel accelerates at a modest rate, and even in weak gravitational fields such as we find at the surface of the earth. We don't notice the non-Euclidean nature of space under ordinary circumstances, because the "warping" is far too slight for us to see.

After Einstein developed his general theory of relativity, light rays from distant stars were observed as they passed close to the sun during a total solar eclipse to find out if the sun's gravitational field, which is strong near its surface, would bend the light rays. The apparent position of a distant star was indeed offset by the presence of the sun near the line of sight. The effect occurred to the same extent as Einstein's general relativity formulas said it should. More recently, a quasi-stellar source (or *quasar*) has been observed whose light has passed close to a suspected black hole on its way to us through intergalactic space. Several images of the quasar appear, indicating that the light rays have been bent by the gravitational field around an invisible object in their path.

Quiz

This is an open book quiz. You may refer to the text in this chapter. A good score is 8 correct. Answers are at the end of the book.

1. The standard unit of length, the meter, is based on known values of the
   - (a) kilogram and the second.
   - (b) second and the speed of light in a vacuum.
   - (c) kilogram and the speed of light in a vacuum.
   - (d) speed of light and the kelvin.
2. In 1 hour (1 h), a beam of light in free space travels approximately
   - (a) 1.08×10^9 m.
   - (b) 1.20×10^5 km.
   - (c) 2.50×10^5 m.
   - (d) 1.08×10^{12} m.

3. According to Isaac Newton's theory, the path of a light ray is bent when the ray passes at an acute angle from air into water because
   (a) magnetic fields surrounding the water atoms cause the light particles to accelerate.
   (b) electric fields surrounding the water atoms cause the light particles to accelerate.
   (c) water molecules cause the wavelength of the light to increase.
   (d) water is denser than air, causing gravitational acceleration of the light particles.
4. Newton's and Einstein's theories of light both led to the conclusion that light particles traveling through space
   (a) can be bent by intense gravitational fields.
   (b) have defined wavelengths based on their color.
   (c) require luminiferous ether to propagate over long distances.
   (d) have speed that depends on the motion of the observer.
5. The results of the Michelson-Morley experiment contributed to Einstein's hypothesis that
   (a) light is a wave phenomenon, and the wavelength depends on the speed of the observer relative to the source.
   (b) the speed of light is lower as seen from an observer retreating from a light source, as compared another observer approaching that same light source.
   (c) the speed of light is the same as observed from all nonaccelerating reference frames.
   (d) the luminiferous either must be stationary relative to the earth.
6. Suppose a ray of light travels through the air and then enters a body of water. Once the ray has crossed the surface boundary, it travels
   (a) more slowly than it did in the air.
   (b) at the same speed as it did in the air.
   (c) faster than it did in the air.
   (d) at a speed that depends on the angle of incidence.

7. Consider a ray of red light that has a wavelength of 750 nm as it propagates through the air. Then it enters a pool of water. Once the ray has crossed the surface boundary, its wavelength
   (a) is longer than 750 nm.
   (b) is still 750 nm.
   (c) is shorter than 750 nm.
   (d) depends on the angle of incidence.
8. Consider a ray of violet light that has a frequency of 7.50×10^{14} Hz as it travels through the air. Then it enters a pool of water. Once the ray has crossed the surface boundary, its frequency
   (a) is lower than 7.50×10^{14} Hz.
   (b) is still 7.50×10^{14} Hz.
   (c) is higher than 7.50×10^{14} Hz.
   (d) depends on the ratio of the propagation speed in the water to the propagation speed in the air.
9. What is the time dilation factor for an object moving through free space at a constant speed of 99.9 percent of the speed of light relative to us?
   (a) 500.
   (b) 22.4.
   (c) 0.999.
   (d) It depends on the wavelength of the light we observe.
10. According to the electromagnetic theory, visible light waves are similar to radio waves, but light waves
   (a) have longer wavelengths.
   (b) have lower frequencies.
   (c) travel faster.
   (d) have shorter periods.

CHAPTER 2

Classical Optics

Classical optics involves the effects of physical substances and objects on reflected and transmitted light. The mathematical analysis of reflection and refraction, in particular, is known as *geometrical optics*.

Reflection

Any smooth surface reflects some of the light that strikes it. If the surface is perfectly flat and shiny such as the face of a pane of glass, a ray is reflected away at the same angle at which it hits. You have heard the expression, "The angle of incidence equals the angle of reflection." This principle (Fig. 2-1) is called the *law of reflection*.

ANGLES

In optics, the angle of incidence and the angle of reflection are defined or measured with respect to a line normal (perpendicular) to the surface at the point where reflection takes place. In Fig. 2-1, these angles are denoted θ, and can range from 0°,

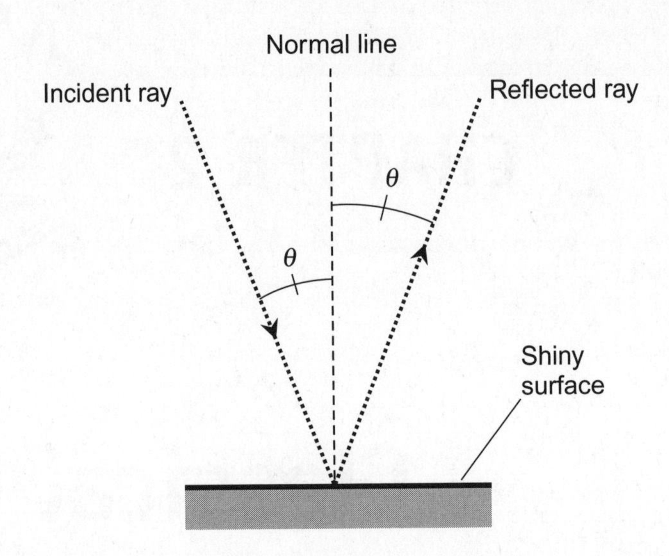

Figure 2-1 When a light ray strikes a shiny surface, the angle of incidence equals the angle of reflection. In this example, both angles are denoted by θ, and are expressed relative to a line normal (perpendicular) to the surface at the point of reflection.

when the light ray strikes at a right angle with respect to the surface, to almost 90°, a grazing angle relative to the surface.

If the reflective surface is shiny but not flat, then the law of reflection applies for any ray of light striking the surface at a specific point. In such a case, the reflection is considered relative to a line normal to a flat plane passing through the point, tangent to the surface at that point. When parallel rays of light strike a curved, shiny surface at many different points, each individual ray obeys the law of reflection, but the reflected rays do not all emerge parallel. In some cases the emerging rays converge; in other cases they diverge.

If a light beam encounters a *matte* surface (one that is not shiny) such as a sheet of paper or a plaster wall, the reflected rays are scattered. On a microscopic scale or for individual photons, the law of reflection holds. On a large scale, however, the law does not hold.

A MIRRORED WALL

Imagine a room that measures 5 m by 5 m, with one mirrored wall. Suppose you stand near one wall (*W* as shown in Fig. 2-2A), and hold a flashlight so that its bulb is 1.000 m away from *W* and 3.000 m away from the mirrored wall. Suppose you aim the flashlight horizontally at the mirrored wall, so that the center of its beam

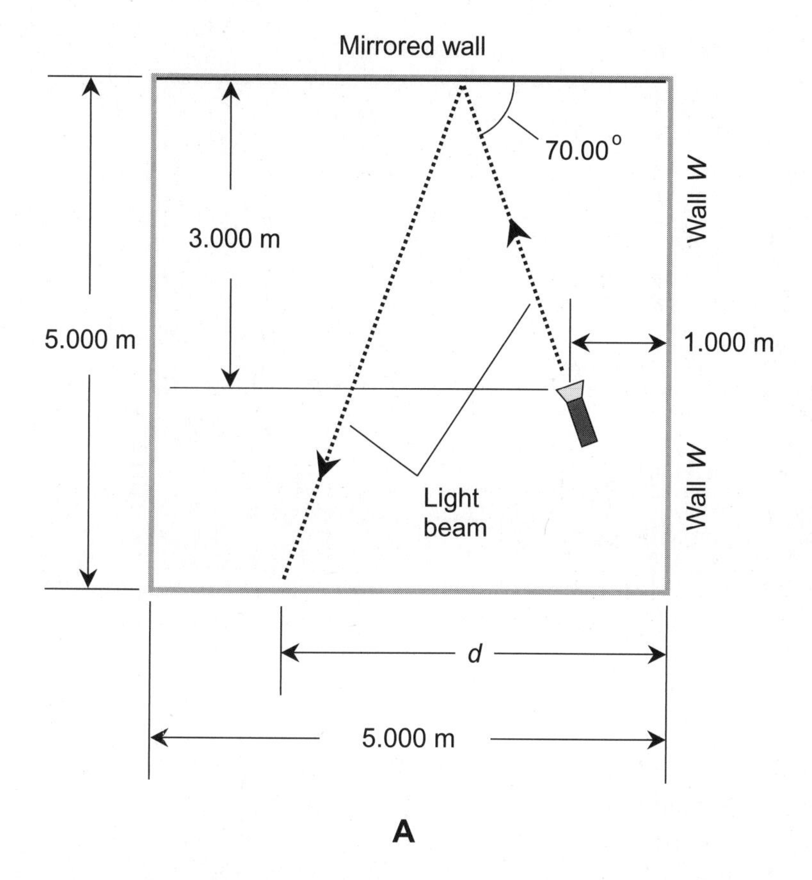

Figure 2-2A A light beam reflects from a mirror and strikes the wall opposite the mirror. The angle of incidence and the angle of reflection are both 20.00°.

strikes the mirror at an angle of 70.00° relative to the mirror surface. The beam reflects from the mirror and encounters the wall opposite the mirror at a certain distance d from wall W.

The light beam travels in a plane parallel to the floor and the ceiling, because the flashlight is aimed horizontally. Figure 2-2B illustrates this situation in geometric detail. The center of the beam strikes the mirror at an angle of 20.00° relative to the normal. According to the law of reflection, it therefore reflects from the mirror at an angle of 20.00° relative to the normal. The path of the light beam forms the *hypotenuses* (longest sides) of two right triangles, one whose base length in meters is e and whose height in meters is 3.000, and the other whose base length in meters is f and whose height in meters is 5.000. Using the trigonometric tangent function, we can calculate e as follows:

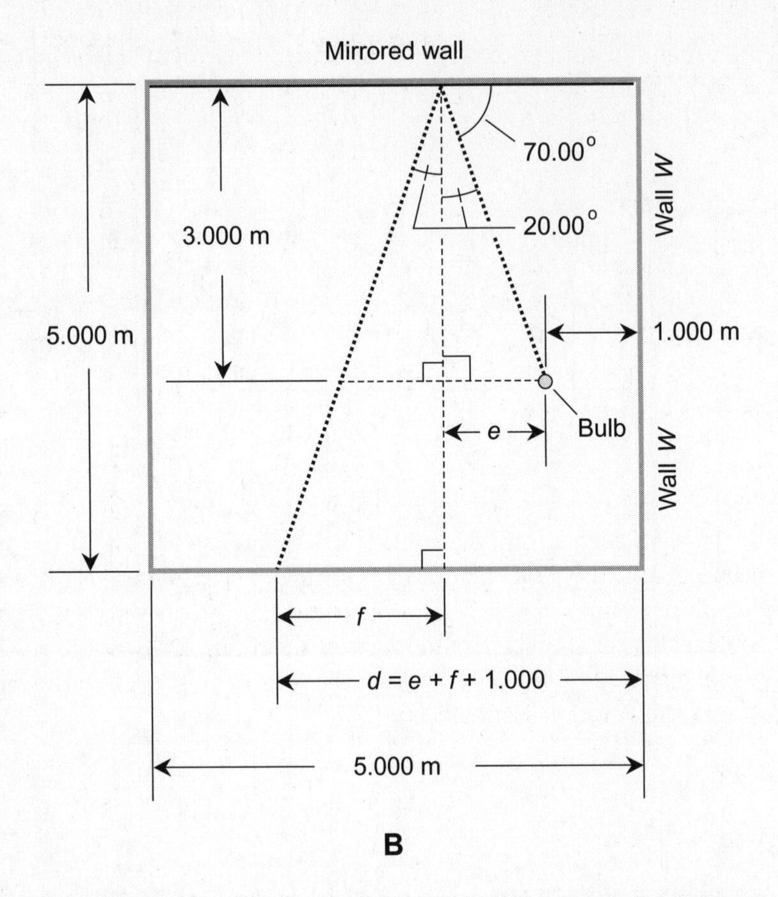

Figure 2-2B Geometrical analysis of the situation shown in Fig. 2-2A.

$$\tan 20.00° = e/3.000$$
$$0.36397 = e/3.000$$
$$e = 0.36397 \times 3.000 = 1.09191 \text{ m}$$

We calculate f in a similar way:

$$\tan 20.00° = f/5.000$$
$$0.36397 = f/5.000$$
$$f = 0.36397 \times 5.000 = 1.81985 \text{ m}$$

Knowing both e and f, we calculate d as follows:

$$d = e + f + 1.000$$
$$= 1.09191 + 1.81985 + 1.000$$
$$= 3.91176 \text{ m}$$

We should round this result off to 3.912 m, because our input data is given to only four significant figures. (The extra figures in the intermediate calculations minimize the risk of cumulative rounding error.)

VIRTUAL IMAGE IN FLAT MIRROR

Suppose you stand in front of a mirror that hangs on a wall, as shown in Fig. 2-3. You see your reflection in the mirror. If you stand at a distance *d* from the wall, then the image of your body, from your point of view, appears to be on the opposite side of the wall, at the same distance *d* from the silvered surface of the mirror as is your real body. When you see the image of a point such as the top of your head or the tip of your right foot, light rays from that point have struck the mirror, reflected from its silvered surface according to the law of reflection, and entered your eyes.

In the example of Fig. 2-3, the collection of reflected rays from all the points on your body, arriving at either of your eyes, forms a *virtual image*. In this context, the term "virtual" refers to the fact that the image cannot be rendered on a screen or on film. If you place a photographic film or photosensitive screen in front of your right eye without any lens or other device to focus the incoming rays, no image will form. You see your reflected image only because the lens in your eye focuses incoming light to points on the *retina* inside the eye.

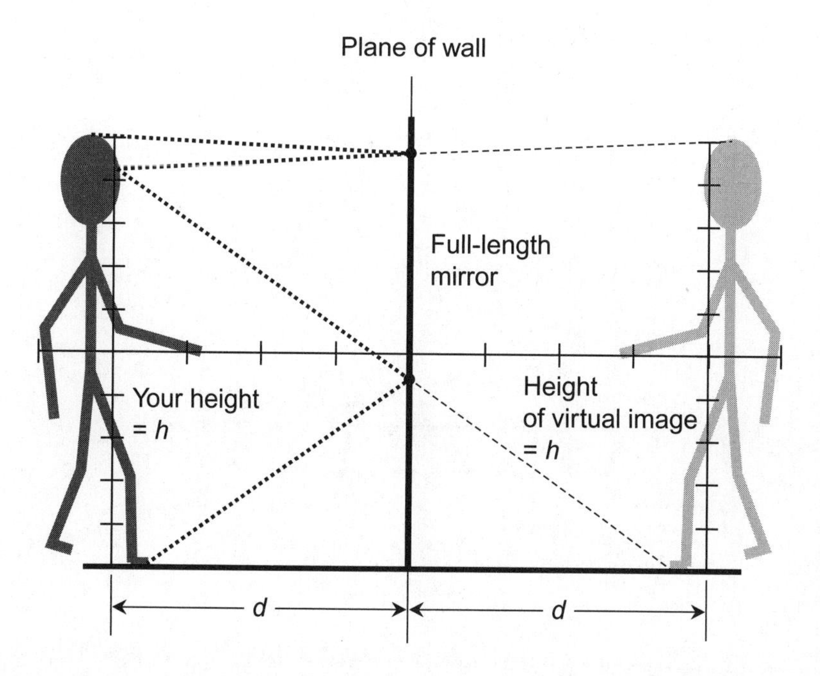

Figure 2-3 When you see your reflection in a flat wall mirror, you see a virtual image.

THE CONVEX MIRROR

A *convex mirror* reflects light so that incident rays, when parallel, are spread out as they leave the surface, as shown in Fig. 2-4A. Converging incident rays, if the *angle of convergence* is just right, are made parallel, or *collimated*, by a convex mirror, as illustrated in Fig. 2-4B. When you look at the reflection of a scene in a convex mirror, the objects all appear reduced in size. The field of vision is enlarged. This phenomenon is used to advantage in some automotive rearview mirrors.

The extent to which a convex mirror spreads light rays depends on the mirror's *radius of curvature*. As the radius of curvature decreases, the extent to which parallel incident rays diverge after reflection increases. The convergence of incoming rays necessary to produce parallel reflected rays also increases. Automotive rearview mirrors have large radii of curvature, so the field of view is not greatly expanded. Because convex mirrors make reflected objects appear more distant than they actually are, some auto makers do not use them, for fear they will cause drivers to misjudge the distances of following vehicles.

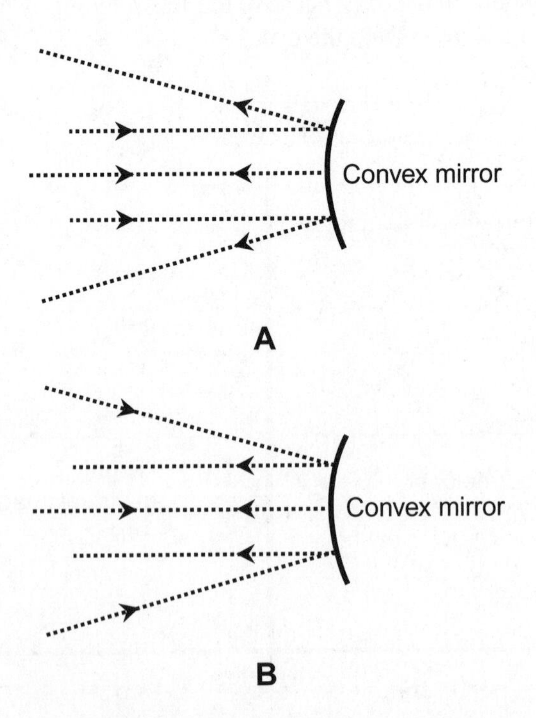

Figure 2-4 At A, a convex mirror spreads parallel incident light rays. At B, the same mirror collimates converging incident light rays.

THE CONCAVE MIRROR

A *concave mirror* reflects light rays in the reverse sense from a convex mirror. When the mirror has the proper contour, and when incident rays are parallel to the axis of the mirror, the rays are reflected so they converge at a *focal point* (Fig. 2-5A) that lies on the axis. A *reflecting telescope* takes advantage of this fact.

When a point source of light is placed at the focal point, a concave mirror reflects the rays so they emerge parallel (Fig. 2-5B). This is the principle by which a flashlight, lantern, or spotlight works. The distance of the focal point from the center of the concave mirror surface is called the *focal length*.

Some concave mirrors have spherical surfaces, but the ideal contour is that of a *paraboloid*, a three-dimensional figure produced by rotating a *parabola* around its axis. When the radius of curvature is large compared with the size of the reflecting surface or when high precision is not required, the difference between a *spherical mirror* and a *paraboloidal mirror* (often imprecisely called a *parabolic mirror*) is negligible. The difference becomes significant if the mirror is intended for use in a precision optical instrument such as a telescope.

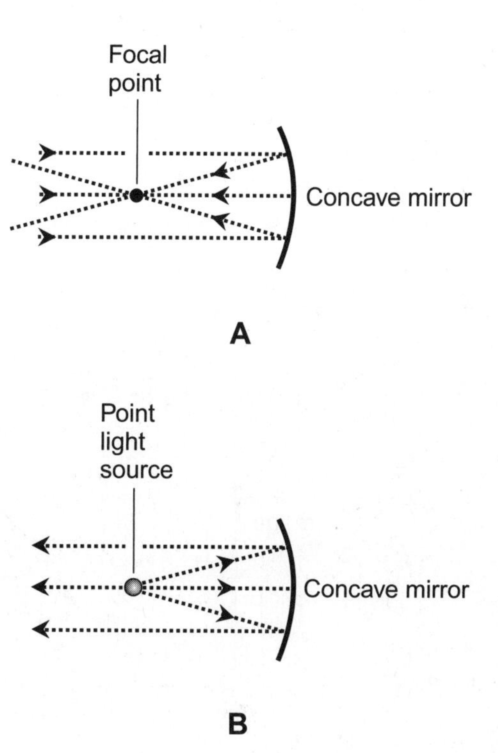

Figure 2-5 At A, a concave mirror focuses parallel light rays to a point. At B, the same mirror collimates light from a point source at the focus.

REAL IMAGE IN CONCAVE MIRROR

Imagine that you stand far away from a spherical or paraboloidal mirror, as shown in Fig. 2-6. Rays of light reflect from your body and reach the mirror surface. At each point where a ray hits the mirror, the angle of incidence is equal to the angle of reflection. The rays converge to form a tiny inverted image of your body near the mirror. If you place photographic film or a photosensitive screen at the location of this image, the image appears clearly. Because of this phenomenon, a focused image from a spherical or paraboloidal mirror is called a *real image*.

The distance of a real image from a concave mirror depends on the distance of the actual object from that same mirror. If your distance from the mirror is "infinity" (many times the focal length), then your real image appears at the focal point. If you are at a lesser distance, then your real image forms at a distance greater than the focal length. Suppose you stand along the mirror axis at a distance r_o from the center of the mirror, where r_o is considerably greater than the focal length f, but not large enough to be called "infinite." Under these circumstances, the distance r_i of your body's real image from the mirror is related to r_o and f like this:

$$1/f = 1/r_i + 1/r_o$$

For this formula to work, all distances must be expressed in the same units. The formula is approximate because, in a practical situation, not every point on an observed object lies at exactly the same distance r_o from the center of the mirror. The discrepancy is small if r_o is many times f. However, as r_o decreases, the discrepancy increases for an object of fixed lateral diameter.

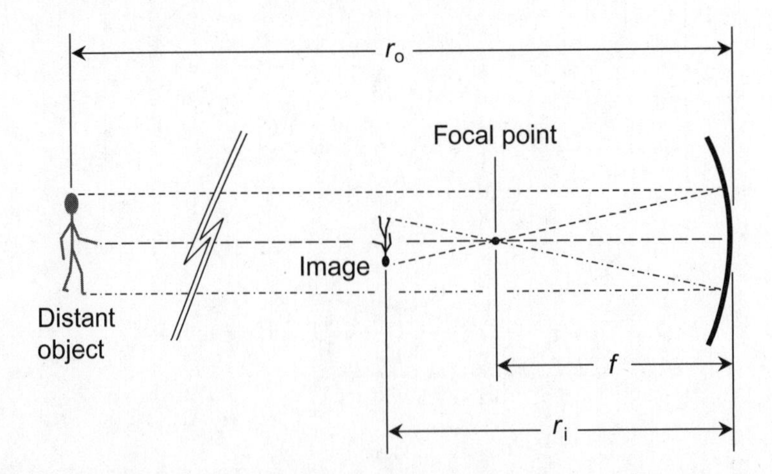

Figure 2-6 A real image forms in a concave mirror when reflected rays come to a focus. The distance r_i of the image from the mirror depends on the distance r_o of the object from the mirror, relative to the focal length f.

If an object is placed at a distance less than f from a concave mirror (that is, closer to the mirror than the focal length), a real image does not exist, because the reflected rays don't come to a focus.

PROBLEM 2-1

In Fig. 2-3, your height is shown as h, and the height of your virtual image is also shown as h. If the wall mirror is positioned at a certain distance from the floor, then that mirror needs to be only half your height to let you see your whole reflection, regardless of how far from the wall you stand. Why is this true?

SOLUTION 2-1

Figure 2-7 is a detailed diagram of the geometry of this situation. Suppose your height is h, the vertical distance from the point E between your eyes to a point H on the top of your head is h_1, and the vertical distance from point E to a point F on the

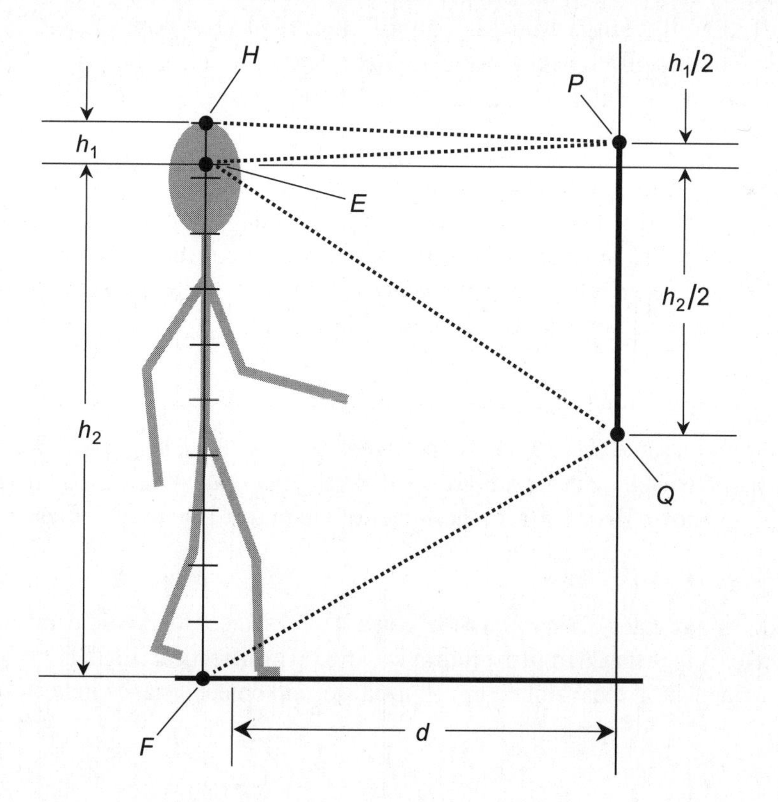

Figure 2-7 Illustration for Problem and Solution 2-1. Your height, h, is equal to $h_1 + h_2$. The minimum necessary height of the mirror is $h_1/2 + h_2/2$, or $h/2$, regardless of the distance d between your eyes and the mirror, and regardless of the distance between your eyes and the top of your head.

floor midway between your feet is h_2. If all three of these distances are expressed in the same units, then

$$h = h_1 + h_2$$

For simplicity, imagine that points H, E, and F all lie along a single vertical line. You look at the image of point H as it is reflected in the mirror. Photons reflect from point H, travel in a straight line, strike the mirror at point P, and reflect back along another straight line to point E. The triangle HPE formed by these three points is an isosceles triangle, because the straight-line distances HP and PE are equal. That means the vertical height difference between points P and E is equal to half the vertical height difference between points H and E. That is $h_1/2$. Similarly, photons reflect from point F, travel in a straight line, strike the mirror at point Q, and reflect back along another straight line to point E. The triangle FQE formed by these three points is an isosceles triangle, because the straight-line distances FQ and QE are the same. It follows that the vertical height difference between points E and Q is half the vertical height difference between points F and E, or $h_2/2$. The vertical distance PQ between points P and Q on the mirror is therefore

$$\begin{aligned} PQ &= h_1/2 + h_2/2 \\ &= (h_1 + h_2)/2 \\ &= h/2 \end{aligned}$$

If the height of the mirror is equal PQ, then the mirror is tall enough to let you see your entire virtual image. We've shown that $PQ = h/2$, which is half your height.

PROBLEM 2-2

Imagine that you stand in front of a concave mirror whose focal length is 200 mm. Your position is such that the mirror axis passes through your body. You are 8.00 m from the center of the mirror. How far from the center of the mirror is your real image?

SOLUTION 2-2

Before doing any calculations, you must convert all distances to the same units. Meters will do here. Your distance from the mirror is $r_o = 8.00$ m. The focal length is $f = 0.200$ m. The formula relating focal length, image distance, and object distance indicates that

$$1/f = 1/r_i + 1/r_o$$

Plugging in the known values, you get

$$\begin{aligned} 1/0.200 &= 1/r_i + 1/8.00 \\ 5.00 &= 1/r_i + 0.125 \end{aligned}$$

$$5.00 - 0.125 = 1/r_i$$
$$4.875 = 1/r_i$$
$$r_i = 1/4.875 \text{ m}$$
$$= 0.205 \text{ m}$$

Your real image is 0.205 m, or 205 mm, from the center of the mirror. This is approximate because your height (unless you are an insect or microbe) is significant compared to your distance from the mirror.

Refraction

A clear pool looks shallower than it actually is because of *refraction*, which occurs when light passes between different media at different speeds. The speed of light is absolute and constant in a vacuum, where it travels at about 2.99792×10^8 m/s. In air, the speed of light is a tiny bit slower than it is in a vacuum; in most cases the difference is not worth worrying about. But in transparent liquid or solid media such as water, glass, quartz, or diamond, the speed of light is significantly less than it is in a vacuum.

INDEX OF REFRACTION

The *refractive index*, also called the *index of refraction*, of a medium is the ratio of the speed of light in a vacuum to the speed of light in that medium. If c is the speed of light in a vacuum and c_m is the speed of light in some transparent medium M, then the index of refraction r_m for the medium M is

$$r_m = c/c_m$$

It's important to use the same speed units, such as meters per second, when expressing c and c_m. According to this definition, the index of refraction of any transparent material is always larger than or equal to 1, because c_m is always less than c.

As the index of refraction for a transparent substance increases, so does the extent to which a ray of light is bent when it strikes the boundary between that substance and air at some angle other than the normal. Various types of glass have different refractive indices. Diamond has a higher refractive index than any glass. The high refractive index of diamond is responsible for the "sparkle" of cut diamonds.

Table 2-1 lists the approximate refractive indices for various transparent materials. Values are for visible light near the middle of the spectrum, at a wavelength of approximately 600 nm.

Table 2-1 Approximate indices of refraction for some materials in the middle of the visible spectrum.

Substance	Value
Air	1.00
Alcohol, ethyl	1.36
Amber	1.55
Cubic zirconia	2.16
Diamond	2.42
Glass, acrylic	1.49
Glass, borosilicate	1.47
Glass, crown	1.52
Glass, flint	1.61
Glycerin	1.47
Ruby	1.76
Salt (sodium chloride), crystal	1.50
Vacuum	1.00
Water, fresh, as ice	1.31
Water, fresh, as liquid	1.33

LIGHT RAYS AT A BOUNDARY

Figure 2-8A illustrates a qualitative example of refraction. In this case, the refractive index of the first medium (below) is higher than that of the second medium (above). A ray striking the boundary at 0° relative to the normal passes through the boundary without changing direction. Any ray that hits at some other angle is bent. As the angle of incidence increases, so does the angle by which the beam is deflected at the boundary. When the angle of incidence reaches a certain *critical angle*, then the light ray is not refracted at the boundary, but is reflected back into the first medium instead. This effect is called *total internal reflection*.

Total internal reflection occurs only when the first medium has a higher index of refraction than the second medium. If a ray of light passes from a substance having a certain refractive index into something with a higher refractive index, total internal reflection does not occur at any angle of incidence. In that situation, there's no critical angle. It is nevertheless possible (and in fact likely) that *some* of the light will be reflected at the boundary regardless of the angle of incidence, just as some of the incident light is reflected from the outside surface of a window on a bright day.

In the atmosphere, the speed of light varies slightly, depending on the density of the air in a particular location. Warm air is less dense than cool air when other factors are

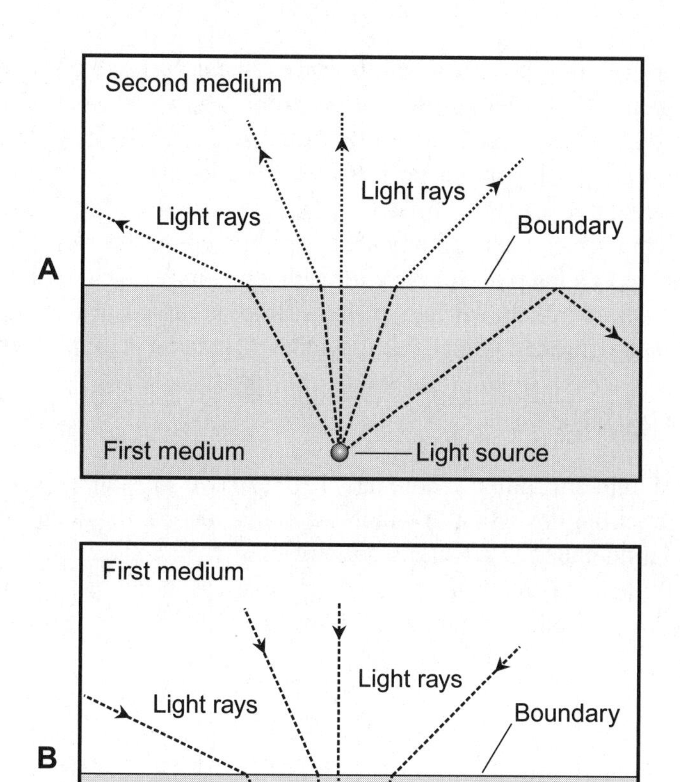

Figure 2-8 At A, light rays strike a boundary where the refractive index decreases. At B, light rays strike a boundary where the refractive index increases.

held constant. Because of this difference in density, warm air has a lower refractive index than cool air at the same altitude. The difference in the refractive index of warm air compared with cooler air can be sufficient to produce total internal reflection if there is a sharp boundary between two air masses whose temperatures are different. This is why, on warm days, you sometimes see "false ponds" over the surfaces of highways or over stretches of hot earth.

Now consider what happens when the directions of the light rays are reversed. This situation is shown in Fig. 2-8B. A ray originating in the first (upper) medium and striking the boundary at a grazing angle is bent downward. This phenomenon causes distortion of landscape images when viewed from underwater. If you have ever been scuba diving, you have witnessed it. The sky, trees, hills, buildings, people, and everything else above the horizon appear within a circle of light that distorts the scene. If you look upward toward the water surface at a near-grazing angle, you see an inverted reflection of the bottom of the body of water, or of other objects in the water with you, just as if the water surface were a mirror.

If the refracting boundary is not flat, the principles shown by Fig. 2-8 still apply for each ray of light striking the boundary at any specific point. The refraction is considered with respect to a flat plane passing through the point, tangent to the boundary at that point. When many parallel rays of light strike a curved or irregular refractive boundary at many different points, each ray obeys the same principle individually.

SNELL'S LAW

When a ray of light encounters a boundary between two substances having different indices of refraction, the extent to which the ray is bent can be determined according to an equation called *Snell's law*.

Figure 2-9 illustrates the principle of Snell's law for a ray of light passing from a medium with a certain refractive index to a medium having a relatively higher

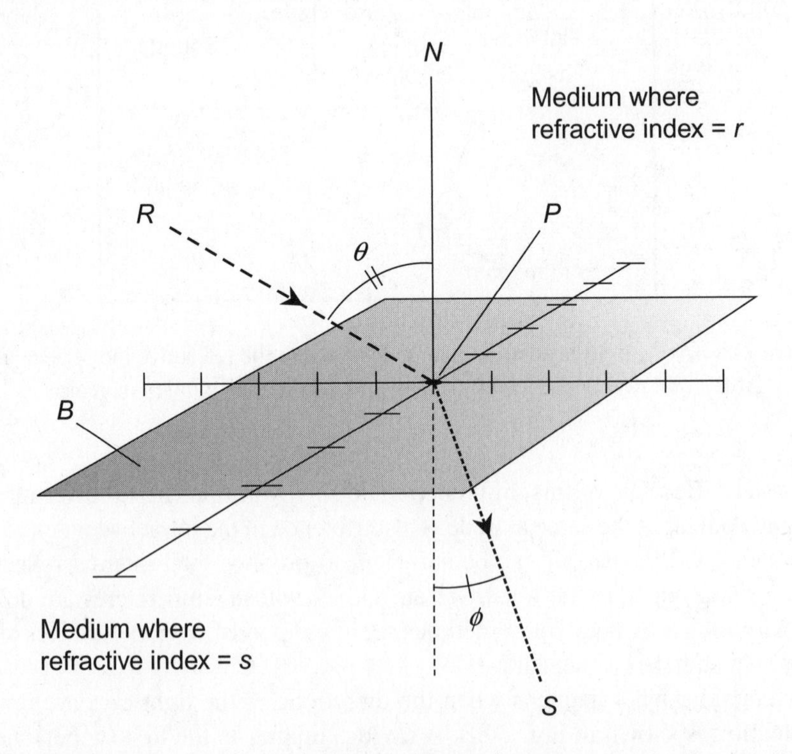

Figure 2-9 Snell's law governs the behavior of a ray of light as it strikes a boundary where the index of refraction increases.

refractive index. Suppose B is a flat boundary between two media M_r and M_s, whose indices of refraction are r and s, respectively. In this case, $r < s$. Imagine a ray of light crossing the boundary at point P, as shown.

Let N be a line passing through point P on plane B, such that N is normal to plane B at point P. Suppose R is a ray of light traveling through M_r that strikes plane B at point P. Let θ be the angle that R subtends relative to the normal line N. Let S be the ray of light that emerges from P into M_s. Let ϕ be the angle that S subtends relative to the normal line N. Then line N, ray R, and ray S all lie in the same plane, and $\phi \leq \theta$. The angles θ and ϕ are equal if and only if the angle of incidence is 0°. The following equations hold for angles θ and ϕ in this situation:

$$\sin \phi / \sin \theta = r/s$$

and

$$s \sin \phi = r \sin \theta$$

Figure 2-10 shows the situation for a ray of light passing from a medium with a certain refractive index to a medium having a relatively lower refractive index. Again, let B be a flat boundary between two media M_r and M_s, whose absolute

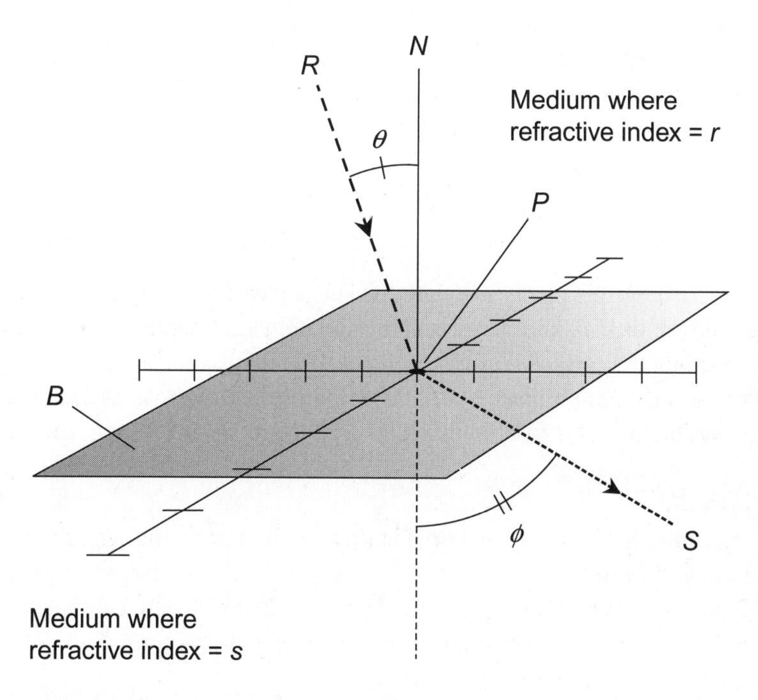

Figure 2-10 Snell's law for a light ray that strikes a boundary where the index of refraction decreases.

indices of refraction are r and s, respectively. In this case, $r > s$. Let N, B, P, R, S, θ, and ϕ be defined as in the previous example. Then line N, ray R, and ray S all lie in the same plane, and $\phi \geq \theta$. As in the previous example (Fig. 2-9), $\theta = \phi$ if and only if $\theta = 0°$. The equations relating the angles and the refractive indices are also the same as in the previous example:

$$\sin \phi / \sin \theta = r/s$$

and

$$s \sin \phi = r \sin \theta$$

DETERMINING THE CRITICAL ANGLE

In the situation shown by Fig. 2-10, the light ray passes from a medium having a relatively higher index of refraction, r, into a medium having a relatively lower index, s. As the angle of incidence θ increases, the angle of refraction ϕ approaches 90°, and ray S gets closer to the boundary plane B. When θ gets large enough, ϕ reaches 90°, and ray S lies exactly in plane B. If θ increases even more, ray R undergoes total internal reflection.

The critical angle is the largest angle of incidence that ray R can subtend, relative to the normal N, without being reflected internally. Let's call this angle θ_c. The measure of the critical angle is the arcsine of the ratio of the smaller index of refraction, s, to the larger index, r, as follows:

$$\theta_c = \arcsin (s/r)$$

PROBLEM 2-3

Suppose a laser device is placed beneath the surface of a liquid freshwater pond. The refractive index of fresh liquid water is approximately 1.33, while that of air is close to 1.00. Imagine that the surface is perfectly smooth. If the laser beam is directed upward so it strikes the surface at an angle of 30.0° relative to the normal, then at what angle, also relative to the normal, will the beam emerge from the surface into the air?

SOLUTION 2-3

Envision the situation in Fig. 2-10 upside down. Then M_r is the water and M_s is the air. The indices of refraction are $r = 1.33$ and $s = 1.00$. The measure of angle θ is given as 30.0°. The unknown is the measure of angle ϕ. We can use the equation for Snell's law, plug in the numbers, and solve for ϕ as follows:

$$\sin \phi / \sin \theta = r/s$$

$$\sin \phi / (\sin 30.0°) = 1.33/1.00$$

$$\sin \phi/0.500 = 1.33$$
$$\sin \phi = 1.33 \times 0.500 = 0.665$$
$$\phi = \arcsin 0.665 = 41.7°$$

PROBLEM 2-4

What is the critical angle for light rays shining from the bottom up toward the surface of a liquid freshwater pond?

SOLUTION 2-4

We can use the formula for critical angle, and envision the scenario of Problem 2-3, where the angle of incidence, θ, can be varied. We plug in the numbers to the equation for critical angle, θ_c, obtaining

$$\begin{aligned}\theta_c &= \arcsin (s/r)\\ &= \arcsin (1.00/1.33)\\ &= \arcsin 0.752\\ &= 48.8°\end{aligned}$$

PROBLEM 2-5

Suppose a laser device is placed above the surface of a smooth, liquid freshwater pool that is of uniform depth everywhere, and aimed downward so the light ray strikes the surface at an angle of 28° relative to the plane of the surface (not a normal to the surface). At what angle, relative to the plane of the pool bottom, will the light beam strike the bottom?

SOLUTION 2-5

This situation is illustrated in Fig. 2-11. The angle of incidence, θ, is equal to 90° minus 28°, the angle at which the laser enters the water relative to the surface. By simple subtraction, we get $\theta = 62°$. We know from Table 2-1 that r, the index of refraction of the air is 1.00, and also that s, the index of refraction of the water is 1.33. We can therefore solve for the angle ϕ, relative to the normal N to the surface, at which the ray travels under the water:

$$\sin \phi/\sin \theta = r/s$$
$$\sin \phi/\sin 62° = 1.00/1.33$$
$$\sin \phi/0.883 = 0.752$$
$$\sin \phi = 0.752 \times 0.883 = 0.664$$
$$\phi = \arcsin 0.664 = 42°$$

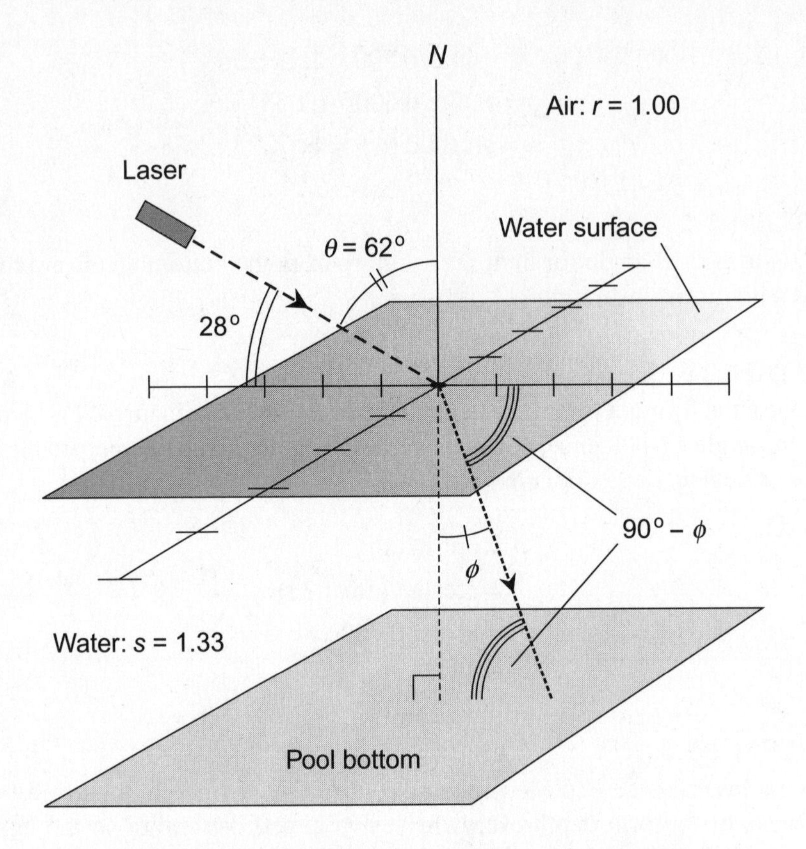

Figure 2-11 Illustration for Problem and Solution 2-5.

We're justified to go to only two significant figures here, because that is the extent of the accuracy of the input data for the angles. The angle at which the laser travels under the water, relative to the water surface, is 90° – 42°, or 48°. Because the pool has uniform depth, the pool bottom is flat and parallel to the water surface. We can conclude that the light beam strikes at an angle of 48° relative to the plane of the pool bottom.

Color Dispersion

The index of refraction for a substance usually depends on the wavelength of the light passing through it. Glass, and virtually any other substance having a refractive index greater than 1, slows down light the most at the shortest wavelengths (blue and violet), and the least at the longest wavelengths (red and orange). This variation of the refractive index with wavelength is known as *color dispersion*. It is the principle by which a prism works.

"RAINBOW" SPECTRA

The more a ray of light is slowed down by the medium from which a prism is made, the more its path is deflected when it passes through the prism. This phenomenon causes a prism to cast a "rainbow" spectrum (Fig. 2-12). The effect is also responsible for the multicolored glitter of clear substances with high indices of refraction such as quartz and diamond. In general, as the index of refraction of a medium increases, so does the extent of the color dispersion when white light passes through it.

Dispersion is important in optics for two reasons. First, a prism can be used to make a *spectrometer*, which is a device for examining the intensity of visible light at specific wavelengths. Second, dispersion degrades the quality of white-light images viewed through simple lenses. It is responsible for the multicolored borders that appear around the images of objects viewed through binoculars, telescopes, or microscopes with low-quality lenses.

PROBLEM 2-6

Suppose a ray of white light, shining horizontally, enters a prism made of a solid transparent material. Suppose the prism cross-section is an exact equilateral triangle with the base oriented horizontally, as shown in Fig. 2-13A. If the index of refraction of the prism is 1.52000 for red light and 1.52500 for yellow light, what is the angle δ between rays of red and yellow light as they emerge from the prism? Assume that the surrounding medium is a vacuum (free space), so its index of refraction is 1.00000 for light of all colors. Find the answer to the nearest hundredth of a degree.

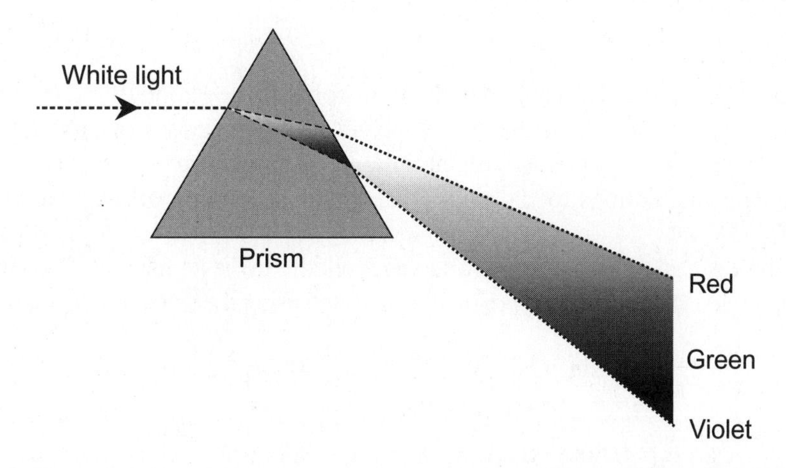

Figure 2-12 Light rays of different colors are refracted at different angles through a prism.

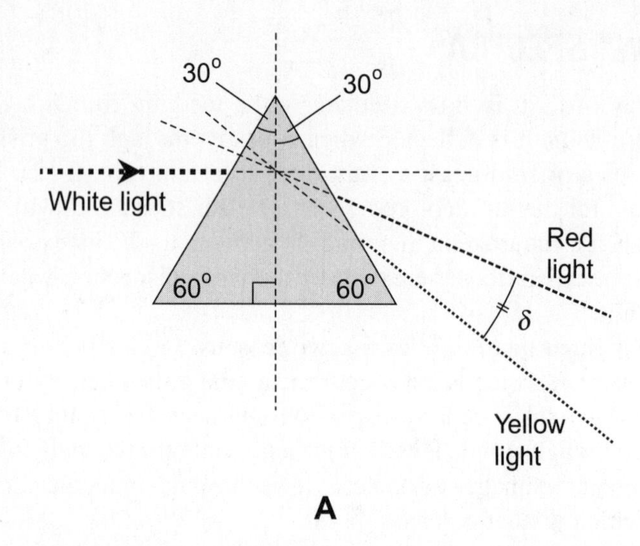

Figure 2-13A Illustration for Problem 2-6. Assume that the prism is made of a transparent material whose index of refraction is 1.52000 for red light and 1.52500 for yellow light.

SOLUTION 2-6

There are at least two ways to approach this problem. Both methods require several steps to complete. Let's do it this way:

- Follow the ray of red light all the way through the prism and determine the angle at which it exits the prism.
- Follow the ray of yellow light in the same way.
- Determine the difference in the two exit angles by subtracting one from the other.

Refer to Fig. 2-13B. The ray of white light comes in horizontally, so the angle of incidence is 30°. Let's consider this figure exact for purposes of calculation. To minimize the risk of problems with cumulative rounding errors, let's carry out our intermediate calculations to six significant figures, rounding off at the end of the process.

The angle ρ_1 that the ray of red light subtends relative to the normal line N_1, as it passes through the first surface into the prism, can be found using the refraction formula:

$$\sin \rho_1/\sin 30.0000° = 1.00000/1.52000$$
$$\sin \rho_1/0.500000 = 0.657895$$
$$\sin \rho_1 = 0.500000 \times 0.657895 = 0.328948$$
$$\rho_1 = \arcsin 0.328948 = 19.2049°$$

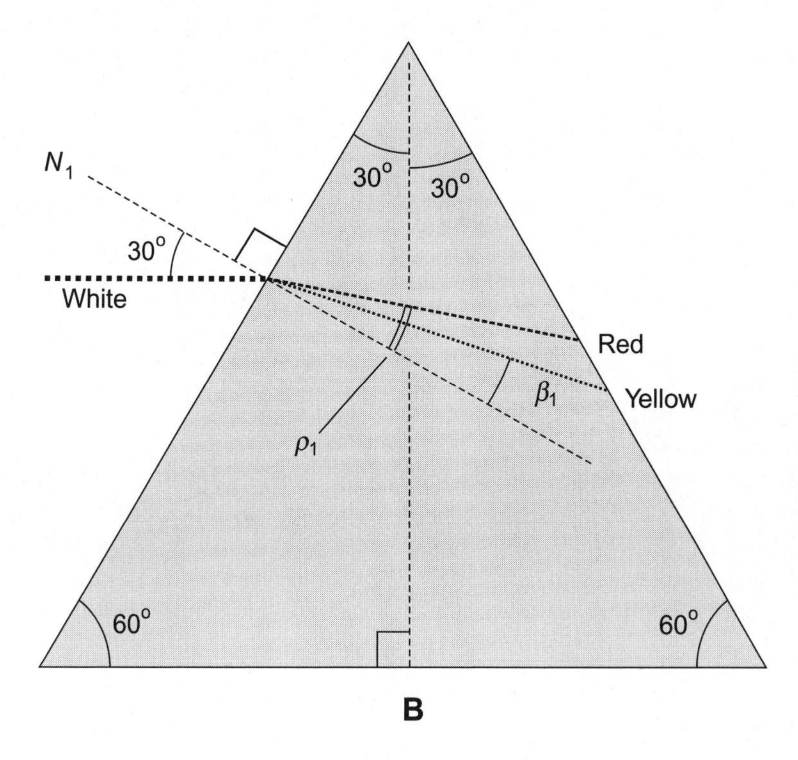

Figure 2-13B Illustration for first part of solution to Problem 2-6.

Because the normal line N_1 to the first surface slants at 30.0000° relative to the horizontal, the ray of red light inside the prism slants at 30.0000° − 19.2049°, or 10.7951°, relative to the horizontal. Now look at Fig. 2-13C and compare it with Fig. 2-13B. The line normal to the second surface for the red ray (call it N_{2r}) slants at 30.0000° with respect to the horizontal, but in the opposite direction from line N_1. Therefore, the angle of incidence ρ_2, at which the ray of red light strikes the inside second surface of the prism, is equal to 30.0000° + 10.7951°, or 40.7951°. Again we use the refraction formula, this time to find the angle ρ_3, relative to the normal N_{2r}, at which the ray of red light exits the second surface of the prism:

$$\sin \rho_3 / \sin 40.7951° = 1.52000/1.00000$$
$$\sin \rho_3 / 0.653356 = 1.52000$$
$$\sin \rho_3 = 0.653356 \times 1.52000 = 0.993101$$
$$\rho_3 = \arcsin 0.993101 = 83.2659°$$

Now we must repeat all of these calculations for the ray of yellow light. Refer again to Fig. 2-13B. The ray of white light comes in horizontally, so the angle of

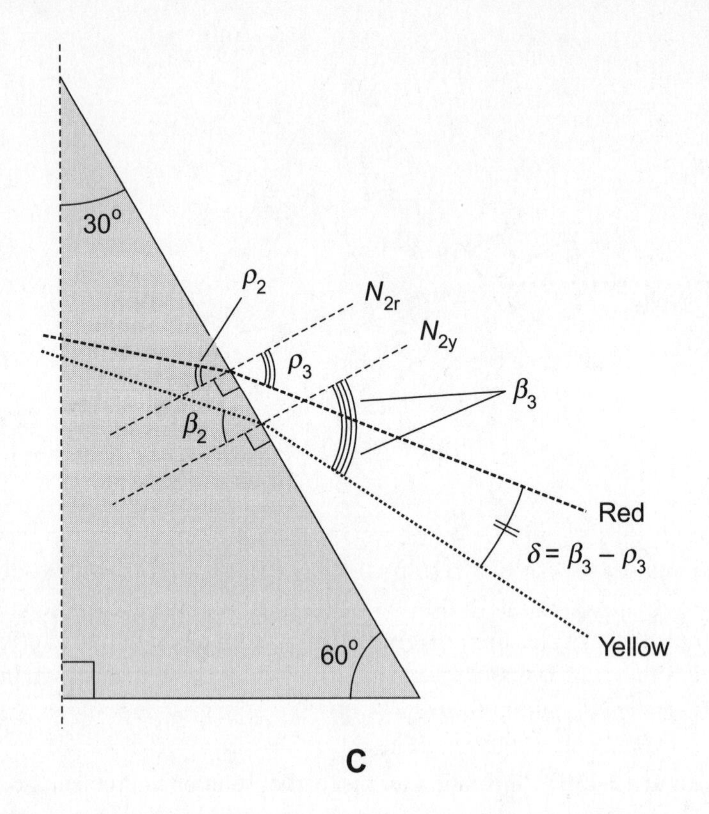

Figure 2-13C Illustration for second part of solution to Problem 2-6.

incidence is 30.0000°, as before. The angle β_1 that the ray of yellow light subtends relative to N_1, as it passes through the first surface into the prism, is

$$\sin \beta_1/\sin 30.0000° = 1.00000/1.52500$$
$$\sin \beta_1/0.500000 = 0.655738$$
$$\sin \beta_1 = 0.500000 \times 0.655738 = 0.327869$$
$$\beta_1 = \arcsin 0.327869 = 19.1395°$$

Because the normal line N_1 to the first surface slants at 30.0000° relative to the horizontal, the ray of yellow light inside the prism slants at 30° – 19.1395°, or 10.8605°, relative to the horizontal. Now, once again, compare Fig. 2-13C with Fig. 2-13B. The line normal to the second surface for the yellow ray (call it N_{2y}) slants at 30.0000° with respect to the horizontal, but in the opposite direction from line N_1, so the angle of incidence β_2, at which the ray of yellow light strikes the inside second surface of the prism, is equal to 30.0000° + 10.8605°, or 40.8605°. Again we use the refraction formula, this time to find the angle β_3, relative to the normal N_{2y}, at which the ray of yellow light exits the second surface of the prism:

$$\sin \beta_3 / \sin 40.8605° = 1.52500/1.00000$$
$$\sin \beta_3 / 0.654220 = 1.52500$$
$$\sin \beta_3 = 0.654220 \times 1.52500 = 0.997686$$
$$\beta_3 = \arcsin 0.997686 = 86.1015°$$

The difference $\beta_3 - \rho_3$ is the angle δ we seek, the angle between the rays of yellow and red light as they emerge from the prism. This, rounded off to the nearest hundredth of a degree, is

$$\beta_3 - \rho_3 = 86.1015° - 83.2659°$$
$$= 2.84°$$

PROBLEM 2-7

Suppose you want to project a "rainbow" spectrum onto a screen, so that the spectrum image measures exactly 10 cm from the red band to the yellow band using the prism as configured in Problem 2-6. At what distance *d* from the screen should the prism be placed? Consider the position of the prism to be the intersection point of extensions of the red and yellow rays emerging from the prism, as shown in Fig. 2-14. Measure the distance *d* along the yellow ray to the nearest centimeter. Assume that the screen is oriented at a right angle to the yellow ray.

SOLUTION 2-7

We know that $\delta = 2.84°$ (accurate to three significant figures) from the solution to Problem 2-6. We are told that the length of the spectrum image from the red band

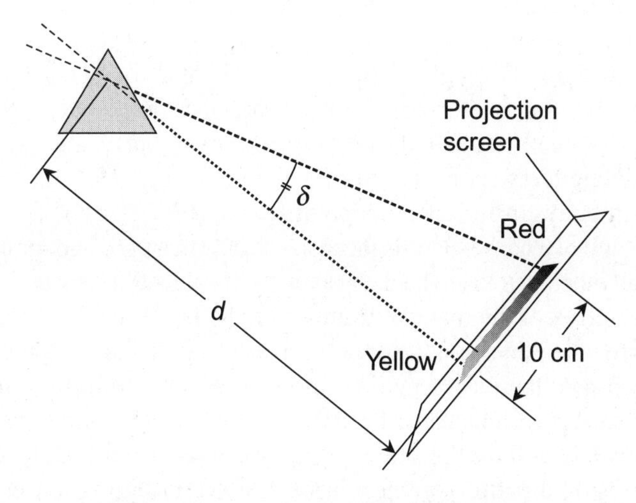

Figure 2-14 Illustration for Problem 2-7.

to the yellow band, as it appears on the screen, is exactly 10 cm, so we can carry that figure to as many significant figures as we want by adding a decimal point followed by zeros. Let's go to some extra significant figures during the calculation process, rounding off at the end. We solve for *d* using right-triangle trigonometry, just as we might do if we were conducting a land survey:

$$\tan 2.84° = 10.0000/d$$
$$0.049608 = 10.0000/d$$
$$d = 10.0000/0.049608$$
$$d = 202$$

The screen should be placed 202 cm from the prism as measured along the yellow ray of light.

Lenses

The refraction of visible light in solid transparent media can be used to advantage. This fact was first discovered when experimenters found that specially shaped pieces of glass can make objects look larger or smaller. Lenses work because they refract rays of light to a greater or lesser extent depending on where, and at what angle, the rays enter or emerge from their surfaces.

THE CONVEX LENS

You can buy a *convex lens* in a novelty store or department store. You should easily be able to find a "magnifying glass" up to 10 or 15 cm in diameter. The term *convex* arises from the fact that one or both faces of the glass bulge outward at the center. A convex lens is sometimes called a *converging lens* because, after passing through the lens, parallel light rays converge to a focal point (Fig. 2-15A). A convex lens can also collimate the rays of light from a point source (Fig. 2-15B).

The focal length of a convex lens is the distance between the center of the lens and the focal point. If all other factors are held constant, the focal length is inversely proportional to the refractive index of the material from which the lens is made. If the difference in thickness between the center and the edges of a convex lens increases but the lens diameter remains constant, the focal length becomes shorter. The effective area of the lens, measured in a plane perpendicular to the axis, is called the *light-gathering area*.

Some convex lenses have the same degree of curvature on each face; others have different degrees of curvature on their faces. Some converging lenses have one flat face; they are called *plano-convex lenses*.

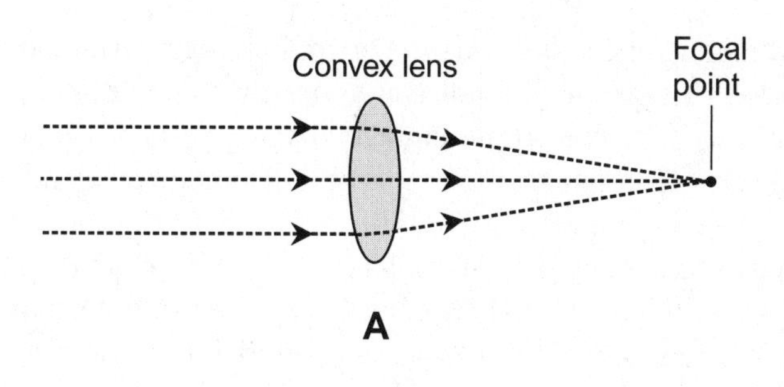

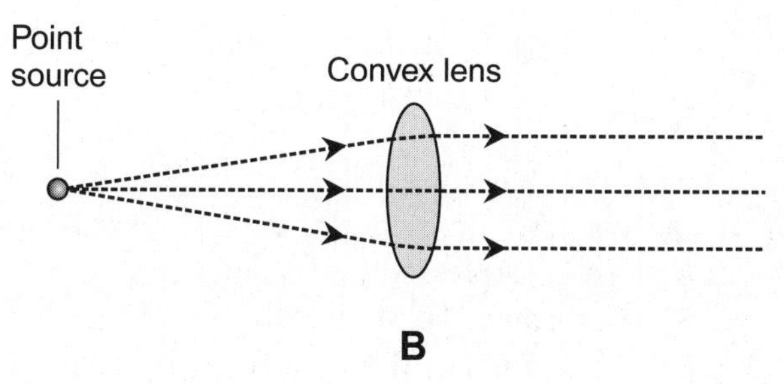

Figure 2-15 At A, a convex lens focuses parallel light rays to a point. At B, the same lens collimates light from a point source at the focus.

REAL IMAGE IN CONVEX LENS

Suppose you stand at some distance from a convex lens, as shown in Fig. 2-16. Rays of light reflect from your body, pass through the lens, and converge to form an inverted real image on the opposite side of the lens. The distance of the real

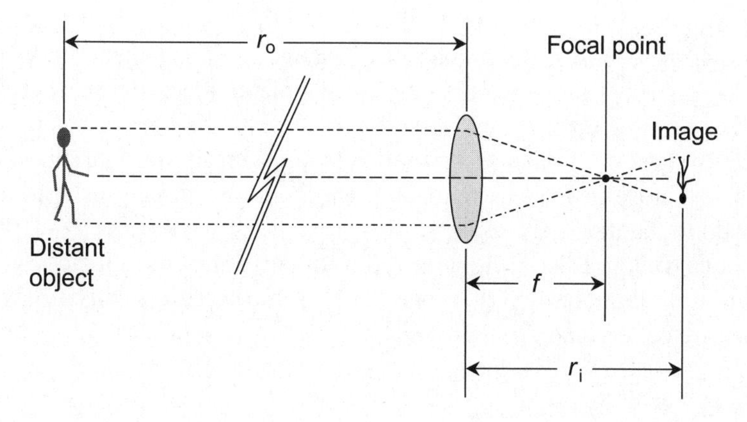

Figure 2-16 A convex lens produces a real image by focusing rays of light as they pass through. The distance r_i of the image from the lens depends on the distance r_o of the object from the lens, relative to the focal length f.

image from the lens depends on how far from the lens you stand, and also on the refractive index of the lens. If you stand at a distance of many times the focal length ("at infinity"), then your real image appears at the focal point. If you are at a lesser distance, then your real image is at a distance somewhat greater than the focal point.

Suppose you stand along the lens axis at a distance r_o from the center of the lens, where r_o is greater than the focal length f. Under these circumstances, the distance r_i of your body's real image from the lens is related to r_o and f in the same way as for a concave mirror:

$$1/f = 1/r_i + 1/r_o$$

This formula is not exact, because not every point on an observed object lies at precisely the same distance r_o from the center of the lens. For the formula to work, the distances f, r_i, and r_o must be expressed in the same units.

If an object is placed at a distance less than f from a convex lens (closer to the lens than the focal length), a real image does not form, because the refracted rays don't come to a focus.

THE CONCAVE LENS

You might have trouble finding a *concave lens* in a department store, but you should be able to order one from a specialty catalog or Web site. The term *concave* refers to the fact that one or both faces of the glass curve inward at the center. This type of lens is also called a *diverging lens*. It spreads parallel light rays outward (Fig. 2-17A). It can collimate incoming rays that converge to the correct extent (Fig. 2-17B).

As with convex lenses, the properties of a concave lens depend on its diameter and on the surfaces' radii of curvature. The greater the difference in thickness between the edges and the center of the lens, the more the lens causes parallel rays of light to diverge, if all other factors are held constant. If you look through a concave lens at a close-in object such as a coin, the object's features appear smaller than they do to the unaided eye.

Some concave lenses have the same radius of curvature on each face; others have different radii of curvature on their two faces. Some diverging lenses have one flat face; these are called *plano-concave lenses*.

PROBLEM 2-8

Suppose a simple convex lens is made from glass with a higher index of refraction for blue light than for red light. How does the focal length for red light (f_{red}) compare with the focal length for blue light (f_{blue})?

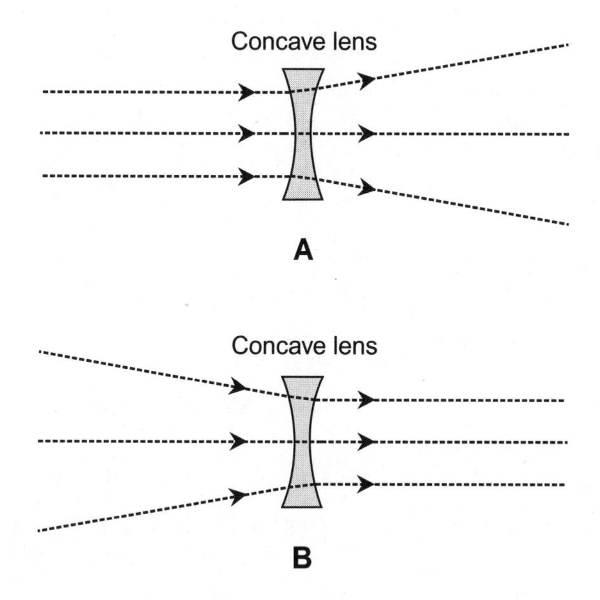

Figure 2-17 At A, a concave lens spreads parallel light rays. At B, the same lens collimates converging light rays.

SOLUTION 2-8

The glass bends blue light more, so f_{blue} is less than f_{red}. Incoming parallel light rays are focused to a point closer to the lens for blue light than for red light, as shown in Fig. 2-18. This phenomenon, called *chromatic aberration*, always occurs with simple lenses. The result is blurring of images in telescopes and cameras, along with "rainbow" halos around distant, white-light point sources such as stars. The effect can be minimized by special lens construction, in which multiple layers of glass having different indices of refraction are laminated to form a *compound lens* or *color-corrected lens.*

PROBLEM 2-9

Suppose you place a point source of light, all of whose energy is concentrated at a single wavelength of 665.78 nm, behind and along the axis of a convex lens whose focal length is 455.33 mm. You want the rays of light to come to a focus at a distance of exactly 20.000 m from the lens. How far from the plane containing the lens should the point source be placed?

SOLUTION 2-9

The precise wavelength in this situation indicates that chromatic aberration is not a factor. (The value of this figure is irrelevant otherwise.) Before doing any calculations, we must convert all distances to the same units. Let's use meters. The distance of the point light source (which can be considered the object) from the plane

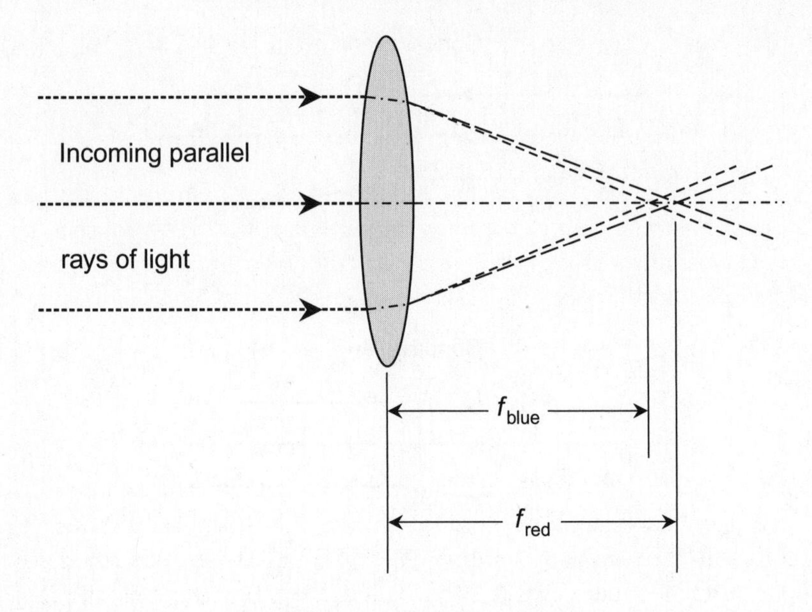

Figure 2-18 Illustration for Problem and Solution 2-8. The focal length for red light is f_{red}, and the focal length for blue light is f_{blue}.

containing the lens is r_o. The focal length is given as $f = 455.33$ mm, which is 0.45533 m. The distance of the image from the lens is given as $r_i = 20.000$ m. The formula relating focal length, image distance, and object distance tells us that

$$1/f = 1/r_i + 1/r_o$$

Plugging in the known values, we get

$$\begin{aligned} 1/0.45533 &= 1/20.000 + 1/r_o \\ 2.1962 &= 0.050000 + 1/r_o \\ 2.1462 &= 1/r_o \end{aligned}$$

From this, we can determine that

$$\begin{aligned} r_o &= 1/2.1462 \\ &= 0.46594 \text{ m} \\ &= 465.94 \text{ mm} \end{aligned}$$

The point light source should be placed 465.94 mm behind the lens to bring the rays to a focus at a distance of 20.000 m. If the point source lies on the axis of the lens, this value can be considered accurate to the number of significant figures given, because the effective height and width of the source are insignificant compared to its distance from the lens.

Quiz

This is an open book quiz. You may refer to the text in this chapter. A good score is 8 correct. Answers are at the end of the book.

1. What is the critical angle for a ray of visible light that starts out in diamond having a refractive index of 2.42 and encounters a boundary with fresh water having a refractive index of 1.33?
   - (a) There is no critical angle in this situation.
   - (b) 57.7°.
   - (c) 33.3°.
   - (d) We need more information to answer this question.
2. Look again at the situation posed in Problem 2-6. Suppose a different kind of glass is used in the prism. Imagine that for all colors, this new glass has a refractive index that is 90 percent of the refractive index of the glass described in Problem 2-6. Under these circumstances, the angle δ between the red and yellow light rays, as they emerge from the prism, is
   - (a) less than 2.84°.
   - (b) equal to 2.84°.
   - (c) more than 2.84°.
   - (d) nonexistent, because the rays are reflected internally at the second surface.
3. Suppose a convex lens has a focal length of exactly 1/3 m. When a particle of sand is placed at a certain distance from the lens along the axis of the lens, the real image of the particle appears at that same distance from the lens, but on the opposite side. What is that distance?
   - (a) 1/3 m.
   - (b) 1/2 m.
   - (c) 2/3 m.
   - (d) We need more information to answer this question.
4. Suppose, in the situation of Question 3, the particle of sand is moved so that it is only 1/6 m from the lens. What happens to the real image of the particle?
   - (a) It gets twice as far away from the lens as it was before.
   - (b) It gets 4 times as far away from the lens as it was before.
   - (c) It gets 8 times as far away from the lens.
   - (d) It vanishes.

5. Suppose that you stand in front of a concave mirror whose focal length is 10 cm. Your position is such that the mirror axis passes through your body. You start out at a distance of several meters from the mirror, and gradually back straight away from it. The real image of your body, produced by the mirror, moves as you move. How?
   (a) It starts out less than 10 cm from the mirror and gradually increases, approaching but never exceeding 10 cm from the mirror.
   (b) It starts out farther than 10 cm from the mirror and gradually decreases, approaching but never becoming less than 10 cm from the mirror.
   (c) It starts out at 10 cm from the mirror and gradually increases, approaching but never exceeding 20 cm from the mirror.
   (d) It starts out at 10 cm from the mirror and gradually decreases, approaching but never becoming less than 5 cm from the mirror.
6. Suppose that you stand in front of a convex lens whose focal length is 1.00 m. Your position is such that the lens axis passes through your body. You start out at a distance of 4.00 m from the lens, and gradually move toward it. The real image of your body, produced by the lens, moves as you move. How?
   (a) It starts out at 1.33 m from the lens and gradually increases.
   (b) It starts out at 1.33 m from the lens and gradually decreases.
   (c) It starts out at 750 mm from the lens and gradually increases.
   (d) It starts out at 750 mm from the lens and gradually decreases.
7. What is the critical angle for a ray of visible light that starts out in crown glass having a refractive index of 1.52 and encounters a boundary with flint glass having a refractive index of 1.61?
   (a) There is no critical angle in this situation.
   (b) 19.2°.
   (c) 70.8°.
   (d) We need more information to answer this question.
8. A virtual image
   (a) is inverted with respect to the object creating the image.
   (b) cannot be rendered directly on photographic film.
   (c) is produced by the diverging rays from a convex mirror.
   (d) cannot be directly observed.

9. Look once again at the situation posed in Problem 2-6. Suppose the prism is made of a transparent substance with a refractive index that, for any given color of light, is twice the refractive index of the glass described in Problem 2-6. Under these circumstances, the angle δ between the red and yellow light rays, as they emerge from the prism, is
   - (a) less than 2.84°.
   - (b) equal to 2.84°.
   - (c) more than 2.84°.
   - (d) nonexistent, because the rays are reflected internally at the second surface.
10. Total internal reflection can occur
    - (a) when a ray of light passes from a medium having a certain refractive index into another medium having a *lower* refractive index, but only if the angle of incidence *exceeds* the critical angle with respect to a line normal to the boundary.
    - (b) when a ray of light passes from a medium having a certain refractive index into another medium having a *higher* refractive index, but only if the angle of incidence *exceeds* the critical angle with respect to a line normal to the boundary.
    - (c) when a ray of light passes from a medium having a certain refractive index into another medium having a *lower* refractive index, but only if the angle of incidence *is smaller than* the critical angle with respect to a line normal to the boundary.
    - (d) when a ray of light passes from a medium having a certain refractive index into another medium having a *higher* refractive index, but only if the angle of incidence *is smaller than* the critical angle with respect to a line normal to the boundary.

CHAPTER 3

Common Optical Devices

The optical instruments that lay people usually encounter are virtual-image-producing devices such as video displays, as well as real-image-producing cameras, projectors, microscopes, and telescopes.

Color Mixing

The wavelength of visible radiant energy has a dramatic effect on the way you perceive it. When most of the energy is concentrated near a single wavelength, you see an intense *hue*. The vividness or purity of a hue is called the *saturation*. The hue and saturation can be considered together as a single variable called *color*. The *brightness* of colored light is a function of how much total energy the light contains at a given wavelength or within a certain range of wavelengths. In video displays, brightness is sometimes called *brilliance* or *intensity*.

THE PRIMARY HUES

Color depends on the dominant wavelength, or the distribution of wavelengths, contained in a beam of radiant light. This light may have passed through filters or translucent materials on its way to your eyes from the source, or it may have been reflected from one or more intermediate surfaces.

When light of three particular hues is combined with equal intensity, the result is *white light* or *gray light*, which has the same brightness at all wavelengths within the visible range. These three hues are called the *primary hues* (or, somewhat less precisely, the *primary colors*). Individually, they appear as *red*, *green*, and *blue*. By mixing radiant light of these three hues in different combinations, all possible colors and brightness levels can be obtained. The maximum brightness is limited only by the total radiant energy in all sources. The deliberate combining of primary hues in this way is known as *additive color mixing*. Red and blue at equal brightness levels combine to produce *magenta* (pink-lavender), red and green in equal brightness levels combine to produce *yellow*, and green and blue in equal brightness levels combine to produce *cyan* (blue-green), as shown in the *Venn diagram* of Fig. 3-1.

In a video display, primary-color combinations result in the various shades and hues that you see. In a *cathode-ray-tube* (CRT) color display (or "color picture tube") three interwoven sets of tiny phosphor dots produce the images. One-third of the dots glow red, one-third glow green, and one-third glow blue. When viewed from a distance, the individual dots merge, and the overall effects of color and brightness appear. More advanced displays such as the liquid-crystal or plasma types work on the same principle, but they employ more efficient components.

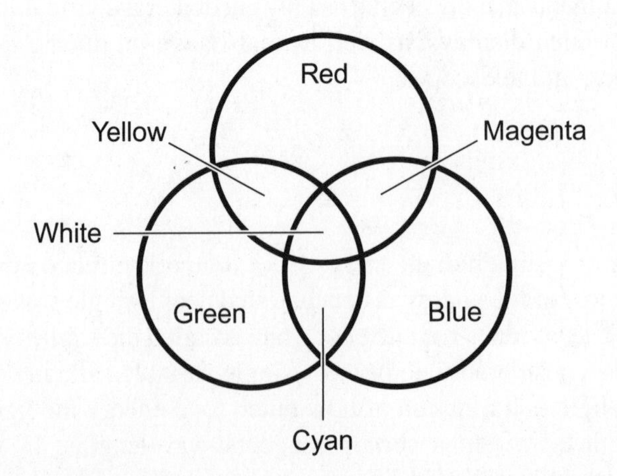

Figure 3-1 Additive mixing of the primary hues red, green, and blue.

THE RGB MODEL

The *red/green/blue* (RGB) *model* takes advantage of the fact that beams or sources of radiant light having the primary hues, emitted in various brightness proportions, are sufficient to produce all colors that the human eye and brain can perceive.

A *digital color palette* is obtained by combining pure red, green, and blue light. We can assign each primary hue an axis in three-dimensional (3D) space as shown in Fig. 3-2, calling the axes R (for red), G (for green), and B (for blue). The brightness of each hue can range from 0 to 255, or binary 00000000 to 11111111. The result is 16,777,216 (256^3) possible colors. Any point within the cube represents a unique color.

Colors in the RGB model can be represented by six-digit *hexadecimal* numbers, such as 005CFF. In this scheme, the first two digits represent the red (R) intensity in 256 levels ranging from 00 to FF. The middle two digits represent the green (G) intensity, and the last two digits represent the blue (B) intensity. Some RGB systems use only 16 levels for each primary hue (binary 0000 through 1111, or hexadecimal 0 through F). This produces 4096 (16^3) possible colors.

THE PRIMARY PIGMENTS

Pigment, in contrast to color, refers to the property of a substance that causes it to absorb light at particular wavelengths. The absorbed energy cannot be reflected or

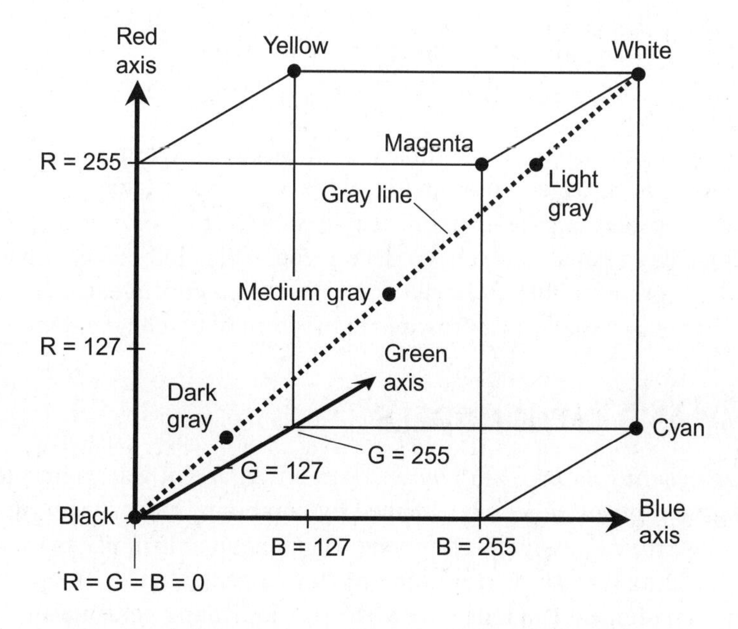

Figure 3-2 Colors in the red/green/blue (RGB) model can be portrayed as points within a cube.

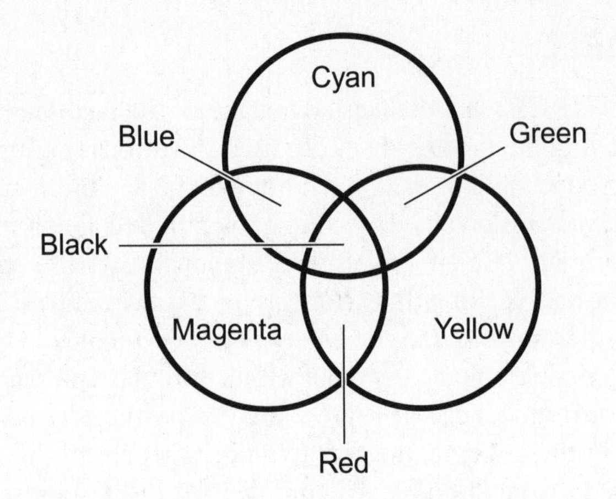

Figure 3-3 Subtractive mixing of the primary pigments cyan, magenta, and yellow.

transmitted; it is usually converted to heat in the medium. The perceived color of light reflected from a pigmented surface differs from the color of the light that strikes it, unless that incident light happens to be the same as the color of the pigmented surface as it appears in ordinary daylight.

When substances such as ink or paint having cyan, magenta, and yellow pigments are combined in equal amounts, the result is gray or black. The same effect occurs when white light is passed through three transparent filters in sequence having these same pigments. For this reason, cyan, magenta, and yellow are called the *primary pigments*.

By mixing media having the primary pigments in different combinations, all possible values of reflected color and brightness can be obtained. The deliberate combining of primary pigments in this way is *subtractive color mixing*. Cyan and yellow pigments in equal amounts produce green. Cyan and magenta pigments in equal amounts produce blue. Magenta and yellow pigments in equal amounts produce red. Figure 3-3 shows this principle in the form of a Venn diagram.

THE CMY AND CMYK MODELS

The *cyan/magenta/yellow* (CMY) *model* takes advantage of way pigmented media mix. A *digital pigment palette* is obtained by combining the primary pigments in various ratios. We can assign each primary pigment an axis in 3D space as shown in Fig. 3-4, calling the axes C (for cyan), M (for magenta), and Y (for yellow). The density of each pigment can range from 0 to 100, indicating percentages. Including the 0 values, this gives us 1,030,301 (101^3) possible pigments. Any point within the cube represents a unique pigment.

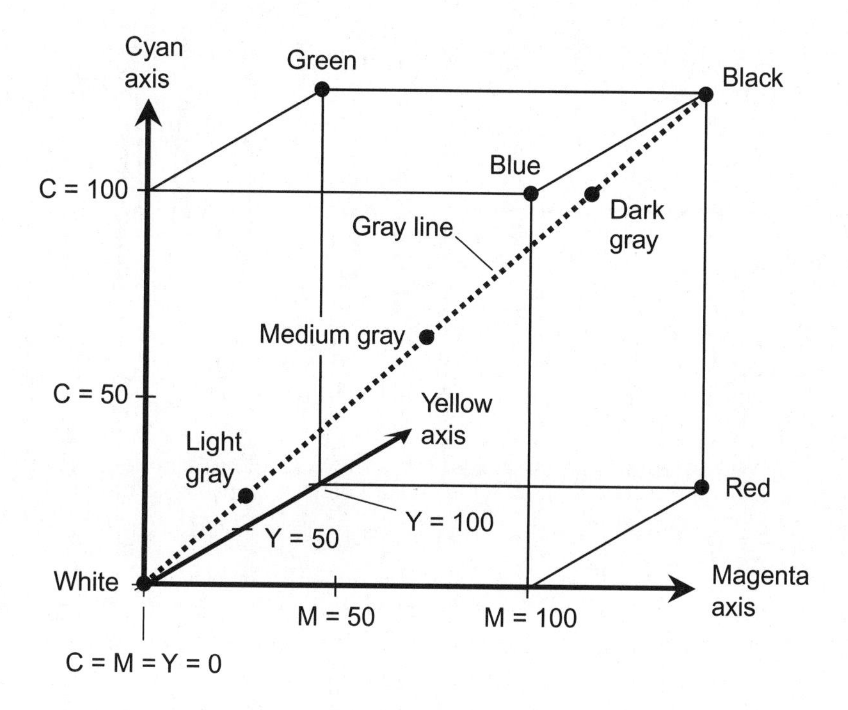

Figure 3-4 Pigments in the cyan/magenta/yellow (CMY) model can be portrayed as points within a cube.

Some pigment schemes include black (K), because it is expensive and wasteful to combine large amounts of cyan, magenta, and yellow ink or paint to get gray or black. This is especially true if printed documents are intended to be published in *grayscale* (such as those in this book). Therefore, you will often read or hear about the *CMYK model.*

PROBLEM 3-1

If white or gray light is defined as radiation of equal intensity throughout the entire range of visible wavelengths (approximately 750 nm to 400 nm), how can light at only three specific wavelengths (red, green, and blue) combine to produce true white or gray?

SOLUTION 3-1

This effect is the result of the fact that perceived illumination is relative, not absolute. Figure 3-5A is a graph of the *spectral distribution* of gray light, which appears white in a dark environment or black in a bright environment. The wavelength is shown on the horizontal axis, and the relative intensity is shown on the vertical axis. Figure 3-5B is a graph of the spectral distribution of red, green, and blue light of high saturation and equal intensity. Clearly, the spectral distributions at A and B differ. But as we see them, the distributions shown at A and B produce the same gray.

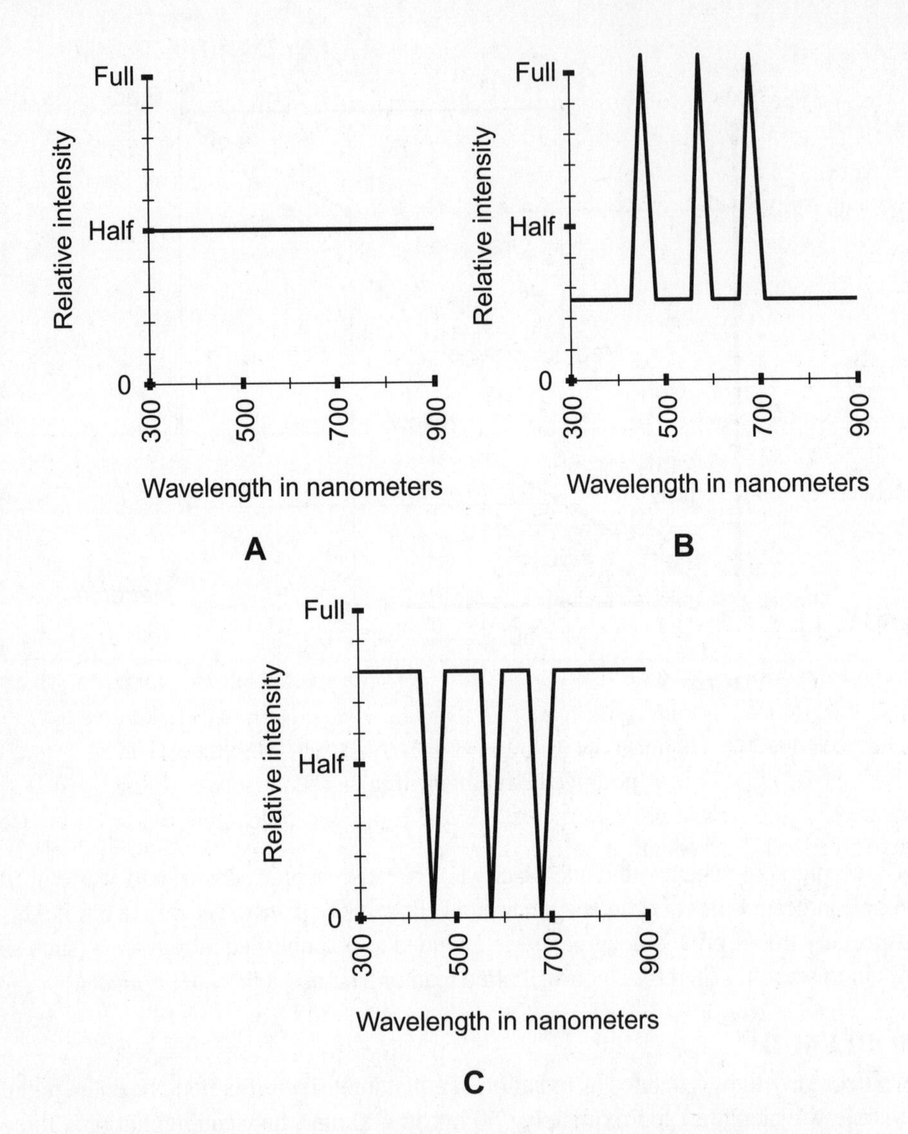

Figure 3-5 At A, spectral distribution of gray light. At B, spectral distribution of the combination of high-saturation red, green, and blue lights. At C, spectral distribution of light that is blocked at the red, green, and blue wavelengths.

PROBLEM 3-2

If black is defined as the complete absence of radiation at any wavelength in the visible range, how can color filters that block light at only three specific wavelengths (red is blocked to obtain cyan, green is blocked to obtain magenta, and blue is blocked to obtain yellow) combine to produce gray?

SOLUTION 3-2

Figure 3-5C is a graph of the spectral distribution of cyan, magenta, and yellow that results from the selective blocking of red, green, and blue respectively. Again, this spectral distribution differs from that shown at A. But the distribution of Fig. 3-5C produces light that appears gray to the human eye.

Cameras

A *camera* converts a visible scene into a form suitable for reproduction on a tangible, two-dimensional medium such as paper or a display screen. Some cameras are designed to capture only single fixed scenes, while others can capture a moving scene and convert it into a sequence of fixed images.

PINHOLE CAMERA

The simplest optical camera is called a *pinhole camera*, because it can be made by literally punching a hole in an opaque box using a pin. Light rays from an external object or scene pass through the hole to form a real image on the inside rear surface of the box, as shown in Fig. 3-6. Refraction does not occur; every light ray travels in a straight line from a point in the external scene to the corresponding point on the image. A piece of photographic film is placed against the back of the box.

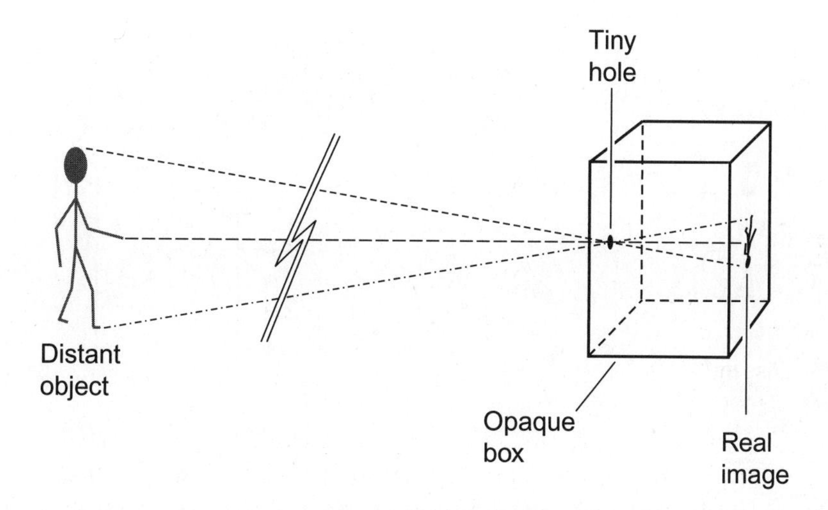

Figure 3-6 Principle of the pinhole camera.

The pinhole camera produces reasonably sharp images only when the hole is tiny, compared with the size of the real image. The *image resolution* (amount of detail portrayed) also improves as the depth of the box increases, if all other factors are held constant. However, as the diameter of the hole decreases or the depth of the box increases, the time required to adequately expose a photographic medium becomes longer, because the real image becomes dimmer per unit area. If motion takes places during the exposure period, a blurred or unrecognizable image will result.

BASIC CAMERA

In the traditional fixed-image camera, a convex lens, rather than a simple hole, is used to produce real images of external scenes and objects on a film or screen. Light rays from the outside are refracted as they pass through the lens, as shown in Fig. 3-7. The camera depth *d* is mechanically adjusted by a *focus control* to provide the sharpest possible real image. This adjustment can be accomplished by moving the lens or by moving the *backplane* where the photosensitive material is attached.

For extremely distant objects, *d* is essentially equal to focal length of the lens. For objects at lesser distances, *d* increases. When the camera is used to photograph close-up scenes, *d* varies considerably as the distance of the external object changes. This effect produces a limited *depth of field*, or minimum-to-maximum range of object distances that generates acceptable images, for any fixed setting of *d*. The image resolution depends on the diameter and quality of the lens, as well as on the accuracy with which *d* is adjusted.

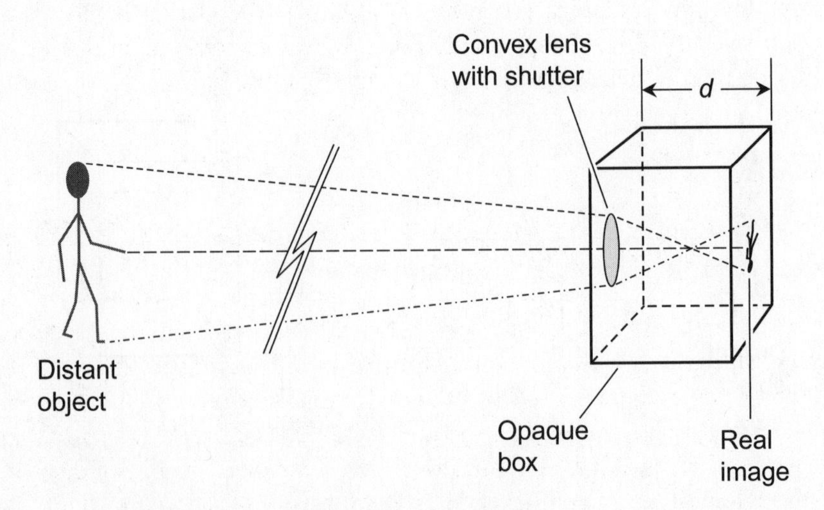

Figure 3-7 Principle of the conventional fixed-image camera. The depth of the box, *d*, is adjusted to produce the sharpest possible real image.

The lens is equipped with a timed *shutter* that ensures proper film exposure. The shutter can also partially obstruct the lens by creating an adjustable, circular opening centered at the center of the lens. The ideal exposure time depends on the brightness of the outside scene, the diameter of the lens, the diameter of the opening allowed by the obstruction, and the sensitivity of the film.

CHARGE-COUPLED DEVICE (CCD)

A *charge-coupled device* (CCD) is a *semiconductor integrated circuit* (also called a *chip*) that converts visible-light images into digital signals. Some CCDs also work at IR or UV wavelengths. Astronomers use CCDs to record and enhance images of celestial objects. Most digital cameras, both fixed-image and full-motion, employ CCDs.

The image focused on the retina of your eye, or on the film of a camera, is an *analog image*. It can have infinitely many configurations, and infinitely many variations in hue, brightness, contrast, and saturation. An analog quantity varies smoothly and continuously. But a digital computer needs a *digital image* to make sense of, and enhance, what it "sees." *Binary* digital signals have only two possible states: on and off. These are also called *high* and *low*, or 1 and 0. An excellent approximation of an analog real image can be assembled from binary digital states or signals. This allows a computer program to process the image, bringing out details and features that would otherwise be impossible to detect.

Figure 3-8 shows a simplified block diagram of a CCD. The image falls on a matrix containing millions of tiny semiconductor-based photoelectric sensors that resemble microscopic "solar cells" or "electric eyes." Each sensor produces one *pixel* (picture element). The computer (not shown) can employ all the tricks characteristic of any good graphics program. In addition to rendering high-contrast or false-color images, the CCD and computer together can detect and resolve images much fainter than is possible with conventional camera film.

VIDICON

The *vidicon* is a full-motion video camera that was once popular in *videocassette recorders* (VCRs). A lens focuses the incoming image onto a screen whose electrical conductivity varies depending on the intensity of light that strikes it. A device called an *electron gun* generates a stream of electrons that sweeps from left to right across the screen, steered by *deflection coils*. Multiple sweeps occur in parallel horizontal lines from the top of the screen to the bottom, forming a *raster scan* of the screen. As the beam scans the photoconductive surface, the screen becomes electrically charged, and then it discharges at different rates in different regions. The rate of discharge in a certain region on the screen depends on the intensity of the visible light striking that region. Figure 3-9 is simplified cutaway view of a vidicon. The vidicon is rarely used today; it has been supplanted by CCD-based video cameras.

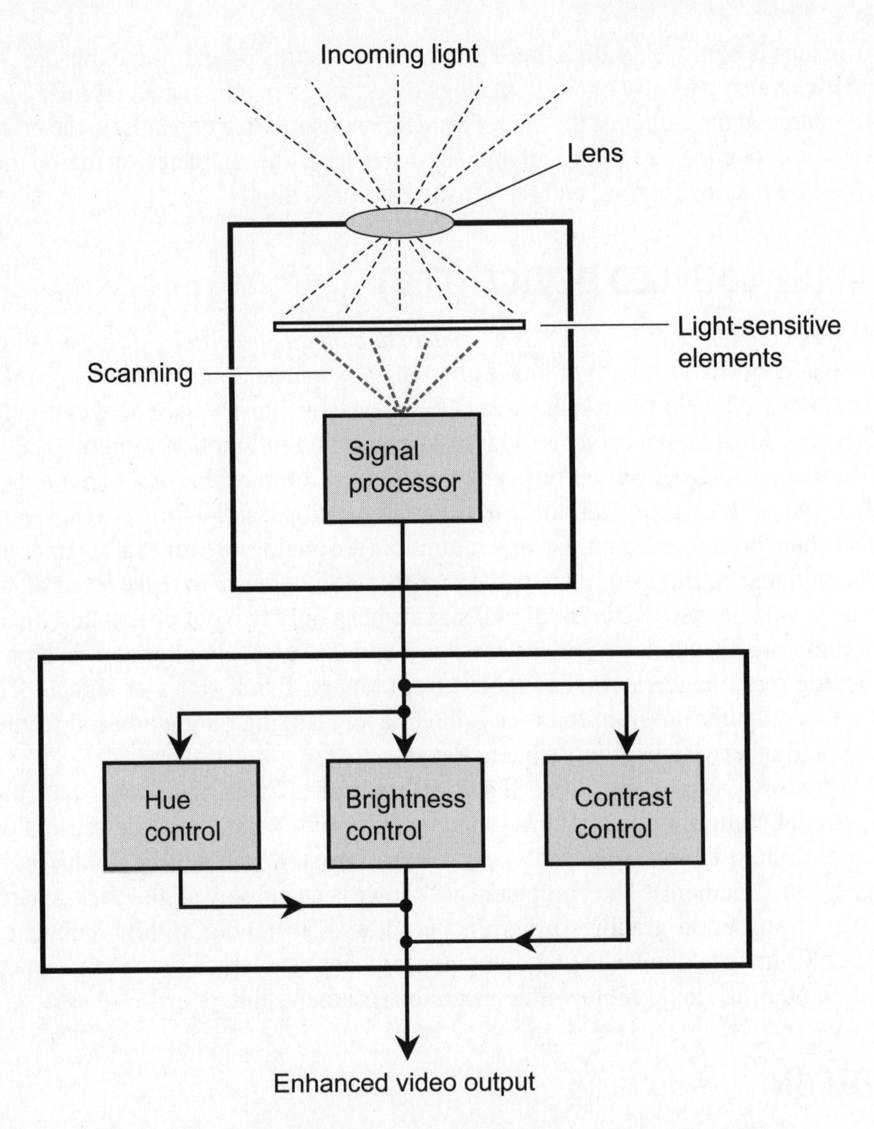

Figure 3-8 Functional diagram of a charge-coupled device (CCD) and basic image-processing system.

IMAGE ORTHICON

The *image orthicon* is another full-motion video camera that was once widely used in commercial television broadcasting. It is constructed much like the vidicon, except that there is a *target electrode* behind the photosensitive electrode, which is also called the *photocathode* (Fig. 3-10). When a single electron from the photocathode

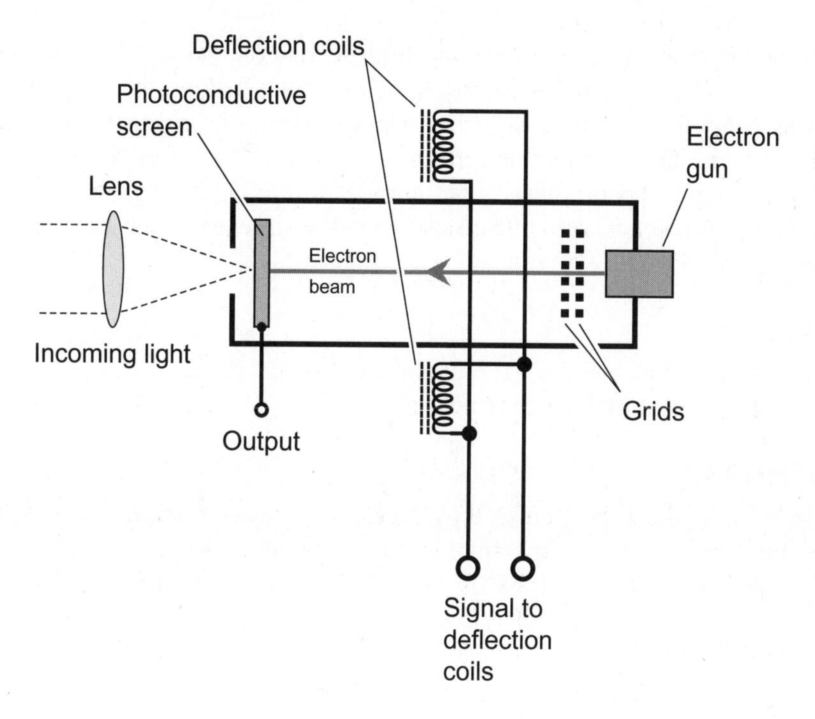

Figure 3-9 Functional diagram of a vidicon, showing one pair of deflection coils. This device is of historical interest.

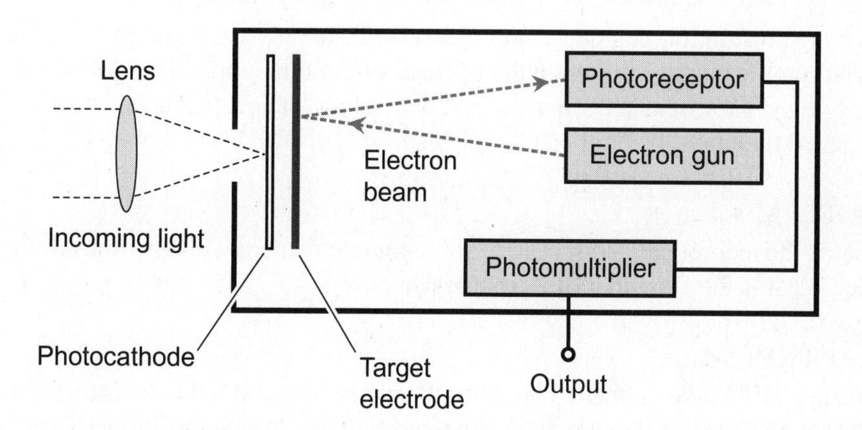

Figure 3-10 Functional diagram of an image orthicon. The deflection coils are not shown. This device is of historical interest.

strikes the target electrode, multiple electrons, called *secondary electrons*, are dislodged. This phenomenon amplifies the video signal, making the image orthicon useful in dimly lit environments. A beam emitted from the electron gun scans the target electrode. The secondary electrons cause some of this beam to be reflected. The greatest return beam intensity corresponds to the brightest parts of the video image. The return beam is modulated as it scans the target electrode and is picked up by a *receptor electrode*.

PROBLEM 3-3

How can the depth of field for a traditional fixed-image camera be increased? What must be sacrificed to obtain this increase?

SOLUTION 3-3

The depth of field can be increased by decreasing the *aperture diameter*, which is the diameter of the circular opening that transmits light through the lens. In effect, this reduces the diameter of the lens without changing its focal length, so the ratio of the focal length to the effective lens diameter increases. This ratio is called the *f-number*, the *f-ratio*, or the *f-stop*, and is symbolized as N in equations. The *f*-number can also be denoted by writing *f*, followed by a forward slash, followed by a number representing the aperture diameter.

If D_a is the aperture diameter and f is the focal length of the camera lens, both measured in the same units, then

$$N = f / D_a$$

As the *f*-number increases, the depth of field also increases. The required film-exposure time gets longer, however, because less light is allowed into the camera. In addition, the image resolution degrades. The reduction in image resolution might or might not be significant, depending on the fineness of the film emulsion or the number of pixels in the photosensitive element, and depending on how much the image will be expanded to obtain the final print or display.

PROBLEM 3-4

What is the theoretical depth of field for a pinhole camera with an arbitrarily small hole? What is the *f*-number of a pinhole camera?

SOLUTION 3-4

In theory, if the hole diameter could be reduced to zero and light could pass through it, the depth of field would be from zero to infinity. In a "real-world" pinhole camera, the effective depth of field is from a few meters to arbitrarily far away. The hole, unlike a lens, has no focal length, so the notion of the *f*-number is meaningless.

Projectors and Displays

A *projector* employs a convex lens to produce a large real image of a small object. A *display* produces a virtual image whose size can vary from a diagonal measure of a few centimeters to several meters.

BASIC PROJECTOR

Functionally, a projector works according to the same principles as a camera. In a basic fixed-image projector, a convex lens produces real images on a screen by focusing the light from a small internal backplane. Light rays from the backplane are refracted as they pass through the lens and land on the screen, as shown in Fig. 3-11. The projector depth d is mechanically adjusted by a focus control to optimize the sharpness of the real image on the screen. The most common way to adjust d is to vary the position of the lens with a mechanical device such as a rack-and-pinion drive.

For extremely distant screens, d is essentially equal to the focal length f of the lens, or slightly greater. For screens close to the projector, d is significantly greater than f. The size of the projected image depends on the ratio of distance D_p of the projection screen from the lens to the distance d of the backplane from the lens when the image is properly focused. In general, if D_o is the distance between any

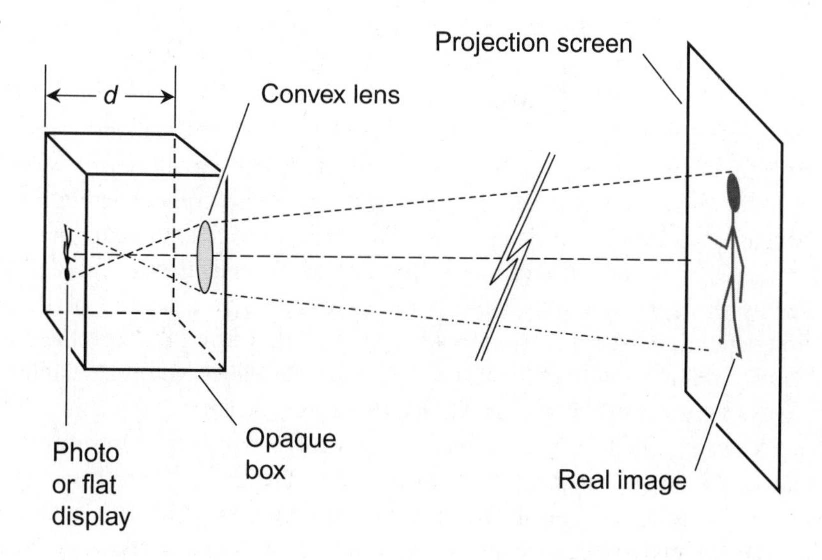

Figure 3-11 Principle of the conventional fixed-image projector. The depth of the box, d, is adjusted to produce the sharpest real image on the screen.

two points on the object on the backplane and D_s is the distance between the corresponding two points on the projection screen, then

$$D_s/D_o = D_p/d$$

If A_o is the total area of the object on the backplane and A_s is the total area of the image on the projection screen, then

$$A_s/A_o = (D_p/d)^2$$

As with a camera, the image resolution obtainable with a projector depends on the diameter and quality of the lens, as well as on the accuracy with which d is adjusted.

CATHODE-RAY TUBE

A cathode-ray tube (CRT) is a display technology that was once universal in television (TV) receivers, oscilloscopes, and computer displays. While the original CRT is not as common today as it once was, derivative technologies have been developed and are currently employed in advanced displays.

In a conventional CRT, an electron gun emits a high-intensity beam of electrons. This beam is focused and accelerated as it passes through electrodes called *anodes* that carry a positive charge. The electrons continue until they strike a screen whose inner surface is coated with *phosphor* that glows visibly when viewed by an external observer. Beam deflection makes video displays possible. This deflection can be accomplished either with electrostatic fields or with magnetic fields.

Figure 3-12 is a simplified cross-sectional rendition of a *monochrome*, or single-color, *electromagnetic CRT*. The device contains two sets of *deflecting coils*, one for horizontal beam motion and the other for vertical beam motion. (For simplicity of illustration, only one set of coils is shown here.) These coils generate magnetic fields because they carry electric currents. The strength of the magnetic field generated by these coils, and therefore the extent of the electron-beam deflection, increases as the current through the coils increases. The direction of the beam deflection depends on the direction in which the current flows through the coils.

The *horizontal deflection coils* receive a certain current waveform, which causes the beam to sweep across the screen. After each sweep, the beam jumps instantly back for the next sweep. The *vertical deflection coils* receive a different current waveform, which causes the beam to move down the screen. All this time, the electron beam varies in intensity in accordance with an analog video signal, so the moving spot changes in brilliance. The end result is a full-motion video image on the screen.

In a color picture tube or computer monitor, there are three electron beams, one each for red, green, and blue colors. Each beam works independently of the other

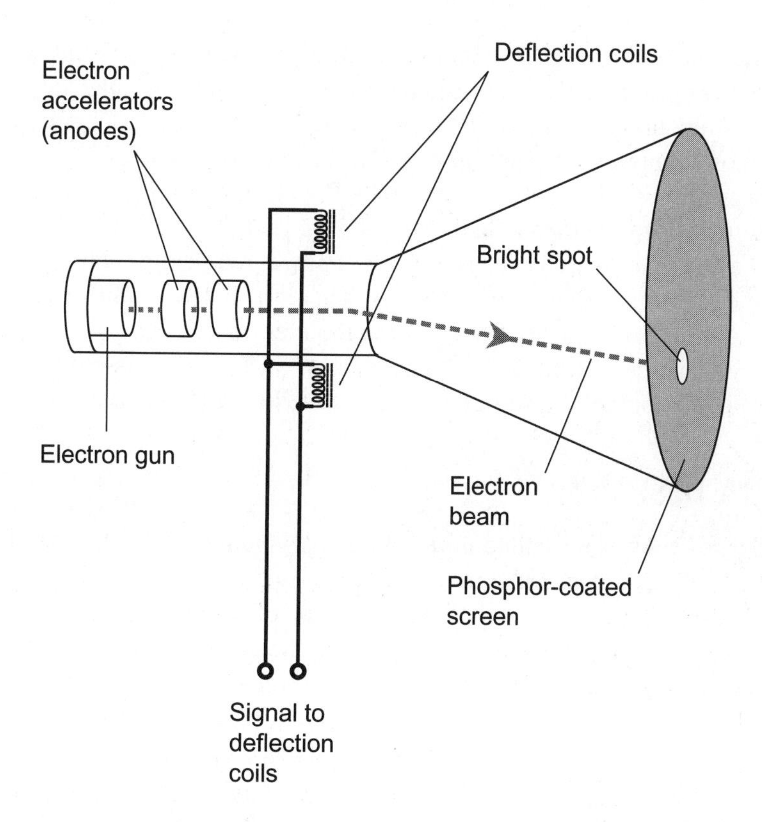

Figure 3-12 Simplified cross-sectional rendition of an electromagnetic CRT, showing one pair of deflection coils. This device is mainly of historical interest.

two. Three images are therefore superimposed on the screen: a red (R) image, a green (G) image, and a blue (B) image. As viewed externally from a distance, these images combine according to the RGB model to produce the full range of colors.

LIQUID-CRYSTAL DISPLAY

A liquid-crystal display (LCD) is a solid-state, flat-screen device that can show geometric shapes and colors. The simplest LCDs are used as alphanumeric displays in calculators, meters, wristwatches, and radios. More sophisticated LCDs are used in computer displays and TV receivers.

The LCD contains a fluid whose light-transmitting and light-reflecting properties depend on the intensity and orientation of an electric field. The fluid is confined between transparent, electrically conductive plates. When a voltage is applied to the plates, the resulting electric field causes a change in the molecules of the liquid, altering the way light passes through the display. The transmitted light can vary in hue, saturation, and brightness. There is a limit to how fast the liquid can change state. In recent years, LCDs have been developed that are fast enough, at room temperature, to function as displays in conventional TV sets. Such LCDs are also suitable for most computer graphics applications. The speed of the LCD is affected by the temperature. Extreme cold causes the LCD to change state slowly.

One of the most significant advantages of the LCD, compared with a CRT display, is the fact that the LCD needs much less power to operate. This makes it ideal for all kinds of portable electronic devices, where the batteries must last as long as possible. Another advantage of the LCD is the fact that it is easy to read in bright sunlight, yet can be backlit for use in darkness. Still another asset of the LCD is image detail comparable or superior to that of the best CRT displays.

PLASMA DISPLAY

In a *plasma display*, also called a *plasma display panel* (PDP), thousands of tiny fluorescent cells are placed in a matrix between two glass panels. The cells can be independently charged and discharged by means of electrodes. The electrodes on the front (viewing) side of the panel are transparent, allowing light to be transmitted to the viewer. Each cell contains the inert gases *neon* and *xenon*, which produce UV when a voltage is applied between the electrodes. The UV radiation strikes phosphors that produce colored light, in the same way as old-fashioned "neon signs" do. Because the PDP is, in effect, a large compound fluorescent light bulb, the power consumption is relatively high for images reproducing bright scenes.

Each pixel in the display consists of three *subpixels*, one consisting of a phosphor that glows red, one consisting of a phosphor that glows green, and one consisting of a phosphor that glows blue. The intensity of each subpixel can be varied independently to produce all of the colors the human eye and brain can perceive according to the RGB model. In a well-engineered and properly operated PDP, the color reproduction is remarkably accurate, and the *contrast ratio* (the intensity of the brightest part of the image divided by the intensity of the darkest part of the image) can be upwards of 10,000 to 1. The entire panel is only a few centimeters thick.

OTHER DISPLAY TECHNOLOGIES

Several other display technologies have been developed or are in the research and development phase. The *surface-conduction electron-emitter display* (SED) is similar to the PDP, except that high-speed electrons, not UV, cause the phosphor to glow. The SED offers contrast ratios as high as 100,000 to 1. Various types of *laser-projection display* have also been developed or proposed. Researchers claim that these displays can reproduce a wider spectrum of colors and have longer useful lives than other types of displays.

PROBLEM 3-5

Consider a projector designed for the display of transparencies 35 mm high. A transparency is placed 20 cm from the lens, and the projected image is 1.4 m tall. Approximately how far from the projector lens is the display screen?

SOLUTION 3-5

We can use the projection formula mentioned previously, based on the fact that the screen is many times more distant from the lens than the transparency. If D_o is the distance between any two points on the transparency (in this case 35 mm or 0.035 m), D_s is the distance between the corresponding two points on the projection screen (1.4 m), and d is the distance of the lens from the transparency (20 cm or 0.20 m), then

$$D_s/D_o = D_p/d$$

where D_p is the distance of the projector lens from the display screen. We can multiply through by d and switch the left and right sides of the above equation to get

$$D_p = dD_s/D_o$$

Converting all distance units to meters and plugging in the numbers gives us

$$\begin{aligned} D_p &= 0.20 \times 1.4/0.035 \\ &= 8.0 \text{ m} \end{aligned}$$

Optical Microscopes

Optical microscopes are designed to greatly magnify the images of objects that are too small to resolve with the unaided eye. The simplest microscope consists of a convex lens that can provide *magnification* (also called *power* and symbolized by the "times sign," ×) of up to about 20×. In laboratory work, the *compound microscope* is preferred because it allows for much greater magnification than a single lens.

FUNDAMENTAL PRINCIPLE

A basic compound microscope employs two lenses. The *objective* has a short focal length, in some cases 1 mm or less, and is placed near the specimen or sample to be observed. A real image of the specimen forms at some distance above the objective, where the light rays come to a focus. The distance (let's call it s) between the objective and this image is always greater than the focal length f of the objective.

The *eyepiece* has a longer focal length than the objective. It magnifies the real image produced by the objective. Illumination is provided by shining visible light up through a translucent specimen or down onto an opaque specimen. Sometimes UV can be shone through or on specimens to cause certain regions to fluoresce. Figure 3-13 is a simplified diagram of a compound microscope, showing how the light rays are focused and how the specimen can be illuminated.

Laboratory-grade compound microscopes have two or more objectives. One of these is selected by manually rotating a wheel to which all the objectives are attached. This feature provides several different levels of magnification with a single eyepiece.

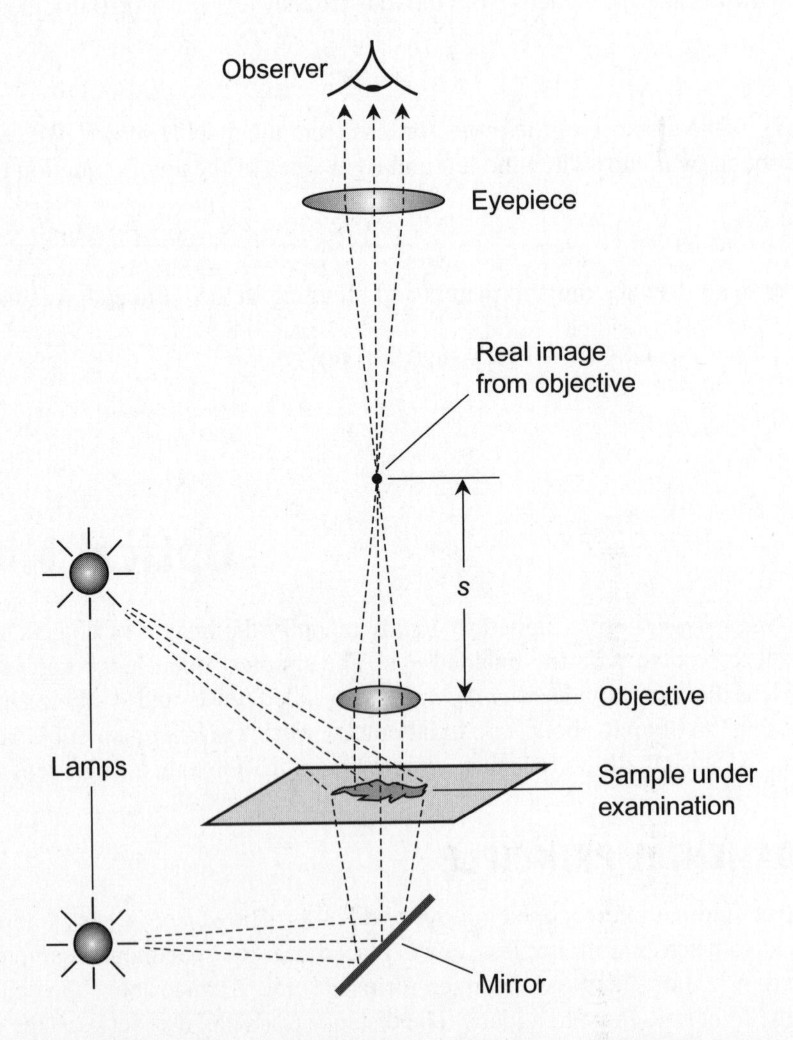

Figure 3-13 Illumination and focusing in a compound optical microscope.

In general, as the focal length of the objective becomes shorter, the magnification of the microscope increases. Some compound microscopes can magnify images up to about 2500×. A hobby-grade compound microscope can provide decent image quality at magnifications up to roughly 1000×.

FOCUSING

A compound microscope is focused by moving the entire assembly, including both the eyepiece and the objective, up and down with respect to the observed sample. This movement must be done with a precision mechanism, because the depth of field (the difference between the shortest and the greatest distance from the objective at which an object remains in good focus) is exceedingly small. As the focal length of the objective decreases, the depth of field also decreases, and the focusing therefore becomes increasingly critical. A high-magnification objective lens may have a depth of field on the order of 2 μm (2×10^{-6} m) or less.

If the eyepiece is moved up and down in the microscope tube assembly while the objective remains in a fixed position, the magnification varies. However, microscopes are usually designed to provide the best image quality for a specific eyepiece-to-objective separation, such as 16 cm (approximately 6.3 in).

If a bright enough lamp is used to illuminate the specimen under examination, and especially if the specimen is transparent or translucent so it can be lit from underneath, the eyepiece can be removed from the microscope and the image projected onto a screen several meters away. A diagonal mirror can reflect this image to a screen mounted on a wall. This technique works best when the focal length of the objective is relatively long.

MICROSCOPE MAGNIFICATION

Suppose f_o is the focal length of the objective lens in a compound microscope, and f_e is the focal length of the eyepiece in the same units. Assume that the objective and the eyepiece are placed along a common axis, and that the distance between their centers is adjusted for proper focus (Fig. 3-14). Let the distance from the objective to the real image of the object be represented by the variable s. The magnification m, a dimensionless quantity, is given by

$$m = [(s - f_o)/f_o]\ [(f_e + 0.25)/f_e]$$

The quantity 0.25 represents the *near point* of the human eye, which is the closest distance over which the eye can focus on an object. For people with average vision, the near point is approximately 0.25 m (10 in).

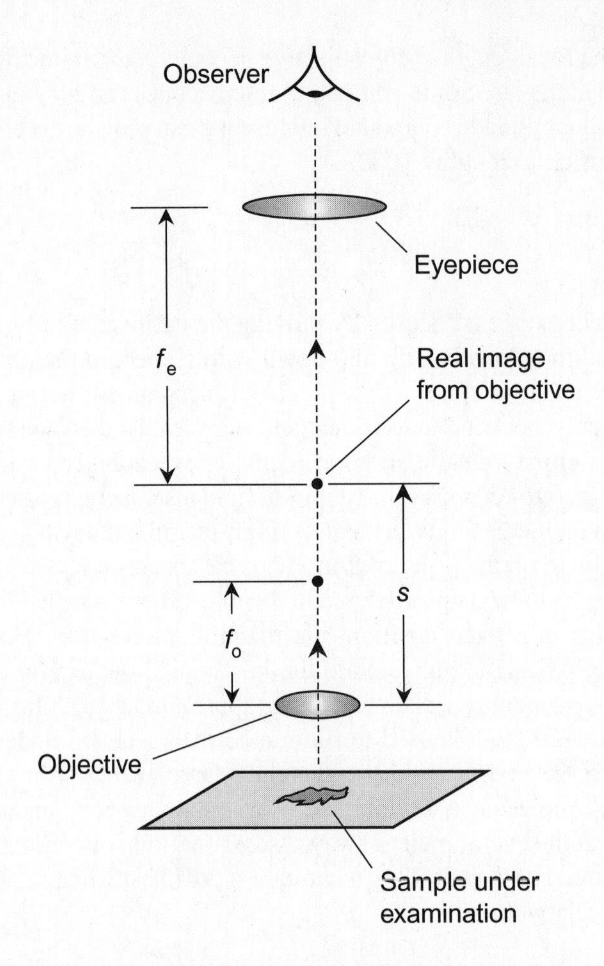

Figure 3-14 Calculation of the magnification factor in a compound microscope.

A less formal method of calculating the magnification of a microscope is to multiply the magnification of the objective by the magnification of the eyepiece. These numbers are provided with objectives and eyepieces, and are based on the use of an air medium between the objective and the specimen, and also on the standard distance between the objective and the eyepiece. If m_e is the magnification of the eyepiece and m_o is the magnification of the objective, then the magnification m of the microscope as a whole is

$$m = m_e m_o$$

PROBLEM 3-6

A compound microscope objective is specified as 10×, while the eyepiece is rated at 5×. What is the magnification m of this instrument?

SOLUTION 3-6

Multiply the magnification factor of the objective by that of the eyepiece:

$$m = 5 \times 10$$
$$= 50\times$$

NUMERICAL APERTURE AND RESOLUTION

In an optical microscope, the *numerical aperture* of the objective is an important specification in determining the resolution, or the amount of detail the microscope can render. In Fig. 3-15, let L be a line passing through a point P in the specimen to be examined, and also through the center of the objective. Let K be a line passing through P and intersecting the circular outer edge of the objective lens opening. Let q be the measure of the angle between lines L and K. Let M be the medium between the objective and the sample under examination. (This medium is not always air.) Let r_m be the refractive index of M. Based on these variables, the numerical aperture of the objective, A_o, is given by

$$A_o = r_m \sin q$$

In general, as A_o increases, the resolution improves. There are three ways to maximize the numerical aperture of a microscope objective having a given focal length:

- Maximize the diameter of the objective
- Maximize the value of r_m
- Minimize the wavelength of the illuminating light

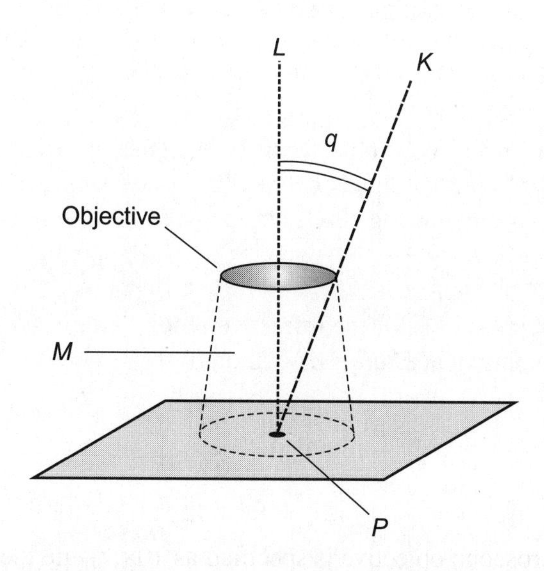

Figure 3-15 Determination of numerical aperture for a microscope objective.

Large-diameter objectives having short focal length, thereby providing high magnification, are difficult to construct. Therefore, when scientists want to examine an object in high detail, they can use blue light, which has a relatively short wavelength. Alternatively, or in addition, the medium *M* between the objective and the specimen can be changed to something transparent with a high index of refraction, such as water, clear oil, or glycerin. This shortens the wavelength of the illuminating beam that strikes the objective, because it slows down the speed of light. (Remember from basic physics the relation between the speed of a wave, the wavelength, and the frequency.) A side-effect of this tactic is a reduction in the effective magnification of the objective, but this can be compensated for by increasing the distance between the objective and the eyepiece.

CHROMATIC ABERRATION

The glass in a simple glass lens refracts the shortest wavelengths of light slightly more than the longest wavelengths. The focal length is therefore shorter for violet light than for blue light, shorter for blue than for yellow, and shorter for yellow than for red. This effect, called *chromatic aberration*, produces rainbow-colored halos or fringes in highly magnified images illuminated by white light.

Chromatic aberration can be eliminated by the use of light having a single, well-defined wavelength. If a full-color image is desired, chromatic aberration can be minimized by the use of a *compound lens* for the objective. A compound lens has two or more sections made of different types of glass, glued together with transparent adhesive. The curvature radii of the lenses are selected so that the lenses individually produce chromatic aberration equal in extent but opposite in sense, so the undesirable chromatic effects essentially "cancel out" over the range of visible wavelengths.

PROBLEM 3-7

Suppose that when a microscope objective is brought near a specimen, the angle q (as illustrated in Fig. 3-15) is 20.0 angular degrees (20.0°). What is the numerical aperture of this objective if air, and nothing else, exists between the specimen and the lens?

SOLUTION 3-7

We plug the numbers $r_m = 1.00$ (the refractive index of air) and $q = 20.0°$ into the formula for the numerical aperture A_o, obtaining

$$
\begin{aligned}
A_o &= r_m \sin q \\
&= 1.00 \sin 20.0° \\
&= 1.00 \times 0.342 \\
&= 0.342
\end{aligned}
$$

PROBLEM 3-8

Suppose that when a microscope objective is brought near a specimen, the angle q (from Fig. 3-15) is 20.0°. What is the numerical aperture of this objective if glycerin with a refractive index of 1.47 is used between the specimen and the lens?

SOLUTION 3-8

We plug the numbers $r_m = 1.47$ and $q = 20.0°$ into the formula for the numerical aperture A_o, getting

$$
\begin{aligned}
A_o &= r_m \sin q \\
&= 1.47 \sin 20.0° \\
&= 1.47 \times 0.342 \\
&= 0.503
\end{aligned}
$$

Optical Telescopes

The first *optical telescopes* were developed in the 1600s. They, like microscopes, used combinations of lenses. Any telescope that enlarges distant images with lenses alone is called a *refracting telescope*. Later, mirrors were employed to build *reflecting telescopes*.

GALILEAN REFRACTOR

The Italian scientist *Galileo Galilei*, who gained notoriety in the early 1600s by daring to suggest that the earth and other planets orbit the sun, devised a telescope consisting of a convex-lens *objective* and a concave-lens *eyepiece*. His first telescope magnified the apparent diameters of distant objects by a factor of a few times. Some of his later telescopes magnified up to 30 times. The so-called *Galilean refractor* (Fig. 3-16A) produces an *erect image*, that is, a right-side up view of external scenes. In addition to appearing right-side-up, images are also true in the left-to-right sense. The magnification factor, defined as the number of times the angular diameters of distant objects are increased, depends on the focal length of the objective, and also on the distance between the objective and the eyepiece.

Galilean refractors are available today, mainly as novelties for terrestrial viewing. Galileo's original refractors had objective lenses only 2 or 3 cm (about 1 in) across; the same is true of most Galilean telescopes these days. Some such telescopes have sliding, concentric tubes, providing variable magnification. When the inner tube is pushed all the way into the outer one, the magnification factor is the lowest; when the inner tube is pulled all the way out, the magnification is highest. The image remains in good focus over the entire magnification-adjustment range. As the distance between the objective and the eyepiece increases, the brightness of the observed image decreases. These instruments are sometimes called "spy glasses."

KEPLERIAN REFRACTOR

A more sophisticated refracting telescope employs a convex-lens objective with a long focal length, and a smaller, convex-lens eyepiece with a short focal length. Unlike the Galilean telescope, the so-called *Keplerian refractor* (Fig. 3-16B) produces an *inverted image*; it is upside-down and backward. (The device is named after *Johannes Kepler*, the astronomer who supposedly invented it several centuries ago.) The distance between the objective and the eyepiece must be exactly equal to the sum of the focal lengths of the two lenses in order for the image to be clear, assuming the distant object is infinitely far away.

The magnification factor of a Keplerian refractor depends on the ratio between the focal lengths of the objective and the eyepiece. This factor can be adjusted by

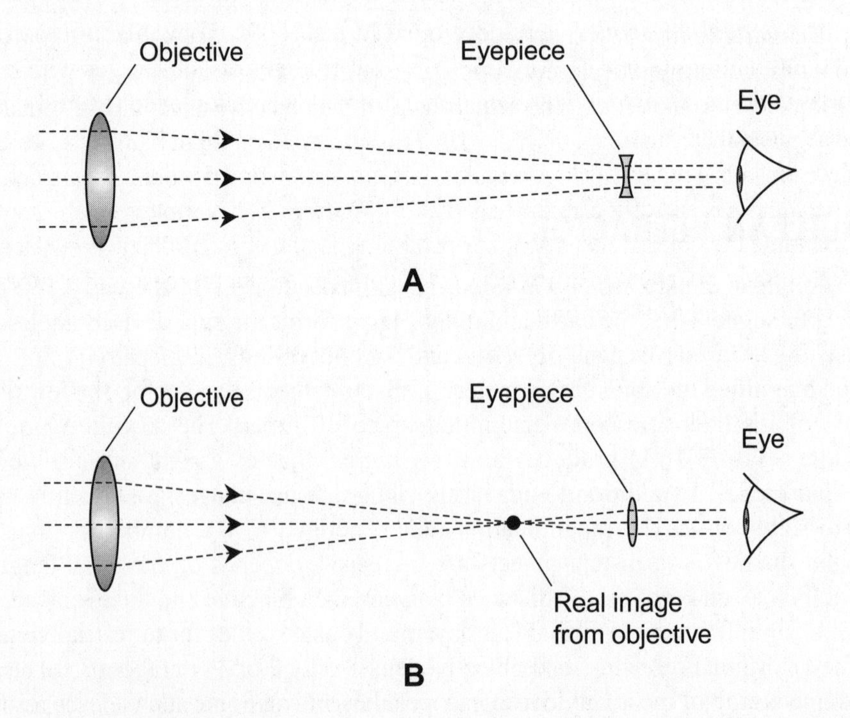

Figure 3-16 The Galilean refractor (A) uses a convex objective and a concave eyepiece. The Keplerian refractor (B) has a convex objective and a convex eyepiece.

using eyepieces with longer or shorter focal lengths. The shorter the focal length of the eyepiece, the greater is the magnification factor, assuming the focal length of the objective lens remains constant.

The Keplerian telescope is preferred over the Galilean type, mainly because the Keplerian design can provide a larger *apparent field of view*, which is the angular diameter, as directly seen by the eye, of the circular region in which objects appear through the telescope. Galilean telescopes have apparent fields of view so narrow that looking through them is an uncomfortable experience.

The largest refracting telescope in the world is a Keplerian refractor at the Yerkes Observatory in Wisconsin. Its objective lens has a diameter of 40 in, or slightly more than 1 m. Keplerian refractors are used by thousands of amateur astronomers worldwide.

LIMITATIONS OF REFRACTORS

A well-designed refracting telescope is a pleasure to use. Nevertheless, there are certain problems inherent in their design. These are known as *spherical aberration*, *chromatic aberration*, and *lens sag*.

Spherical aberration results from the fact that spherical convex lenses don't bring parallel light rays to a perfect focus. Thus, a refracting telescope with a spherical objective will focus a ray passing through its edge a little differently than it will treat a ray passing closer to the center. The actual focus of the objective is not a point, but a short line segment that runs along the lens axis. This effect causes slight blurring of images of objects that have relatively large angular diameters, such as the moon, the sun, interstellar *nebulae* (gas and dust clouds), and "nearby" galaxies. The problem can be corrected by grinding the objective lens so it has paraboloidal, rather than spherical, surfaces.

Chromatic aberration in a telescope occurs for the same reason as it does in a microscope, and the solution is the same.

Lens sag occurs in the largest refracting telescopes. When an objective is larger than approximately 1 m (about 40 in) across, it becomes so massive that its own weight distorts its shape. Glass is not perfectly rigid, as you have noticed if you have seen the reflection of the landscape in a large window on a windy day. There is no truly effective way to get rid of this problem with a refractor, except to put the telescope in space where the force of gravitation is near zero.

NEWTONIAN REFLECTOR

The problems inherent in refracting telescopes, particularly chromatic aberration and lens sag, can be largely overcome by using mirrors instead of lenses as objectives. A *first-surface mirror*, with the silvering on the outside so the light never passes through glass, can be manufactured so that it brings light to a focus that doesn't change as the wavelength changes. Mirrors can be supported from behind, so they can be made larger than lenses without running into the sag problem.

Isaac Newton designed a reflecting telescope that was free of chromatic aberration. His design is still used in many reflecting telescopes today. The *Newtonian reflector* employs a concave objective mounted at one end of a long tube. The other end of the tube is open to admit incoming light. A small, flat mirror is mounted at a 45° angle near the open end of the tube to reflect the focused light through an opening in the side of the tube containing the eyepiece, as shown in Fig. 3-17A.

The flat mirror obstructs some of the incoming light, slightly reducing the effective surface area of the objective mirror. As a typical example, suppose a Newtonian reflector has an objective mirror 20 cm in diameter. The total surface area of this mirror is approximately 314 cm^2. If the square eyepiece mirror measures 3 cm by 3 cm, then, its total area is 9 cm^2, which is about 3 percent of the total surface area of the objective.

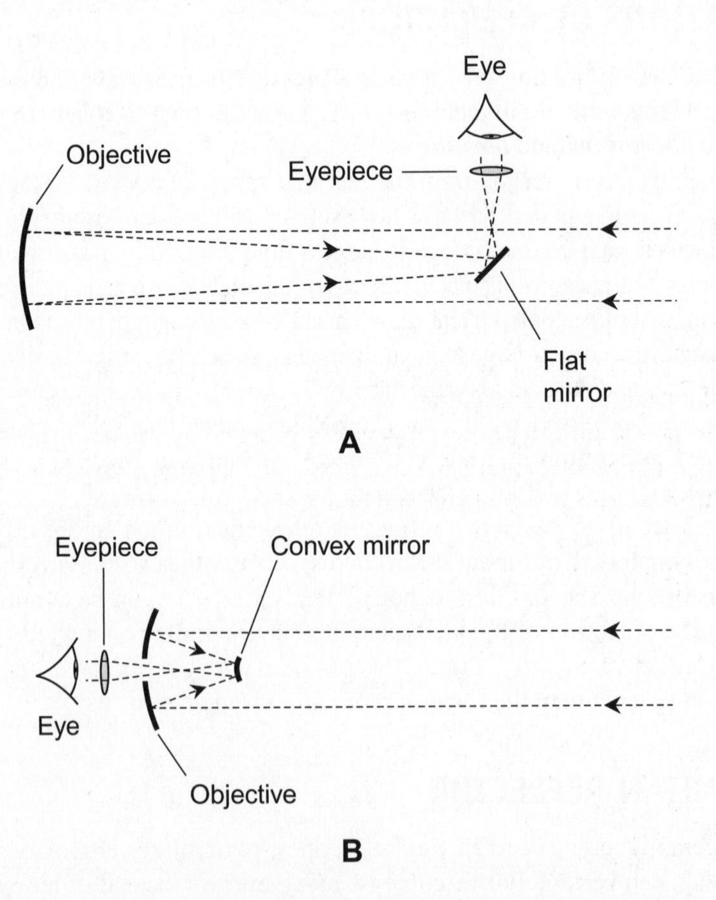

Figure 3-17 In a Newtonian reflector (A), the eyepiece is in the side of the telescope tube. In a Cassegrain reflector (B), the eyepiece is in the center of the objective mirror.

Newtonian reflectors have limitations. Some people find it unnatural to "look sideways" at distant objects. If the telescope has a long tube, a ladder is needed to view objects at high elevations. These annoyances can be overcome by using a different way to get the light to the eyepiece.

CASSEGRAIN REFLECTOR

Figure 3-17B shows the design of the *Cassegrain reflector*. The eyepiece mirror is closer to the objective than in the Newtonian design. It is not angled, but convex. The convexity of the eyepiece mirror increases the effective focal length of the objective mirror. Light reflects from the convex mirror and passes through a small hole in the center of the objective, containing the eyepiece.

The Cassegrain reflector can be made with a physically short tube and an objective mirror having a smaller radius of curvature than that of a Newtonian telescope having the same diameter. Therefore, a Cassegrain telescope is less massive and less bulky than an equivalent Newtonian telescope. Cassegrain reflectors with heavy-duty mountings are physically stable, and they can be used at low magnification to obtain wide-angle views of the sky.

MAGNIFICATION

In a telescope, magnification (×) is the factor which the instrument increases the observed angular diameters of distant objects. A 20× telescope makes the moon, whose disk subtends about 0.5° of arc as seen with the unaided eye, appear 10° in angular diameter. A 180× telescope makes a crater on the moon with an angular diameter of only 1 minute of arc, or (1/60)°, appear 3° in angular diameter.

Magnification is calculated in terms of the focal lengths of the objective and the eyepiece. If f_o is the focal length of the of the objective and f_e is the focal length of the eyepiece (in the same units as f_o), then the magnification factor, *m*, is given by

$$m = f_o / f_e$$

For a given eyepiece, as the effective focal length of the objective increases, the magnification of the whole telescope also increases. For a given objective, as the effective focal length of the eyepiece increases, the magnification of the telescope decreases.

RESOLUTION LIMIT

The *resolution limit* of a telescope is an expression of its ability to separate two objects that are close to each other in the sky. The resolution limit is measured in an angular sense, usually in seconds of arc, or units of (1/3600)°. The smaller the number, the better the resolution. The resolution limit is sometimes called *resolving power*, although that usage can be misleading because it suggests that the factor would increase as the objective diameter increases.

The best way to measure a telescope's resolution limit is to scan the sky for known pairs of stars that appear close to each other in the angular sense. Astronomical data charts can determine which pairs of stars to use for this purpose. Another method is to examine the moon, and then use a detailed map of the lunar surface to ascertain how much detail the telescope can render.

The resolution limit increases with magnification, but only up to a point. In practice, the resolution limit also depends on the acuity of the observer's eyesight if direct viewing is contemplated, or the coarseness of the grain of the photographic or detecting surface if a camera is used.

When all other factors are held constant, the resolution limit of a telescope varies approximately according to the *inverse* of the objective diameter. For example, if telescope A has an objective twice the diameter of the objective in telescope B, and if telescope A can resolve two objects separated by 0.2 seconds of arc (0.2″), then telescope B can only be expected to resolve objects down to an angular separation of 0.4″.

LIGHT-GATHERING AREA

The *light-gathering area* of a telescope is a quantitative measure of its ability to collect light for viewing or photographing faint objects. It can be defined in centimeters squared (cm^2) or meters squared (m^2), the effective surface area of the objective lens or mirror as measured in a plane perpendicular to its axis. Sometimes, light-gathering area is expressed in inches squared (in^2).

For a refracting telescope, given an objective radius of r, the light-gathering area A can be calculated according to the following formula:

$$A = \pi r^2$$

where π is approximately equal to 3.14159. If r is in centimeters, then A is in centimeters squared. If r is in inches, then A is in inches squared. If r is in meters, then A is in meters squared.

In a reflecting telescope, given an objective radius of r, the light-gathering area A is given by

$$A = \pi r^2 - B$$

where B is the area obstructed by the secondary mirror assembly. If r is in centimeters and B is in centimeters squared, then A is in centimeters squared. If r is in inches and B is in inches squared, then A is in inches squared. If r is in meters and B is in meters squared, then A is in meters squared.

ABSOLUTE FIELD OF VIEW

When you look through the eyepiece of a telescope, you can see anything within a cone-shaped region whose apex is at the telescope, as shown in Fig. 3-18. The *absolute field of view* is the angular diameter q of this cone. The angle q can be specified in degrees, minutes, and/or seconds of arc. Sometimes the angular radius is specified instead of the angular diameter.

The absolute field of view depends on several factors. The magnification of the telescope is important. When all other factors are held constant, the absolute field of view is inversely proportional to the magnification. If you double the magnification, you cut the absolute field of view in half. If you reduce the magnification by a factor of 4, you increase the absolute field of view by a factor of 4.

Another factor that affects the absolute field of view is the *focal ratio* or *f-ratio*: the objective's focal length divided by its actual diameter as measured in the same units. A telescope's *f*-ratio is denoted by writing *f*, followed by a forward slash, followed by the ratio expressed as a number. Thus, for example, if the focal length of the objective is 200 cm and its actual diameter is 20 cm, the *f*-ratio is *f*/10. If all other factors are held constant, the maximum absolute field of view that can be obtained with the telescope decreases as the *f*-ratio increases. Long, narrow telescopes tend to have small fields of view. Short, wide telescopes have large fields of view.

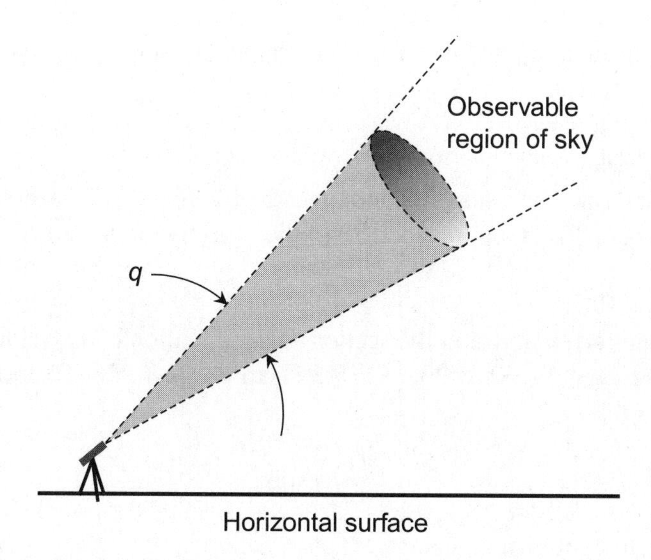

Figure 3-18 A telescope's absolute field of view, q, is measured in angular degrees, minutes, and/or seconds of arc.

APPARENT FIELD OF VIEW

The viewing angle provided by the eyepiece—the *apparent field of view*—is also an important consideration for telescope viewing. Some types of eyepieces have a wide apparent field, such as 60° or even 90°. Others have narrower apparent fields, in some cases less than 30°. Galileo's first refracting telescope had an apparent field of view only a few degrees wide. The apparent field of view depends on the construction of the eyepiece. It is also limited to a certain extent by the construction of the telescope.

PROBLEM 3-9

How much more light can a refracting telescope with a 250-mm-diameter objective gather, compared with a refracting telescope having an objective whose diameter is 50 mm? Express the answer as a percentage.

SOLUTION 3-9

Light-gathering area is proportional to the square of the objective's radius. Therefore, the ratio of the larger telescope's light-gathering area to the smaller telescope's light-gathering area is proportional to the square of the ratio of their objectives' diameters. Let's call the ratio of the objective diameters *k*. Then we have

$$\begin{aligned} k &= 250/50 = 5.0 \\ k^2 &= 5.0^2 \\ &= 25 \end{aligned}$$

The larger telescope gathers 25 times, or 2500 percent, as much light as the smaller one.

PROBLEM 3-10

Suppose a telescope has a magnification factor of 220× with an eyepiece of 0.500 cm focal length. What is the focal length of the objective in meters?

SOLUTION 3-10

Use the formula given above in the section "Magnification." The value of f_o in this case is the unknown. We know that f_e = 0.500 cm and m = 220. Therefore

$$\begin{aligned} m &= f_o/f_e \\ 220 &= f_o/0.500 \\ f_o &= 220 \times 0.500 \\ &= 110 \text{ cm} \\ &= 1.10 \text{ m} \end{aligned}$$

PROBLEM 3-11

Suppose the absolute field of view provided by a telescope is 20 arc minutes. If a 10-mm eyepiece is replaced with a 15-mm eyepiece that provides the same viewing angle as the 10-mm eyepiece, what happens to the absolute field of view provided by the telescope?

SOLUTION 3-11

The 15-mm eyepiece provides 10/15, or 2/3, as much magnification as the 10-mm eyepiece. Therefore, the absolute field of view of the telescope using the 15-mm eyepiece is 3/2, or 1.5, times as wide as that using the 10-mm eyepiece.

Quiz

This is an open book quiz. You may refer to the text in this chapter. A good score is 8 correct. Answers are in the back of the book.

1. Suppose you have purchased a hobby microscope for your grade-school child. The most powerful objective lens in this device is rated at "60 power," while the most powerful eyepiece is rated at "15 power." What is the approximate maximum magnification that this device can provide?
   (a) 16×
   (b) 75×
   (c) 900×
   (d) More information is necessary to answer this question.
2. In a Cassegrain telescope, the small convex mirror
   (a) reduces the magnification factor of the eyepiece.
   (b) increases the effective focal length of the objective.
   (c) increases the light-gathering power of the telescope.
   (d) produces a virtual image that is always in clear focus.

3. A refracting telescope with an objective 60.00 mm in diameter has
   (a) 2.250 times the light-gathering power of a refracting telescope with an objective 40.00 mm in diameter.
   (b) 1.500 times the light-gathering power of a refracting telescope with an objective 40 mm in diameter.
   (c) 1.225 times the light-gathering power of a refracting telescope with an objective 40 mm in diameter.
   (d) the same light-gathering power as a refracting telescope with an objective 40 mm in diameter.
4. A telescope with an objective 4 in across has a resolution limit, expressed in arc seconds, equal to approximately
   (a) 4 times that of a telescope with an objective 2 in across.
   (b) twice that of a telescope with an objective 2 in across.
   (c) half that of a telescope with an objective 2 in across.
   (d) 1/4 that of a telescope with an objective 2 in across.
5. Suppose you have several cans of paint in various pigments along with white paint, but no black or gray paint. You can obtain light gray by mixing white paint with equal quantities of
   (a) red, yellow, and blue paint.
   (b) red, blue, and green paint.
   (c) cyan, magenta, and yellow paint.
   (d) cyan, magenta, and green paint.
6. In a conventional fixed-image camera using a convex lens, what happens to the *f*-number if the aperture diameter is doubled?
   (a) It increases by a factor of 4.
   (b) It doubles.
   (c) It is cut in half.
   (d) It decreases by a factor of 4.

7. An overhead projector is designed for the display of transparencies 12 in high. A transparency is placed 18 in from the lens, and the projected image is 24 feet (ft) tall. Approximately how far is the display screen from the projector lens?
   (a) 72 ft
   (b) 36 ft
   (c) 24 ft
   (d) 18 ft
8. As the diameter of the hole in a pinhole camera decreases, assuming all other factors are held constant,
   (a) the image on the backplane gets sharper.
   (b) the required film exposure time decreases.
   (c) the image on the backplane gets brighter.
   (d) the required film exposure time does not change.
9. Suppose you want to increase the numerical aperture of a microscope without changing the magnification factor of either the eyepiece or the objective. What can you do?
   (a) Move the eyepiece farther from the objective.
   (b) Move the objective farther from the sample under observation.
   (c) Move the objective closer to the sample.
   (d) Increase the refractive index of the medium between the objective and the sample.
10. A CCD camera produces
    (a) a UV output signal.
    (b) an analog output signal.
    (c) a visible-light output signal.
    (d) a digital output signal.

Common Optical Effects

This chapter is an overview of well-known effects and phenomena that have been observed with visible light, IR, UV, and other forms of EM energy.

Radiation Pressure

The particle model of radiant energy suggests that EM fields can exert force on matter. This force is called *radiation pressure*. Isaac Newton and Johannes Kepler both suggested that radiation pressure from the sun causes the deflection and curvature of comet tails. *James Clerk Maxwell* hypothesized a quantitative relationship between radiation pressure and EM field intensity.

THE CROOKES RADIOMETER

Figure 4-1 is a drawing of a *Crookes radiometer*. Most high-school and college physics students have seen this gadget. It consists of a partially evacuated glass bulb with a set of vanes mounted on a bearing that allows the vanes to rotate in a horizontal plane. Each vane is polished or painted white on one side, and is blackened on the other side. Most Crookes radiometers have four vanes.

When a source of visible or IR radiation is brought near the device, the vanes rotate, with the white sides leading and the black sides trailing. If the device is cooled down after having been heated up, the vanes reverse their sense of rotation, with the black sides leading and the white sides trailing. Because of this windmill-like rotation, the Crookes radiometer is sometimes called a *light mill*.

Evidence indicates that the movement of the vanes is not caused by the direct impact of photons against the surfaces. If that were the case, the device would work even if the bulb were completely evacuated. In addition, the sense of rotation would be reversed from what is commonly observed. Photons striking the white or polished surfaces would "bounce back," producing greater pressure on those surfaces than on the black faces where photons would "stick." But when the vanes are placed in a total vacuum, they do not rotate.

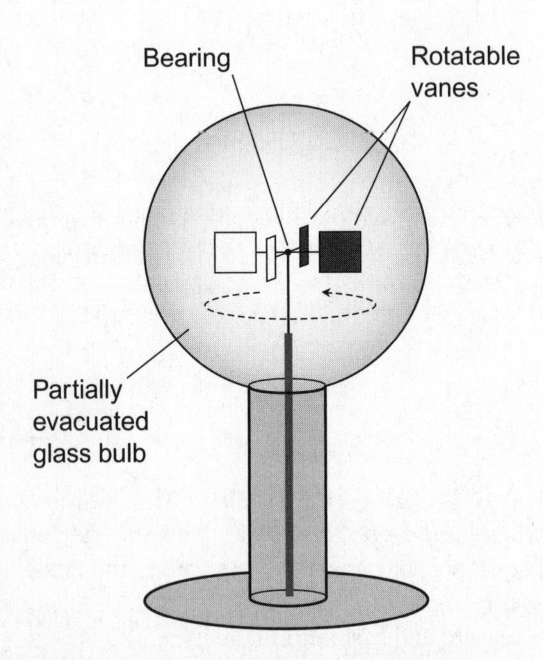

Figure 4-1 The Crookes radiometer can be used to qualitatively determine the intensity of visible light, IR, or UV radiation.

The most common explanation for the vane rotation is that, when exposed to visible, IR, or UV radiation, the dark surfaces absorb more of the incident energy than the white surfaces. Therefore, the black faces become warmer, so the air molecules near the black faces move faster. Some of the kinetic energy of these air molecules is mechanically imparted to the vanes, pushing harder against the dark faces. This energy causes the assembly to spin on its bearing with the white sides leading.

Thermodynamics also offers an explanation as to why the vanes rotate the opposite way (with the dark sides leading) if the whole device is rapidly cooled. In that case, the dark sides radiate IR energy more quickly than the white sides, becoming cooler than the white faces. Under these conditions, the white faces impart more kinetic energy to the adjacent air molecules than the black faces, so the net air pressure is greater against the white faces.

PHOTON MOMENTUM

Photons lack mass in the ordinary sense, so it is difficult to imagine how they could impose mechanical force or pressure in the same way as, say, the wind. But as we learned in Chap. 1, the momentum p_{ph} of a photon can be defined; it is related to the wavelength and the frequency by the formulas

$$p_{\text{ph}} = h/\lambda$$

and

$$p_{ph} = hf/(3.00 \times 10^8)$$

where p_{ph} is in kilogram meters per second, h is Planck's constant in joule-seconds, λ is the wavelength in meters, and f is the frequency in hertz.

Because photons have momentum, it is reasonable to suppose that, if a barrage of photons strikes an object or surface, some or all of that momentum will be transferred to the object or surface. This effect has been observed and measured. Serious experiments to measure radiation pressure began near the end of the 19th century, and they have become increasingly refined and sophisticated since that time.

MEASURING TRUE RADIATION PRESSURE

A Crookes radiometer of ordinary design fails to rotate in a complete vacuum, but not because the incident photons exert no pressure. The problem is that in a light mill of typical construction, the bearing has too much friction, the vanes are too small, and are not made of the proper material. Unless the EM intensity is extreme, the radiation pressure is too weak to overcome these obstacles. The difficulty of measuring radiation pressure is similar to the challenge encountered by physicists trying to measure the gravitational force among objects in a laboratory environment.

In the early 1900s, a Russian physicist, *Pyotr Lebedev*, built a device similar to the Crookes radiometer, but with an evacuated chamber and more sensitive hardware. He used lenses to focus the incoming light, increasing its intensity. He used a *torsion balance* instead of a common mechanical bearing, and the chamber was completely evacuated to eliminate any possibility of interaction with gas molecules.

Figure 4-2 illustrates the principle of a radiometer designed to measure true radiation pressure. This drawing is not an actual rendition of Lebedev's apparatus, and the relative sizes and positions of the components have been modified for clarity. The two vanes are identical in size, shape, and mass. (They appear different in Fig. 4-2 because they are drawn in perspective.) The vane assembly is housed in a vacuum, but the rest of the device can be outside the evacuated chamber.

When EM radiation strikes the *Lebedev radiometer*, the photon pressure on the reflective vane exceeds the photon pressure on the dark vane. In the ideal case, where the reflective vane absorbs none of the incident energy and the dark vane

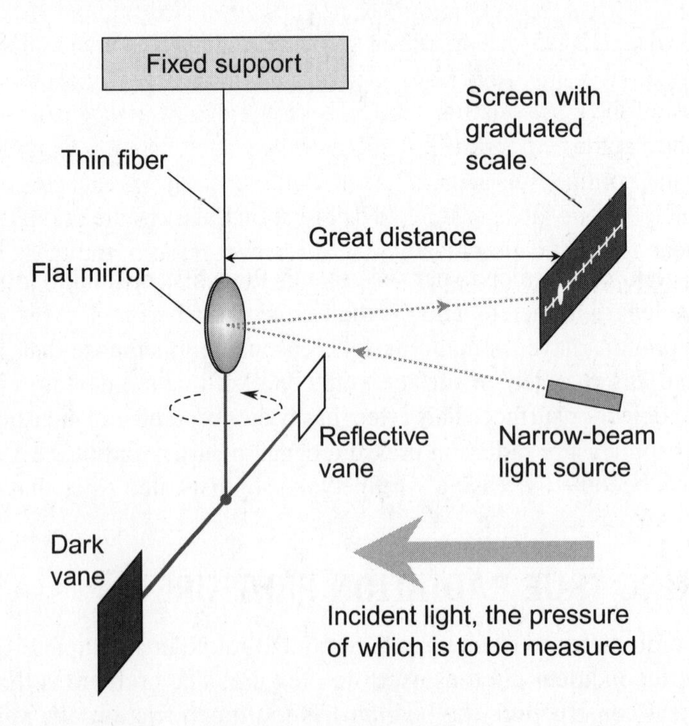

Figure 4-2 Functional diagram of a radiometer designed for measuring true radiation pressure.

absorbs all of it, the force caused by light pressure on the reflective vane is twice the force exerted on the dark vane. This force differential produces torque on the fiber, slightly altering the orientation of the lever between them. An *optical lever*, in which a narrow beam of light is reflected from a mirror to a distant screen, enhances the sensitivity of the device.

RADIATION PRESSURE IN OUTER SPACE

In 1910, Lebedev conducted experiments to determine the effects of radiation pressure on gases. The results of these experiments contributed to theories involving the behavior of nebulae near and between stars in our galaxy. On the scale of interstellar space, where distances are measured in *light years* (units of approximately 10^{13} km), radiation pressure can literally blow cosmic gas and dust around. Photons from stars create physical currents in the galaxy that exhibit patterns similar to those in the earth's atmosphere. This similarity is apparent when photographs of interstellar nebulae, taken through telescopes, are compared with photographs of weather systems taken by earth-orbiting satellites.

The theory of radiation pressure explains why there is a limit to the mass that a star can attain. Stars form because of gravitational attraction among molecules of gas and dust in space. Once a certain amount of matter has accumulated within a region of space, gravitation pulls more gas and dust in, until a dense, massive "ball" is formed. Most of the matter in this "ball" is hydrogen gas. Increasing pressure raises the temperature at the center of the "ball" until hydrogen fusion begins, and a star is born. If the star continues to accumulate matter, it eventually reaches a critical brilliance at which the outward pressure of its radiation balances the gravitational force on matter near the star. This prevents any further increase in the mass of the star, because gas and dust in its vicinity is "blown away" rather than "falling in."

OPTICAL TWEEZER

Radiation pressure is significant on a microscopic scale. A device called an *optical tweezer* takes advantage of radiation pressure, allowing scientists and medical researchers to trap and manipulate cells, bacteria, viruses, and small particles of matter.

Figure 4-3 illustrates how the optical tweezer works. A laser beam is focused by a convex lens so that its parallel rays converge. Drawing A is a side view; drawing B is a view looking directly into the laser beam. The objective lens of a lab microscope can be used to focus the laser beam. In more advanced devices, special lenses called *Fresnel zone plates* are used. The focusing of the laser produces a "fuzzy sphere" of optical energy called a *Gaussian energy distribution* with the greatest intensity at a central point, and steadily decreasing intensity outward in all directions.

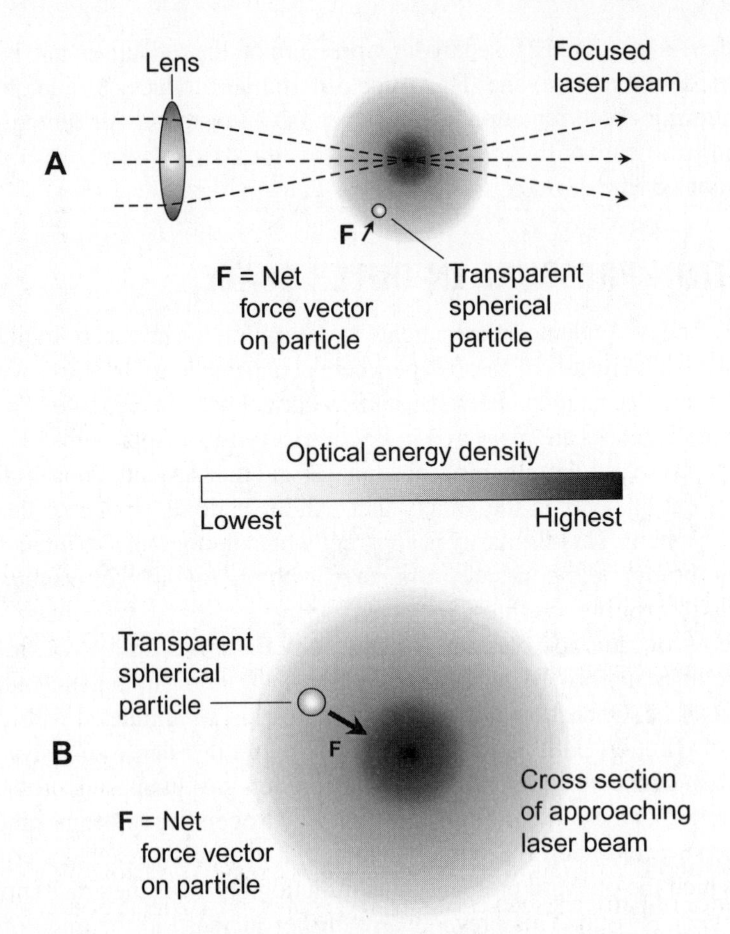

Figure 4-3 Principle of the optical tweezer. At A, side view. At B, looking into the laser beam. The force vector **F** is the result of refraction and scattering effects.

When a transparent or semitransparent particle is brought into the Gaussian region, the particle refracts, reflects, and scatters the light in such a way that a net *restoring force* vector **F** is produced on the particle, pulling the particle toward the point of greatest optical energy density. The force vector **F** occurs both laterally (perpendicular to the direction in which the laser beam travels) and longitudinally (along the line of laser beam propagation), and is a reaction to changes in photon momentum as photons are refracted, reflected, and scattered by the particle.

The optical tweezer works better with some particles than with others. The ideal particle is spherical, transparent, and has an index of refraction significantly higher than that of the surrounding medium.

PROBLEM 4-1

Suppose that solar radiation produces radiation pressure of 4.7×10^{-6} N/m^2 on a flat, perpendicular, reflective surface at a distance of 1.5×10^8 km from the sun. That distance is the approximate radius of the earth's orbit. If a solar sail measuring 1 km^2 is deployed to propel a spacecraft at the same distance from the sun as the earth, how much of the force against this sail is the result of true radiation pressure?

SOLUTION 4-1

We multiply the force per square meter by the surface area of the solar sail in meters, based on the knowledge that the sail is flat, and is oriented so that the sun's rays strike at a right angle. The sail measures 1 km^2, so its surface area is $10^3 \times 10^3 = 10^6$ m^2. The force on the sail is therefore

$$4.7 \times 10^{-6} \text{ N/m}^2 \times 10^6 \text{ m}^2 = 4.7 \text{ N}$$

A practical solar sail would experience greater force than this as a result of bombardment by high-speed subatomic particles constantly emanating from the sun. This particle stream is called the *solar wind*. The pressure or force imposed by the solar wind is different than photon pressure. Solar wind consists of particles with finite, measurable rest mass, moving at less than the speed of light.

Scattering

Visible, IR, and UV radiation can be scattered as a result of various physical effects. The presence of particles such as atoms, molecules, dust grains, bacteria, or viruses can scatter incident radiation. Strong electric or magnetic fields can also deflect EM radiation, producing a scattering effect.

RAYLEIGH AND RAMAN SCATTERING

When EM radiation passes through a particle cloud or gaseous medium in which the individual particles, atoms, or molecules have radii smaller than 10 percent of the wavelength, the energy is scattered in all directions as shown in Fig. 4-4. The effect is similar to that of an elastic collision when a moving particle (in this case a photon) strikes an object. The phenomenon was discovered by the English physicist *Lord Rayleigh* around the year 1900.

In nearly all cases, photon energy is not increased or decreased by an encounter with a material particle, regardless of the extent to which the photon direction changes. Therefore, the wavelength of a reflected photon is almost never altered by a Rayleigh type encounter. However, the extent of the scattering does depend on the wavelength of the photon. In the earth's atmosphere, Rayleigh scattering takes place

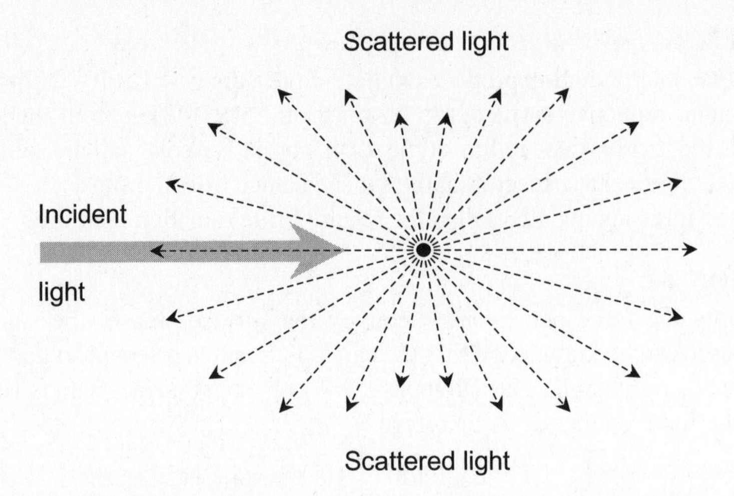

Figure 4-4 Rayleigh scattering. The central dot represents a parcel of particles, atoms, or molecules with radii smaller than 10 percent of the wavelength.

to a markedly greater extent at the shorter visible wavelengths than at the longer wavelengths. Mathematically, the relation is in direct proportion to the fourth power of the frequency, and in inverse proportion to the fourth power of the wavelength. That is,

$$I \propto f^4$$

and

$$I \propto 1/\lambda^4$$

where I is the radiation intensity of the scattered light, f is the frequency, and λ is the wavelength. (The symbol $\propto$ indicates proportionality.) Rayleigh scattering of light in clear air is more than 9 times as great at a wavelength of 400 nm than at a wavelength of 700 nm. This difference explains why a clear daytime sky looks blue from the earth's surface. (At night, the blue is a darker shade, but the sky is never totally black.)

Rayleigh scattering does not occur to exactly the same extent in all directions. However, the *forward scatter* and *backscatter* patterns are symmetrical; they are "mirror images" of each other as shown in Fig. 4-4. The greatest intensity of scattered light is along the axis of the incident ray. The lowest intensity is at right angles to the incident ray.

In approximately one out of every 10^7 Rayleigh-type encounters, the energy of the atom or molecule changes. The photon leaves with different energy (usually less)

than it had before the encounter. The photon frequency decreases and the wavelength increases. The encounter has the characteristics of an inelastic collision. This effect is called *Raman scattering* after the Indian physicist *C. R. Raman* who is credited with its discovery in 1928.

MIE SCATTERING

Rayleigh scattering is rarely observed all by itself, because the earth's atmosphere contains particles with radii much larger than 0.1 wavelength. In some locations, such as the polar regions, there are few such particles; in desert regions or where haze is common, there are many. Larger particles include dust, water droplets, and particulate matter from pollutants such as smoke. When particles are larger in radius than approximately 1 wavelength, most of the incident light is scattered in the forward direction, as shown in Fig. 4-5. This phenomenon is known as *Mie scattering*.

Unlike Rayleigh scattering, the extent of Mie scattering is independent of the wavelength. As a result, light scattered in the Mie mode tends to appear white or gray. This is why clouds usually appear white or gray when illuminated by sunlight in an unpolluted environment during the middle of the day. If the particles themselves are pigmented, the scattered light is tinted. This effect is sometimes observed in deserts or polluted areas. Mie-scattered light also appears tinted if the incident light has been filtered before it strikes the scattering particles.

In the earth's atmosphere, the actual scattering of visible light occurs in a combination of modes. When there is little Mie scattering, the sky has a highly saturated blue or turquoise appearance, and there is little variation in hue or saturation as the

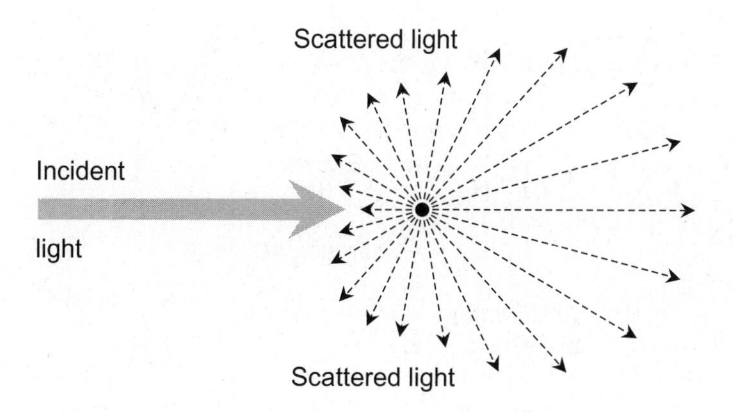

Figure 4-5 Mie scattering. The central dot represents a parcel of particles with radii larger than a wavelength.

viewing angle varies. When there is significant Mie scattering in the absence of clouds, the sky seems to acquire a white glow around the sun. Sunlight filtered through unpolluted fog or clouds appears solid gray. Similar effects occur in moonlight or starlight, although the overall brightness is far lower.

THOMSON SCATTERING

In the preceding examples, it's easy to imagine that the scattering results from "collisions" between photons and material particles (although that view is an oversimplification). An entirely different effect can scatter EM energy in the presence of charged subatomic particles. This phenomenon, called *Thomson scattering*, does not involve "collisions" in any sense, but only interactions between fields and particles. Thomson scattering is observed in cosmic radiation, in the solar corona, and in devices intended to produce controlled hydrogen fusion as a source of useful energy.

In the Thomson scattering mode, it is convenient to imagine the EM energy as a wave phenomenon. An EM wave consists of an electric-field (E) component and a magnetic-field (M) component. The flux lines of these two components are mutually perpendicular at every point. The EM wave travels in the direction perpendicular to both the E-field flux lines and the M-field flux lines (Fig. 4-6).

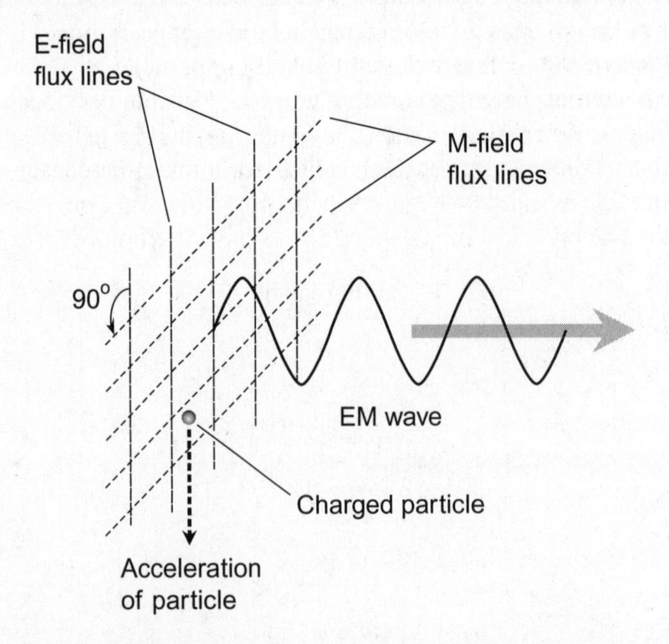

Figure 4-6 In an EM wave, the electric flux lines are perpendicular to the magnetic flux lines. A charged particle caught in the EM field is accelerated parallel to the electric lines of flux.

When an EM wave encounters a charged subatomic particle, especially one with low mass such as an electron, that particle is accelerated by the electric and magnetic components of the EM field. The acceleration occurs parallel to the E-flux lines and perpendicular to the M-flux lines of the original EM field. The acceleration of the particle causes it to emit a secondary EM field with components in all directions except directly in line with the motion of the particle. The strongest EM radiation from the charged particle occurs in the plane perpendicular to its acceleration.

BRILLOUIN SCATTERING

When the refractive index of a medium changes rapidly, EM rays are bent to a variable extent. The most common cause of rapid changes in refractive index is a fluctuation in density produced by acoustic compression waves. *Brillouin scattering* is the scattering of EM rays, particularly IR, visible light, and UV, as a result of such density fluctuations.

Brillouin scattering can occur in gases, liquids, and solids. Often, it is too small to be noticed. For example, the noise from a loud siren or explosion causes significant acoustic compression in the surrounding air, in turn producing irregularities in the refractive index of that air. However, an observer near the source of the noise does not see the direct optical scattering effects. Theoretically, Brillouin scattering takes place in such a situation, but it is not great enough to see without sensitive equipment designed to detect and measure it.

Brillouin scattering is more pronounced in solids, particularly crystalline substances that have indices of refraction significantly larger than 1. When sound waves travel through such a medium along with IR, visible, or UV rays, the acoustic vibrations can produce noticeable *modulation* (changes in the amplitude, wavelength, phase, or polarization) of the EM rays. The modulation follows the waveform of the sound.

COMPTON SCATTERING

High-energy photons can lose or gain energy when they interact with electrons in atoms. The result is a change in the wavelength. An energy loss increases the wavelength of a photon, while an energy gain reduces it. Energy loss is more commonly observed than energy gain. These phenomena are called *Compton scattering*, and the extent of the wavelength change is called *Compton shift*, named after the physicist *Arthur H. Compton* who first observed and explained it.

In Compton scattering, the change in photon energy is attended by a transfer of energy to or from an electron in the matter, so that momentum is conserved during the encounter. This energy exchange can give rise to a new photon that travels in a different direction from the original one. The observed result is both directional and wavelength scattering.

The existence of Compton scattering lends support to the particle theory of EM radiation, because the effects can't be explained on the basis of wave theory alone. In particular, Compton scattering suggests that the inherent wavelength of each particle is inversely proportional to the energy it contains. The phenomena occur most often with X rays and gamma rays. Compton scattering has applications in radiology, where X rays are used to treat certain types of cancer.

IONOSPHERIC EFFECTS

High-speed subatomic particles, UV rays, and X rays from the sun cause *ionization* of the rarefied gases in the earth's upper atmosphere. This ionization occurs at specific altitude ranges in the *ionosphere*, causing absorption, refraction, and scattering of EM waves at some radio frequencies. The refraction and scattering effects make long-distance communication or reception possible in the "shortwave radio band" of the EM spectrum.

The lowest ionized region, the *D layer*, exists at an altitude of about 50 km (31 mi), and is ordinarily present only on the daylight side of the earth. This layer does not contribute to long-distance communication, and in fact sometimes impedes it by absorbing EM energy. The *E layer*, about 80 km (50 mi) above the surface, also exists mainly during the day, but nighttime ionization is occasionally observed. The uppermost layers are called the *F_1 layer* and the *F_2 layer*. The F_1 layer, normally present only on the daylight side of the earth, forms at about 200 km (125 mi) altitude. The F_2 layer exists at about 300 km (187 mi) during daylight and darkness.

At certain radio frequencies, the E layer occasionally returns signals to the earth. Because this mode of propagation is intermittent, it is called *sporadic-E propagation*. It is most often observed at frequencies between 20 MHz and 150 MHz. The propagation range is a few hundred kilometers. The standard frequency-modulation (FM) broadcast band is sometimes affected by sporadic-E propagation.

Under normal conditions, communication by F_1-layer or F_2-layer propagation can be accomplished between any two points on the earth at some frequency or frequencies between 5 MHz and 30 MHz. Amateur radio operators have made two-way contacts with other amateur operators on the opposite side of the earth in this mode using only a few watts of radiated EM power.

PROBLEM 4-2

Suppose a beam of visible light with a wavelength of 521 nm is transmitted through a cloud in which the particles have radii of 22 to 34 nm. Will Rayleigh scattering take place in this situation? If so, and if the wavelength of the beam increases to 638 nm, to what extent will the Rayleigh scattering change? If Rayleigh scattering is not significant in this situation, what is the reason?

SOLUTION 4-2

The particle radii are less than 10 percent of the visible-light wavelength, so Rayleigh scattering will occur at 521 nm. The intensity of the scattered radiation in any particular direction is inversely proportional to the fourth power of the wavelength. The scattered-light intensity therefore changes by a factor of $(521/638)^4$, or 0.445, when the wavelength increases to 638 nm. The intensity of a scattered ray in any particular direction is only 44.5 percent as great at a wavelength of 638 nm as it is at a wavelength of 521 nm.

Diffraction

When a wave disturbance encounters an aperture, a slit, or an obstruction with a well-defined edge, "phantom" sources appear at the same wavelength, complicating the observed effects. This phenomenon is called *diffraction*. Long waves diffract more readily than shorter waves, if all other factors are held constant.

CIRCULAR-APERTURE DIFFRACTION

Suppose that a monochromatic, point source of light shines through a single circular aperture in an opaque barrier. It seems reasonable to suppose that when projected on a screen, the light will produce a well-defined circular spot. Under casual observation, this appears to happen when the aperture is large compared with the wavelength of the light. But when the radius of the hole is less than a few wavelengths, the spot becomes diffuse, and concentric rings appear around it.

The central spot in the projection of a light beam through a small aperture is called an *Airy disk*. It has a bright center and a blurred border. The concentric rings are dimmer. Figure 4-7 shows the general effect of *circular-aperture diffraction*. Drawing A is an edgewise view of the projection apparatus, with cross-sectional views of the light beams after passing through the aperture. Drawing B shows the nature of the pattern on the projection screen. (The brightness is inverted in drawing A; the darkest shades represent the greatest light intensity.) If the aperture becomes smaller or the wavelength of the light increases, the effect becomes more pronounced.

Circular-aperture diffraction occurs with holes of all sizes and with EM radiation at all wavelengths, but we don't notice it with large apertures unless we scrutinize the projected image closely. Nevertheless, circular aperture diffraction limits the maximum obtainable image resolution that can be rendered with an optical instrument. The blurring of the Airy disk and the existence of the external rings degrade the quality of projected images in cameras, microscopes, and telescopes. The effect can also occur with radio telescopes and electron microscopes.

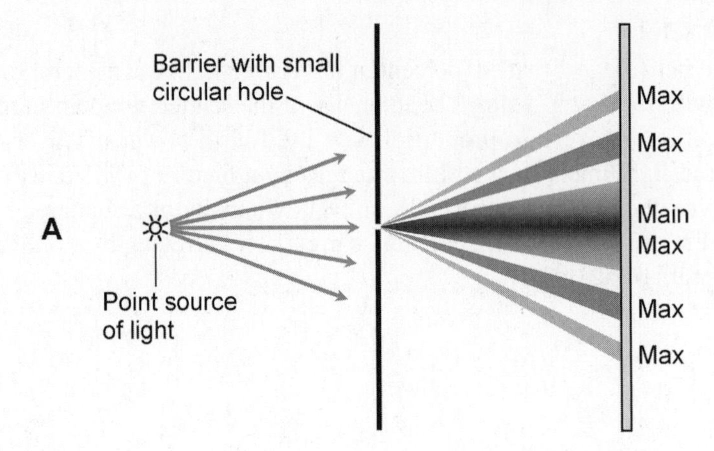

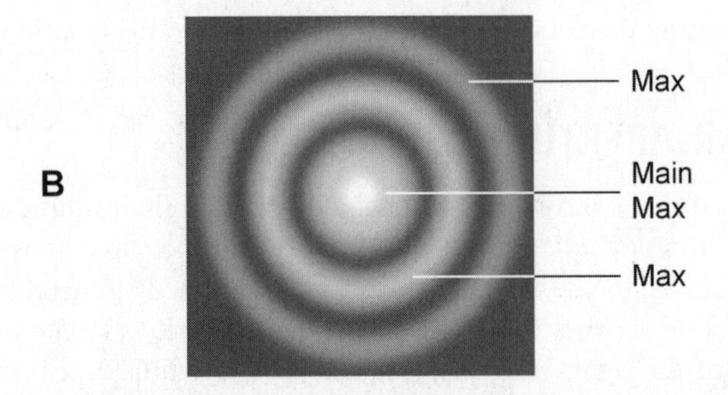

Figure 4-7 Circular-aperture diffraction. The projected image has a disk-shaped main maximum surrounded by ring-shaped minor maxima. At A, an edge-on view with shading inverted. At B, the screen image as viewed face-on.

SINGLE-SLIT DIFFRACTION

Now imagine that a slit is cut horizontally through an otherwise opaque barrier. When a point source of monochromatic light is placed on one side of the barrier and a projection screen is set up on the other side as shown in Fig. 4-8A, a bright horizontal band appears on the screen if the slit is wide compared with the wavelength. To the casual observer, this band has a sharp edge; but on closer observation, some blurring and multiple dimmer bands can be seen. As the slit becomes narrower, the bright central band becomes diffuse, and the dimmer bands on either side become more apparent. When the slit is narrow enough, the resulting pattern appears similar to the image in Fig. 4-8B. This effect is called *single-slit diffraction*.

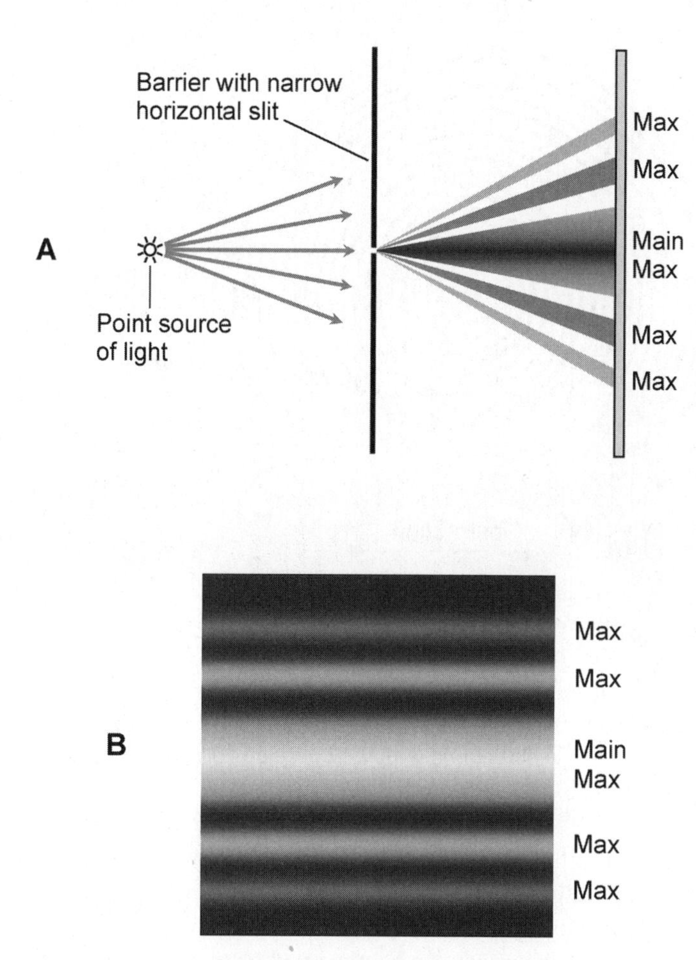

Figure 4-8 Single-slit diffraction. The projected image has a bright central bar with parallel, dimmer bars on either side. At A, an edge-on view, with shading inverted. At B, the screen image as viewed face-on.

CORNERS AND OBSTRUCTIONS

Diffraction can cause EM wave disturbances to propagate around sharp corners. The corner behaves as a new source of energy at the same wavelength, as shown in Fig. 4-9. This effect can theoretically occur with any type of wave energy. Longer waves, corresponding to lower frequencies, diffract more readily than shorter waves at higher frequencies. As a corner becomes sharper relative to the wavelength, diffraction occurs more efficiently.

Corner diffraction happens with water waves, as any surfer knows. It happens with radio waves, especially on the standard amplitude-modulation (AM) commercial

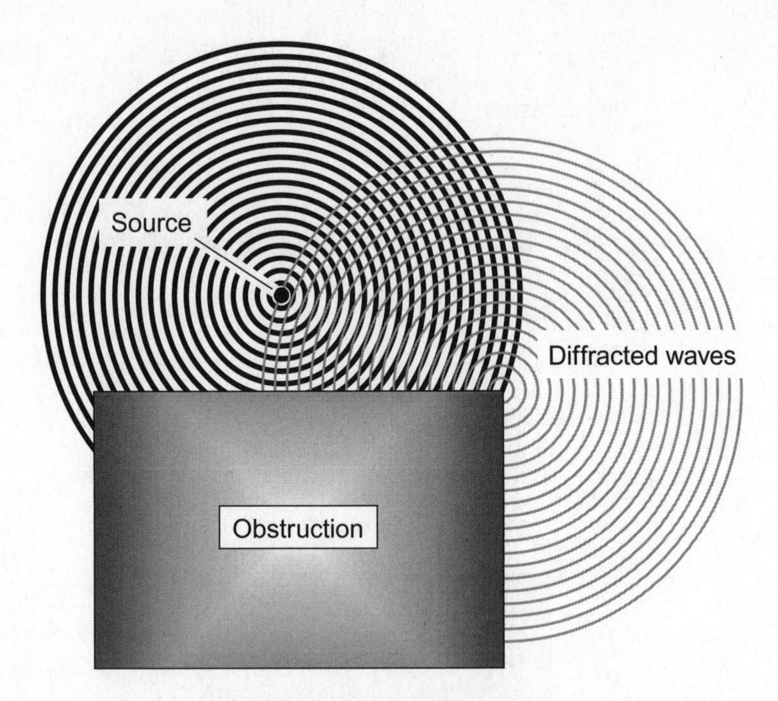

Figure 4-9 Diffraction makes it possible for waves to propagate around obstructions with well-defined corners.

broadcast band where the EM waves are hundreds of meters long. It happens with visible light as well, although the effect can be observed only under idealized conditions. One of the tests by which scientists ascertain the wave nature of a disturbance is to see whether or not the effect can be observed from around a corner.

When an obstruction is tiny in comparison with the wavelength of a disturbance, the waves diffract so well that they pass the object as if it's not there. A flagpole, for example, has no effect on low-frequency sound waves. The pilings of a pier, likewise, are ignored by large ocean swells. Particles of matter are rendered invisible in optical microscopes as a result of this same effect, when those particles are small in diameter compared with the wavelength of the light from the illumination source.

DIFFRACTION GRATINGS

When light waves pass through a barrier having multiple, narrowly spaced, parallel, transparent slits, each slit produces a diffraction pattern similar to the one shown in Fig. 4-8B. The patterns are displaced relative to each other. With white light, the pattern overlap produces color dispersion that can be seen as a "rainbow" on a projection screen. The observed effect is similar to, but more pronounced than, the

dispersion that occurs with glass prisms. The same thing happens when light waves reflect from a flat mirror with multiple dark lines or grooves. Devices of this sort are called *diffraction gratings*.

In an ideal diffraction grating, the spacing between slits, lines, or grooves should be comparable with the wavelength range of visible light, but no less than half the wavelength of visible red light. Assuming red light has a wavelength of 750 nm, the grating should have transparent or reflective lines spaced at least 375 nm apart, which corresponds to approximately 2.67×10^6 lines per meter, or 2670 lines per millimeter. If the lines are more closely spaced than this, dispersion does not occur. The largest workable spacing for a practical diffraction grating is less well-defined, and can range up to several tens of wavelengths before the medium behaves like an ordinary glass or mirror.

Diffraction gratings have existed for centuries. In fact, the principle was discovered at about the same time Isaac Newton demonstrated color dispersion with glass prisms. The first diffraction grating intended for producing color dispersion was built in 1785 by an American experimenter named *David Rittenhouse*. A simple diffraction grating can be manufactured by drawing a set of evenly spaced dark lines on a large sheet of white paper, photographing the sheet, and then developing the film to produce a transparency.

You can observe the color-dispersion effect of closely spaced reflective grooves by holding a blank *recordable compact disc* (CD-R), of the sort used to copy or back up computer data, in bright sunlight and allowing the light to reflect onto a white wall, as shown in Fig. 4-10. If you hold the disc at a certain angle, and if you

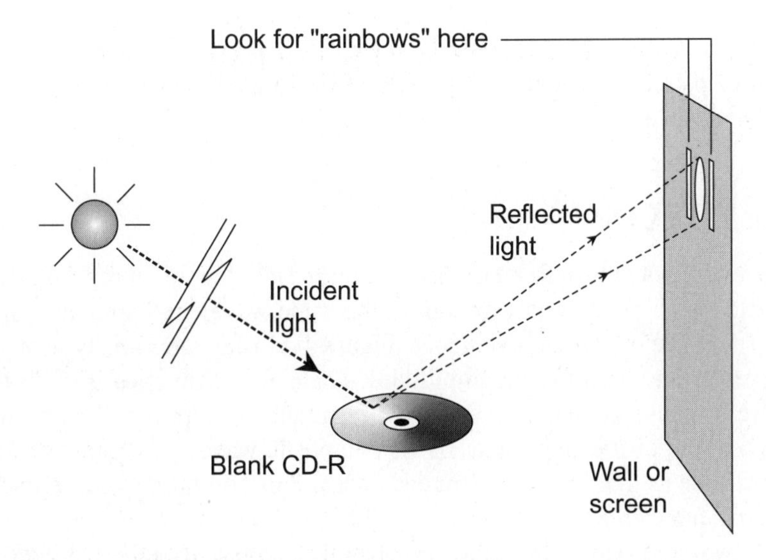

Figure 4-10 A blank recordable compact disc (CD-R) can act as a reflective diffraction grating to produce rainbow spectra.

make sure that the sunlight reflects from the "business side," you'll see "rainbow" spectra on either side of the image. The effect is produced by the alternate reflection and scattering of light among the closely spaced pits on the disc surface.

PROBLEM 4-3

If photons had no wavelike properties, what would happen to a ray of light passing through a circular aperture or a single narrow slit in an opaque barrier? What would happen to a ray of light encountering a sharp corner? What would happen to a ray of light passing through a fine grating?

SOLUTION 4-3

If photons were mere particles, having no characteristics of wavelength or frequency, then a ray of light passing through an aperture would produce a uniform spot with well-defined edges on a projection screen. A ray passing through a narrow slit would produce a single band. No diffraction would be observed at a sharp corner. A ray passing through a fine grating would produce a series of alternating bright and dark bands on a projection screen, with no color dispersion. If the aperture, slit, or grating spacing were smaller than the diameter of a single photon, then no light would get through the barriers, and the projection screens would appear uniformly dark.

Phase and Interference

Two or more waves can combine to produce interesting effects and, in some cases, remarkable patterns. Amplitudes can be exaggerated, waveforms altered, and entirely new waves generated.

PHASE EFFECTS

The *phase* of a wave is defined in *degrees*, with each degree representing 1/360 of a complete cycle. When two waves have the same wavelength and are "lined up," they are said to be in *phase coincidence*. Figure 4-11 illustrates this type of situation for two waves having different amplitudes. The phase difference is 0°. If two sine waves are in phase coincidence, then the resultant is a sine wave with amplitude equal to the sum of the amplitudes of the composite waves. The phase of the resultant is the same as that of the constituent waves. This condition is sometimes called *constructive interference*.

When two sine waves of equal wavelength oscillate exactly 1/2 cycle (180°) apart in time, they are said to be in *phase opposition*. An example is shown in Fig. 4-12. A situation of this sort is also known as *destructive interference*. If two

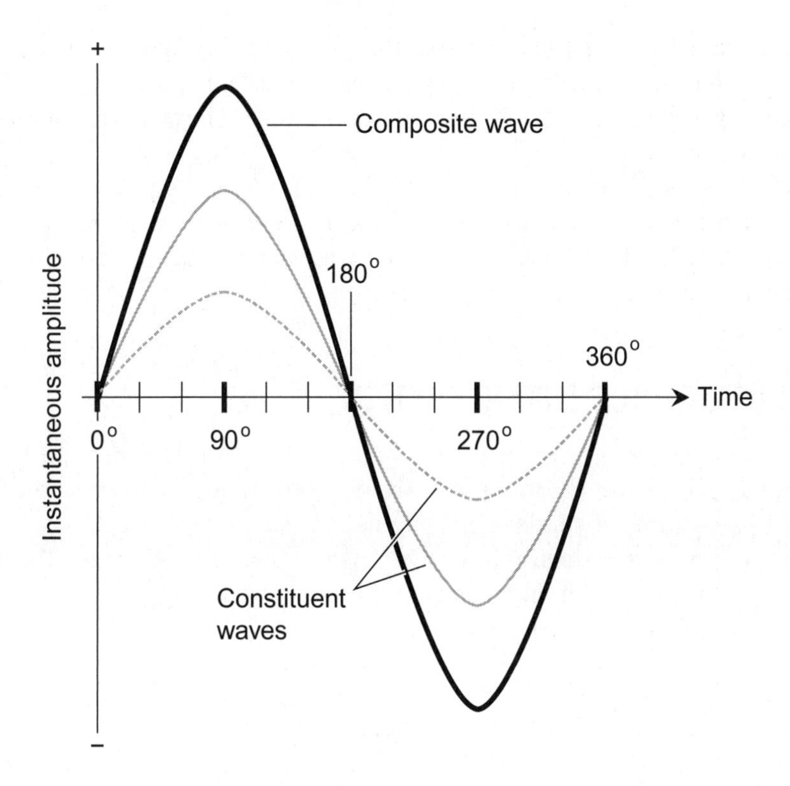

Figure 4-11 Two waves in phase coincidence. Their amplitudes add at every instant in time.

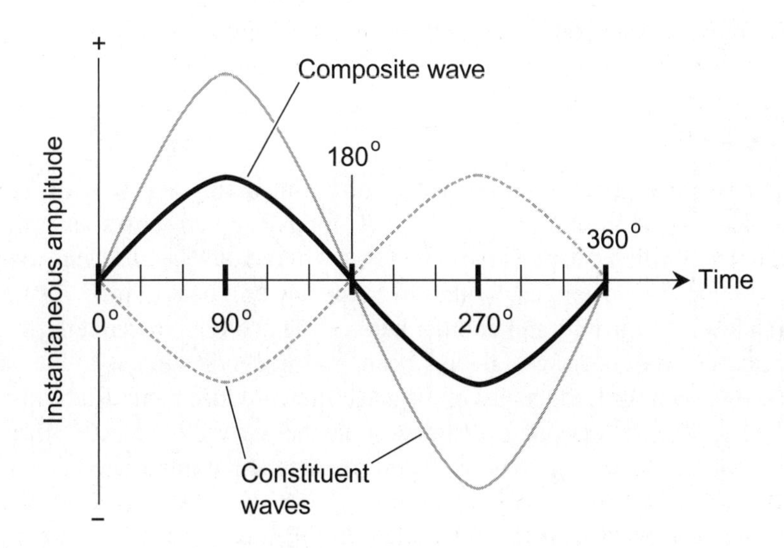

Figure 4-12 Two waves in phase opposition. Their amplitudes subtract at every instant in time.

sine waves have the same frequencies and amplitudes and combine in phase opposition, they completely cancel each other because the instantaneous amplitudes of the two waves are equal and opposite at every point in time. The amplitude of the resultant wave is zero.

Two sine waves having equal wavelength can differ in phase by any amount from 0° (phase coincidence), through 90° (*phase quadrature*, meaning a difference of 1/4 cycle), 180° (phase opposition), 270° (phase quadrature again), to 360° (phase coincidence again).

DOUBLE-SOURCE INTERFERENCE

Imagine you are a surfer who spends every winter on the North Shore of the Hawaiian island of Oahu. In the maritime sub-Arctic, storms constantly cross the Pacific Ocean. If two storms of similar size and intensity are separated by a vast distance, complex swell patterns span millions of square kilometers. Between the storms, swells alternately act with and against each other, producing huge zones of constructive and destructive interference.

Patterns created by interference between multiple wave sources appear at all scales, from storms at sea to orchestra instruments in a concert hall, from radio transmitters to X-ray machines. The slightest change in the relative positions or wavelengths of two sources can make a profound difference in the resulting *interference pattern*. Examples are shown in Figs. 4-13 and 4-14. The *wave trains* are shown as sets of concentric circles. The interference patterns can be clearly seen, having a much different form than that of the wave trains themselves.

HETERODYNING

No matter what the mode, and regardless of the medium, waves of different frequencies can mix to produce new waves. When this happens with sound, the effect is called *beating*; when it happens with EM waves, it is called *heterodyning* or *mixing*. Two waves that are close to each other in frequency will beat to form a *heterodyne* at a much lower frequency, and another heterodyne at a higher frequency.

Beat and heterodyne waves always occur at frequencies equal to the sum and difference between the frequencies of the waves that produce them. Radio-frequency heterodyning was discovered by engineers in the early 1900s. At optical wavelengths, a similar effect can be used to produce frequency-modulated beams of IR, visible light, and UV. In theory, heterodyning can even take place at X-ray and gamma-ray wavelengths. Such effects may be of interest to astronomers as they encounter strange new objects in deep space and try to formulate theories to explain their behavior.

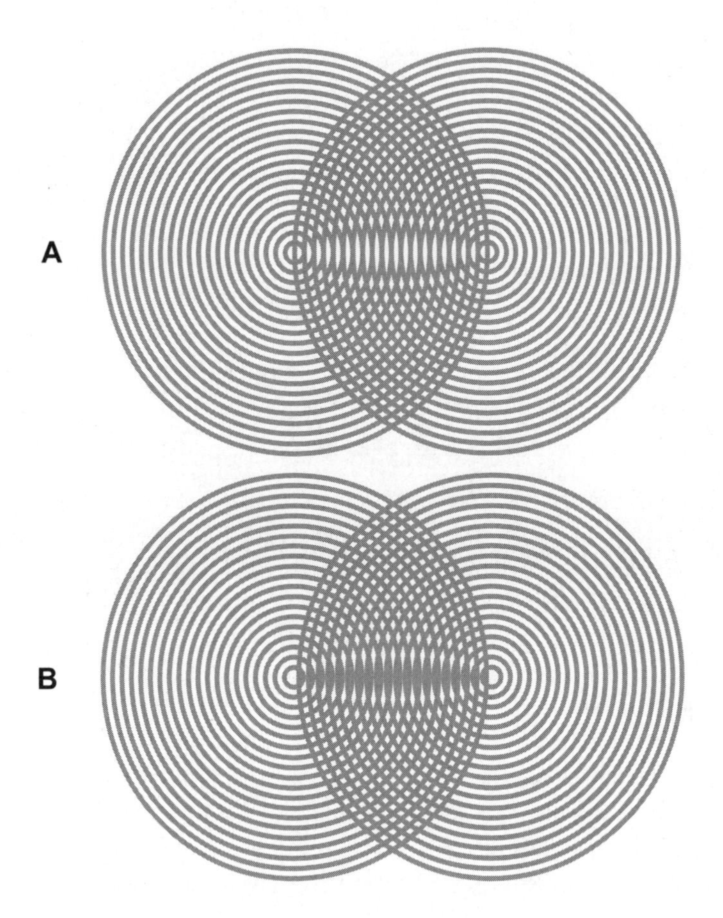

Figure 4-13 A small source displacement can change an interference pattern. Compare the interference patterns in drawings A and B.

Given two waves having different frequencies f and g (expressed in the same units) where $g > f$, they beat together to produce new waves, called *mixing products*, at frequencies x and y as follows:

$$x = g - f$$

and

$$y = g + f$$

Figure 4-15 shows examples of heterodyning in which the lower-frequency mixing products are visually evident. In the top example, the waves, shown by sets of vertical lines, differ in frequency by 10 percent (f and $1.1f$); in the middle example, by 20 percent (f and $1.2f$), and in the lower example, by 30 percent (f and $1.3f$).

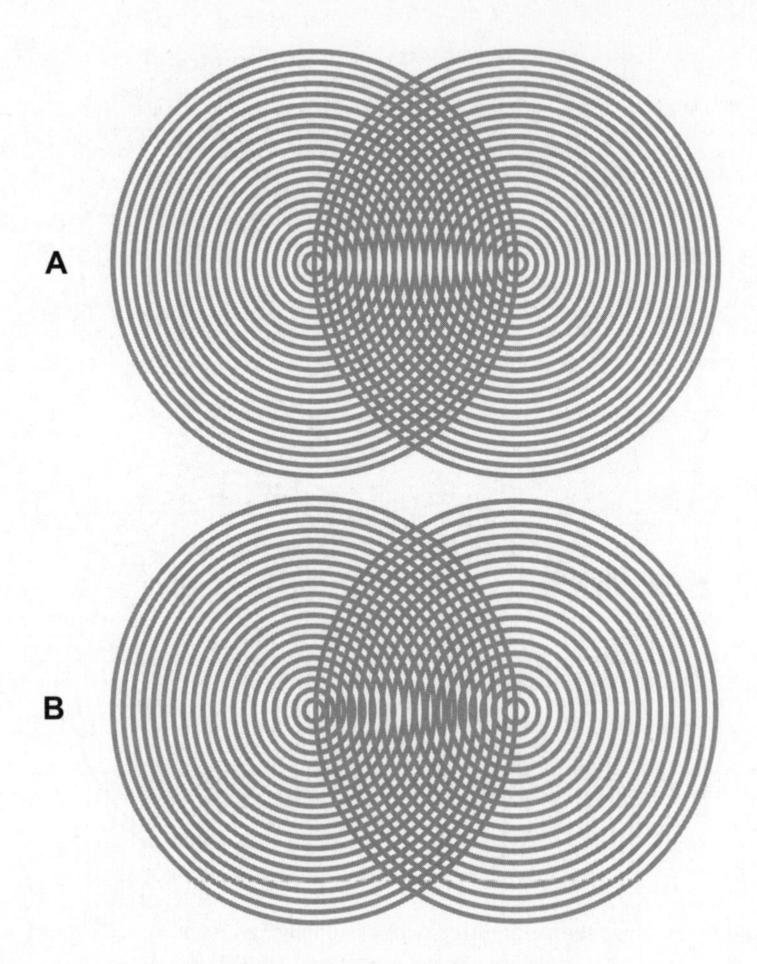

Figure 4-14 Identical wavelength (A) versus a 10 percent difference in wavelength (B). Notice the difference in the interference pattern caused by the wavelength change.

PARALLEL-REFLECTOR INTERFERENCE

Imagine that a ray of light with all its energy concentrated at a single wavelength strikes a pair of flat, parallel surfaces at a right angle. Suppose that the upper surface is semitransparent and partially reflective, the lower surface is totally reflective, the surfaces are separated by 1/4 wavelength ($\lambda/4$) as shown in Fig. 4-16A, and the index of refraction of the medium between the surfaces is the same as the index of refraction of the medium outside them.

The incident wave strikes the upper surface at point *X* and the lower surface at point *Y*. At each point, some light is reflected. Imagine that the upper surface reflects

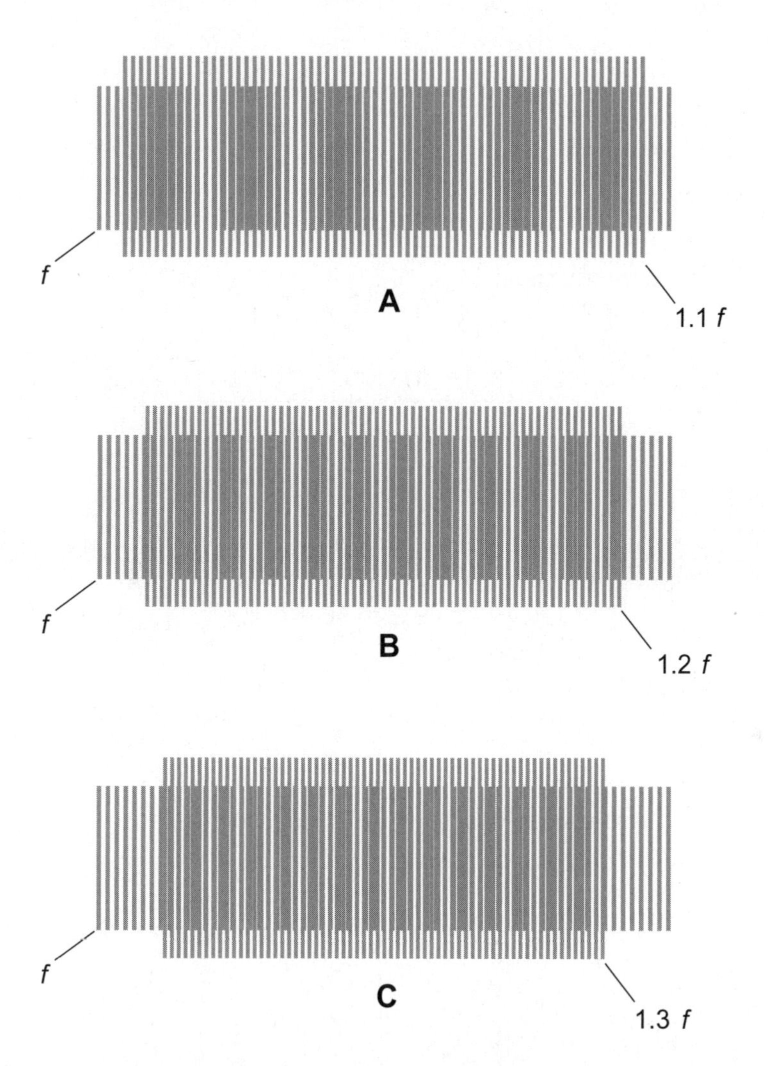

Figure 4-15 Visual rendition of wave heterodyning. At A, the two waves differ in frequency by 10 percent; at B, by 20 percent; at C, by 30 percent. The lower-frequency mixing products are clearly visible.

and transmits the incident light in a certain proportion, so that the reflected waves *X* and *Y* have equal amplitudes when they emerge. The total path difference between the two rays is 1/4 + 1/4 wavelength, or 1/2 wavelength, so the emerging waves *X* and *Y* are in phase opposition (Fig. 4-16B). Because they have equal intensities, these two waves cancel out. If there are no other light sources, an observer near the line of the incident ray will see the pair of reflectors as a black surface.

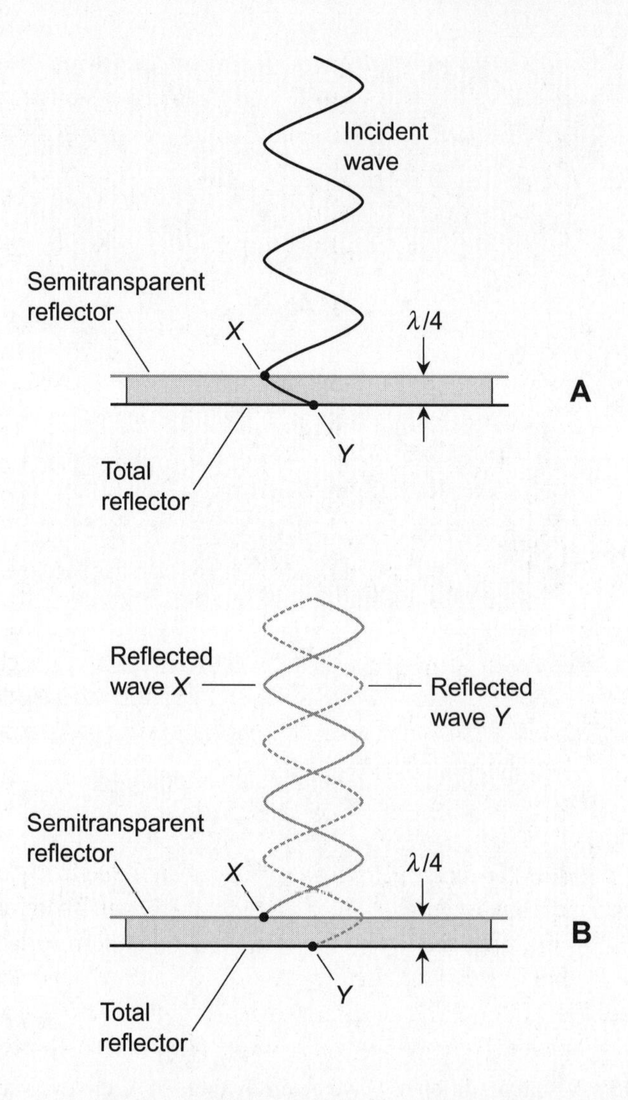

Figure 4-16 Wave interference from reflectors separated by 1/4 wavelength. At A, the incident wave strikes the surfaces at points *X* and *Y*. At B, the reflected waves emerge in phase opposition because the overall path lengths differ by 1/2 wavelength (twice the spacing between reflectors).

If the wavelength of the incident light changes, the phase cancellation will no longer be complete, and the pair of surfaces will seem partially reflective to the observer. If the source wavelength remains constant but the observer moves away from the line of the incident ray so the reflection angle is no longer 90°, then again, the destructive interference will no longer be total, and the pair of surfaces will appear partially reflective.

PROBLEM 4-4

Suppose we have a pair of parallel surfaces such as those shown in Fig. 4-16, separated by a distance of 150 nm, with a vacuum between the surfaces and a vacuum outside them. At what wavelength will a visible-light wave arriving perpendicular to the surfaces experience destructive interference when reflected? If the wavelength is increased by 10 percent, can the pair of surfaces still produce destructive interference between the reflected waves? If so, how? If not, why not?

SOLUTION 4-4

A wave arriving perpendicular to the surfaces will undergo destructive interference when the separation is 1/4 wavelength. In this case, the visible wavelength is 600 nm (4 times the spacing). If the wavelength increases by 10 percent to 660 nm, the apparatus can still produce destructive interference, but the incoming beam must arrive at a certain slant, so that the total path difference between the two reflected waves is 1/2 wavelength.

PROBLEM 4-5

Look again at the system shown in Fig. 4-16. Suppose that the light has a wavelength of 660 nm, and a ray strikes the surface at an angle of incidence so that the reflected waves cancel. What is this angle of incidence with respect to a line normal to the surfaces?

SOLUTION 4-5

For wave cancellation to occur, the total path length that the light ray travels between the surfaces must be 1/2 wavelength, which is $660/2 = 330$ nm. Therefore, the ray must travel $330/2 = 165$ nm from the top surface down to the bottom surface, and another 165 nm from the bottom surface back up to the top surface. The actual separation between surfaces is 150 nm. The necessary angle of incidence θ is therefore equal to the arccosine of the ratio of the actual separation to the distance the ray must travel:

$$\begin{aligned}\theta &= \arccos 150/165 \\ &= \arccos 0.909 \\ &= 24.6°\end{aligned}$$

Polarization

The *polarization* of an EM field is defined as the orientation of the electric-flux lines. Polarization-related effects occur at all wavelengths, from the very-low radio frequencies to the gamma-ray spectrum.

HORIZONTAL, VERTICAL, AND LINEAR POLARIZATION

When the electric lines of flux in an EM wave are oriented parallel to the earth's surface, the field has *horizontal polarization*. In contrast, *vertical polarization* is a condition in which the electric lines of flux are perpendicular to the earth's surface. Light having horizontal polarization reflects well from a horizontal surface such as a pool of water, while vertically polarized light reflects to a variable extent off the same surface. Simple experiments, conducted with *polarized sunglasses*, can illustrate the effects of visible-light polarization. In outer space, the terms "horizontal" and "vertical" are meaningless, but if the electric lines of flux are straight and maintain a constant orientation, then an EM wave is said to have *linear polarization*.

ELLIPTICAL AND CIRCULAR POLARIZATION

The polarization of an EM wave may vary as the EM field propagates through space or through a transparent medium. If the flux lines rotate, the wave has *elliptical polarization*. An elliptically polarized wave can rotate either clockwise or counterclockwise as it approaches an observer. If it rotates counterclockwise, it is said to have *right-hand elliptical polarization*. If it rotates clockwise, it is said to have *left-hand elliptical polarization*.

Elliptically polarized light is the equivalent of two linear-polarized waves that differ by 90° (1/4 cycle) in phase. If the amplitudes of the waves are equal, we have a special case of elliptical polarization known as *circular polarization*. In elliptical and circular polarization, the flux lines go through one full rotation per wave cycle. The angular rate, in radians per second, at which the lines of flux rotate is 2π times the wave frequency in hertz.

BREWSTER EFFECT

In Chap. 2, we learned that when a ray of light crosses a defined, flat boundary from a medium with a certain refractive index r to a medium with a higher refractive index s, the ray is bent at the boundary unless the angle of incidence is 0° (the ray strikes perpendicular to the boundary at the point of incidence). Not all of the light penetrates the boundary, because some of the light is reflected back at the same angle as the incidence angle.

If the incident ray has random polarization so that the wave amplitude is equal in all planes containing the incident path, the transmitted and reflected waves are both polarized. Figure 4-17 illustrates this phenomenon. At some angles of incidence, the effect is small; at other angles it is pronounced. The transmitted wave amplitude is always greatest for waves in the plane containing the transmitted path and a line normal to the surface, and smallest in the plane perpendicular to that. For the reflected wave, the amplitude is smallest in the plane containing the reflected path

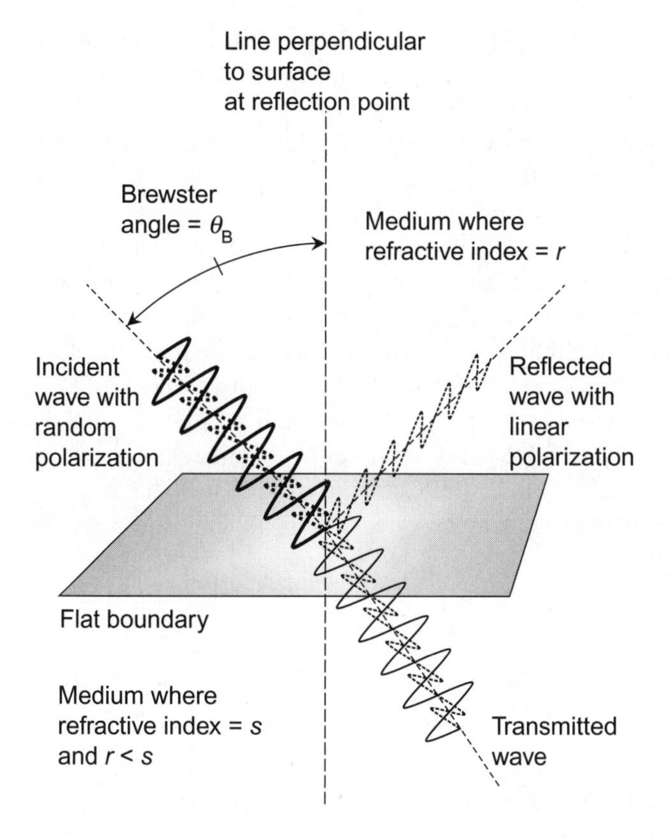

Figure 4-17 When randomly polarized light passes from a medium with a low refractive index r to a medium with a higher refractive index s, the reflected wave is linearly polarized at right angles to the plane of reflection when the angle of incidence equals the Brewster angle θ_B.

and a line normal to the surface, and greatest in the plane perpendicular to that. This phenomenon is called the *Brewster effect*.

When the incident ray arrives at a certain angle called the *Brewster angle* or *polarization angle*, the Brewster effect reaches a maximum, and the reflected wave is linearly polarized in the plane perpendicular to the plane containing the reflected path and a line normal to the surface. If we let θ_B represent the Brewster angle, then

$$\theta_B = \arctan (s/r)$$

where r is the index of refraction in the medium containing the incident and reflected waves, and s is the index of refraction in the medium containing the transmitted wave.

The fact that the reflected wave is linearly polarized causes some polarization of the transmitted wave as well, but the effect is not as pronounced as with the reflected wave. Because the index of refraction of a substance almost always varies with the wavelength of the light traveling through it, the Brewster angle for a boundary between two media is usually wavelength-dependent.

You can observe the Brewster effect by doing an experiment with polarized sunglasses. If you look at the reflection of the sky from the surface of a calm lake, you can rotate the sunglasses and see differences in the amount of reflected light that gets through the lenses. The effect can also be seen with light reflected from glass windows. Brewster effect does not occur at mirrors or shiny metallic surfaces, because they do not allow the light to penetrate the boundary. For Brewster effect to take place, the light must be transmitted, refracted, *and* reflected.

Figure 4-17 shows the Brewster effect for situations where $r < s$. The phenomenon can also occur when $r > s$, in which case the light passes from a medium with a relatively higher refractive index to one with a relatively lower refractive index. The same formula applies. Under these conditions, the Brewster angle is always smaller than the critical angle θ_c, which is given by

$$\theta_c = \arcsin (s/r)$$

When $r > s$, the angles θ_B and θ_c approach each other as the refractive indices of the two media become increasingly different. As the indices of refraction of the two media approach each other, θ_c approaches 90° while θ_B approaches 45°.

POLARIZATION BY RAYLEIGH SCATTERING

When randomly polarized light waves, such as those from the sun, undergo Rayleigh scattering in the atmosphere, polarization occurs in the scattered waves. The greatest polarization effect is in planes perpendicular to the path along which the light travels. When the sun is at the zenith, waves scattered in a plane parallel to the earth's surface tend to exhibit horizontal polarization. When the sun is on the western horizon, waves scattered from the zenith appear polarized in a vertical plane running north and south.

Polarization caused by Rayleigh scattering is not as pronounced as polarization caused by the Brewster effect. It can be noticed by doing an experiment with polarized sunglasses. The best results are obtained on clear days when the sun is midway between the horizon and the zenith. Look through the sunglasses at a point in the sky 90° away from the sun. Then rotate the glasses and observe the variations in the lightness of the sky. The effect will be greatest in the plane perpendicular to the line between you and the sun.

DICHROIC MATERIALS

Certain transparent crystals known as *dichroic materials* absorb light to an extent that varies depending on the polarization. Such crystals are used to manufacture

polarizing filters. One of the most common dichroic materials is a type of boron silicate called *tourmaline*. Polarizing materials can be artificially manufactured by heating, stretching, and staining specially treated plastics. This is how the lenses in polarized sunglasses are made.

When light having circular, elliptical, or random polarization strikes a sheet or plate of dichroic material at a right angle to the surface, the emerging rays have linear polarization that is always in the same plane. When linear-polarized light encounters such a sheet or plate at a right angle to the surface, the intensity of the emerging rays depends on the angle between the plane of the incident waves and the *polarization axis* of the dichroic medium.

You can conduct an experiment to observe the polarizing behavior of dichroic materials if you have access to two pairs of polarized sunglasses, or are willing to take the lenses out of a single pair. Close one eye and hold one lens between your open eye and a dim light source such as a low-wattage light bulb. When you rotate the lens in its own plane, you won't see any change in the intensity of the emerging light, although its polarization will rotate along with the polarization axis of the lens. If you place the other lens in front of the first one and then rotate either lens (but not both), the emerging light varies in intensity (Fig. 4-18). The maximum amount of light energy passes through the pair of lenses when their polarization axes are aligned. The minimum amount of light energy passes through when the polarization axes are perpendicular to each other. When the axes are at an intermediate angle, some, but not all, of light energy gets through the pair of lenses.

MALUS EFFECT

The variation in intensity of the light that gets through a pair of polarizing filters is known as the *Malus effect*. Imagine a pair of ideal polarizing filters, both of which produce perfect linear polarization in light arriving with random polarization. If we know the intensity x of the linear-polarized light passing through the first filter, and if we also know the angle ϕ between the polarization axes of the two filters, then we can calculate the intensity y of the light that emerges from the second filter as shown in Fig. 4-19. The formula is

$$y = x \cos^2 \phi$$

The ratio of the output intensity to the input intensity for the second filter (the one closer to the observer) is

$$y/x = \cos^2 \phi$$

These formulas, taken together, are called *Malus's law* or the *law of Malus*. From the second formula, we can see that when $\phi = 0°$, we have the ratio $y/x = 1$, so the

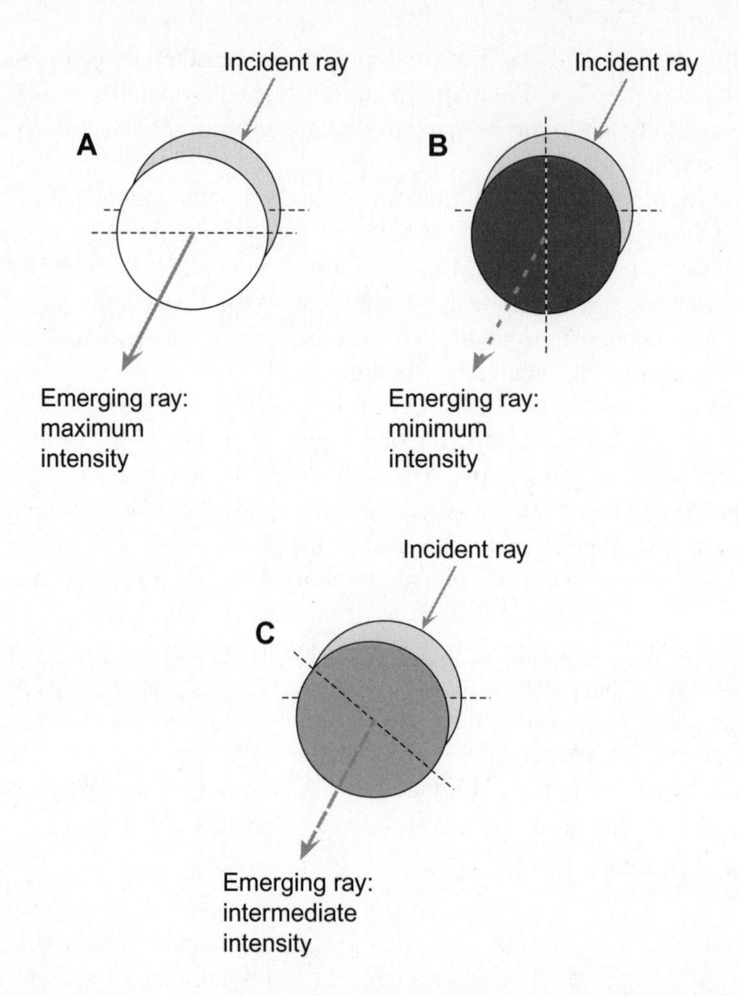

Figure 4-18 When two polarizing filters are placed in front of each other and you look at a light source through them, the intensity of the emerging beam depends on the angle between the polarization axes (dashed lines). At A, the axes are parallel. At B, the axes are perpendicular. At C, the axes are at an intermediate angle.

second filter transmits all of the light energy it receives. It's also apparent that when $\phi = 90°$, we have the ratio $y/x = 0$, so none of the light energy reaching the second filter emerges from it.

PROBLEM 4-6

Imagine a ray of light entering a calm pond of water whose index of refraction is 1.33. Consider the refractive index of the air to be 1.00. At what angle relative to the plane of the pond's surface is the Brewster effect greatest?

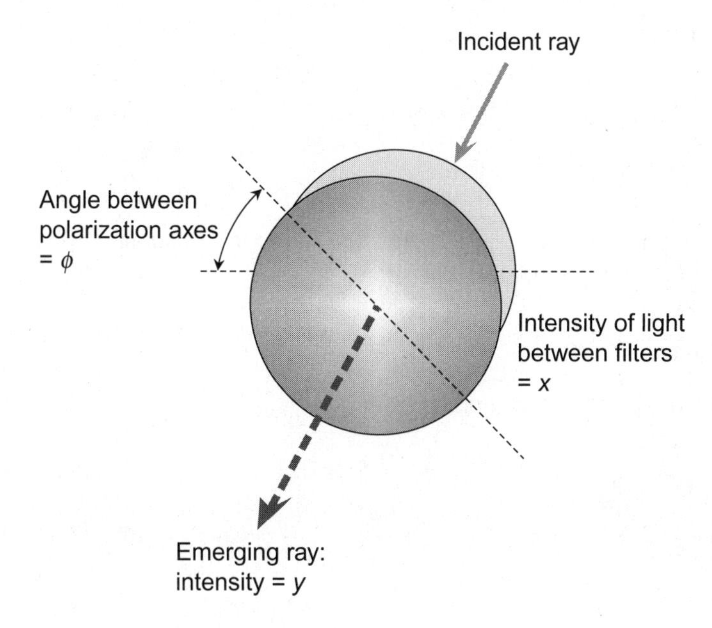

Figure 4-19 The law of Malus. The intensity y of the emerging light can be calculated if the intensity x between the filters, and the angle ϕ between the polarization axes, are known.

SOLUTION 4-6

We use the Brewster formula, using the input values $r = 1.00$ and $s = 1.33$. That gives us the Brewster angle θ_B, as follows:

$$\begin{aligned}\theta_B &= \arctan (s/r) \\ &= \arctan (1.33/1.00) \\ &= \arctan 1.33 \\ &= 53.1^\circ\end{aligned}$$

We must keep in mind the fact that the angle of incidence for a surface or boundary is defined as the angle relative to a normal (perpendicular) line at the point of entry. To obtain the angle θ_s relative to the plane of the surface, we must subtract θ_B from 90°. When we do that, we get

$$\theta_s = 90^\circ - 53.1^\circ = 36.9^\circ$$

PROBLEM 4-7

Consider a pair of ideal polarizing filters, one in front of the other, placed near a source of light so the rays arrive perpendicular to the planes of the filters as shown in Fig. 4-19. At what angle should the polarization axes be placed so that the second filter transmits exactly 1/4 of the light energy it receives?

SOLUTION 4-7

We must work the Malus formula "backward." The ratio of the output energy to the input energy for the second filter, y/x, is given as 1/4 or 0.25. Therefore, we have

$$0.25 = \cos^2 \phi$$

Taking the positive square root of both sides and then transposing the left-hand and right-hand sides of the equation, we obtain

$$\cos \phi = 0.5$$

We can take the arccosine of both sides to get

$$\begin{aligned} \phi &= \arccos 0.5 \\ &= 60° \end{aligned}$$

Quiz

This is an open book quiz. You may refer to the text in this chapter. A good score is 8 correct. Answers are in the back of the book.

1. Consider a system similar to the one shown in Fig. 4-16. Suppose that the surfaces are surrounded by air, and the medium between them is also air. If the surfaces are 149 mm apart, at what wavelength will a visible-light wave arriving perpendicular to the surfaces experience destructive interference when reflected?
   (a) 149 nm
   (b) 298 nm
   (c) 596 nm
   (d) 888 nm
2. Round-the-world "shortwave" communications is possible because of
   (a) Mie scattering.
   (b) Rayleigh scattering.
   (c) ionospheric effects.
   (d) polarization effects.

3. When a light ray passes from a medium having a high index of refraction to one with a lower index of refraction, the Brewster angle is always
   (a) smaller than the critical angle.
   (b) equal to the critical angle.
   (c) larger than the critical angle.
   (d) larger than 45°.
4. Suppose you want to make a diffraction grating for IR energy having a wavelength of 1800 nm. The narrowest workable spacing between the grooves is approximately
   (a) 450 nm.
   (b) 900 nm.
   (c) 1800 nm.
   (d) There is no limit as to how close the grooves can be.
5. Suppose you have a pair of ideal polarizing filters. You place one in front of the other and rotate them so the angle between their polarization axes is 45°. Then you hold the two filters near a source of light so the rays arrive perpendicular to the filters. What percentage of the light energy input to the second lens emerges from it?
   (a) 25 percent
   (b) 50 percent
   (c) 70.7 percent
   (d) 86.6 percent
6. Rayleigh scattering increases as
   (a) the wavelength increases.
   (b) the wavelength decreases.
   (c) the frequency decreases.
   (d) the photon energy decreases.
7. Consider two wave disturbances that mix by heterodyning. If one wave has frequency of f and the other wave has a frequency of $1.5f$, both measured in the same units, then the mixing products have frequencies of
   (a) $0.5f$ and $2f$.
   (b) $0.5f$ and $2.5f$.
   (c) $0.75f$ and $1.25f$.
   (d) $1.5f$ and $2.5f$.

8. Mie scattering is largely responsible for
   (a) the white appearance of the sky on a hazy afternoon.
   (b) the blue color of the sky at noon on a clear day.
   (c) the red color of the sky at sunset on a clear evening.
   (d) All of the above.
9. Sporadic-E effects occur at
   (a) all EM wavelengths.
   (b) IR wavelengths.
   (c) visible wavelengths.
   (d) radio wavelengths.
10. True radiation pressure can
    (a) cause stars to collapse when they burn out.
    (b) turn the vanes of a Crookes radiometer.
    (c) exert force on a reflective surface exposed to sunlight.
    (d) cause charged particles to stream toward the sun.

Laser Fundamentals

The word *laser* is an acronym that stands for *light amplification by stimulated emission of radiation*. Lasers have diverse applications. In this chapter, we'll learn how various types of lasers work, and see how they're used.

What Is Laser Light?

Laser light differs from ordinary light in two ways. First, all of its energy (in theory) is concentrated at a single wavelength, and therefore at a single frequency. Second, all of the "wave packets" (that is, photons) in a laser beam propagate in phase coincidence.

SPECTRAL DISTRIBUTION AND COHERENCE

The visible light from a conventional source such as the sun or a house lamp is not a clean, sine-wave EM disturbance. There are components from the red, or longest visible wavelength, through the violet, or shortest visible wavelength. Usually, there are components outside this range as well. If we look at a graph of light intensity as a function of wavelength for a particular light source, we get a *spectral distribution* for that source.

The spectral distribution for an *incandescent lamp* is biased toward the red end of the visible spectrum, as shown in Fig. 5-1A. A common *fluorescent tube* produces energy oriented toward the middle wavelengths (Fig. 5-1B). Some lamps emit light

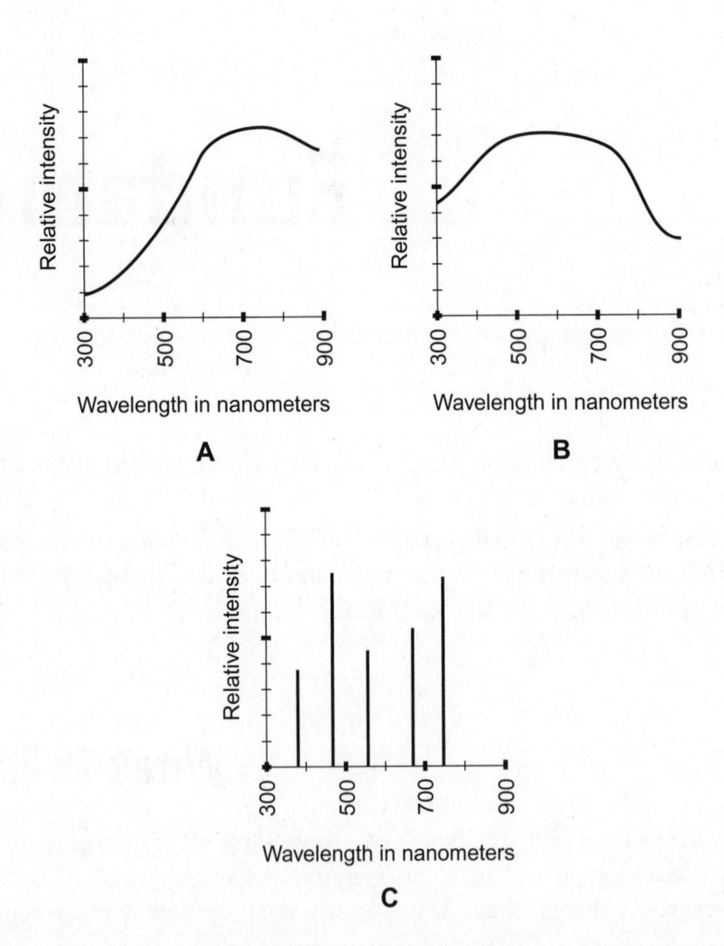

Figure 5-1 At A, the spectral distribution of light from an incandescent lamp. At B, the spectral distribution of light from a fluorescent lamp. At C, the spectral distribution of light from a hypothetical lamp that emits energy at multiple discrete wavelengths.

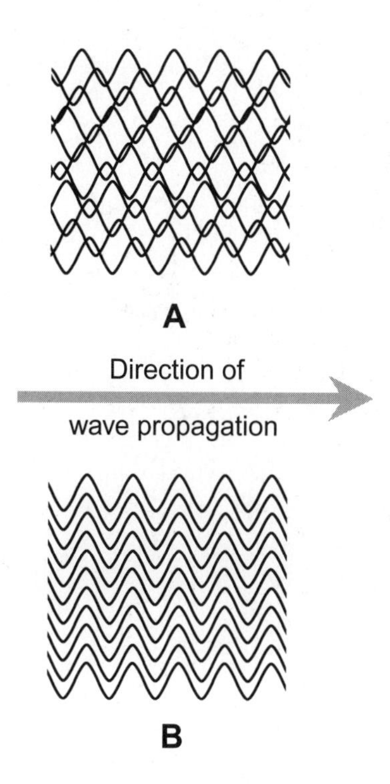

Figure 5-2 At A, waves of monochromatic with waveforms aligned at random. At B, waves of coherent monochromatic light.

at several discrete wavelengths (Fig. 5-1C), producing a distribution that shows well-defined *spectral lines*.

Even a lamp that emits light at discrete wavelengths, such as a *mercury-vapor lamp*, *sodium-vapor lamp*, or *neon-gas lamp*, produces myriad waves in random phase for each specific wavelength, as shown in Fig. 5-2A. Because of the random phase, such light is often called *incoherent*, whether it is monochromatic or not. In contrast, the visible light or other EM energy from a source that radiates at a single wavelength and only one phase component is said to be *coherent* (Fig. 5-2B) because all of the wavefronts are aligned. No ordinary lamp produces coherent light, but lasers do.

THE BOHR ATOM

When an electron is bound by an atomic nucleus, that electron can normally exist only at certain discrete energy levels. Figure 5-3 illustrates a simplified model of an atom, known as the *Bohr atom* after its inventor, physicist *Niels Bohr* who hypothesized it in the early 1900s. The circles represent cross sections of spheres called

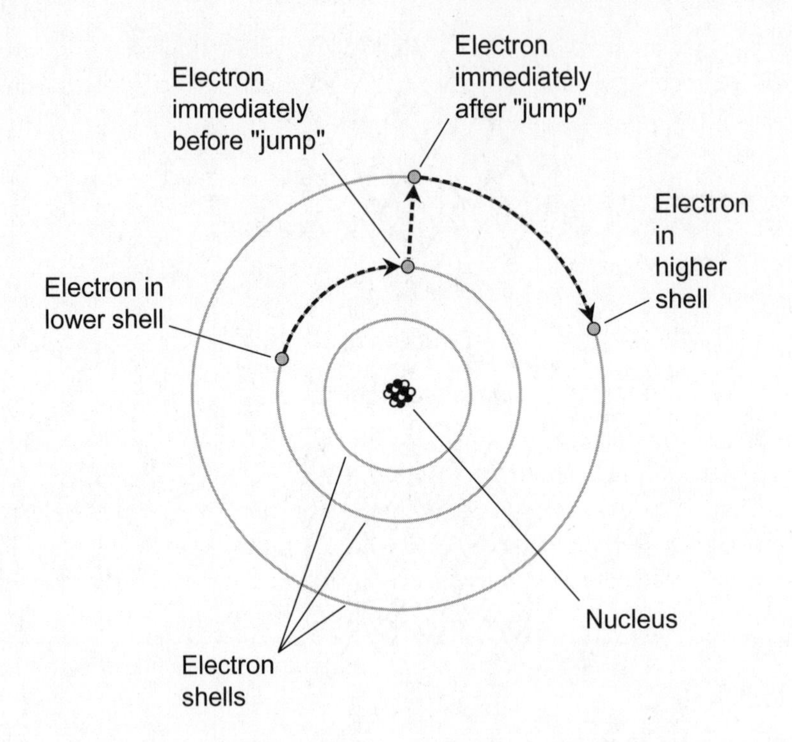

Figure 5-3 The Bohr model of the atom, showing how an electron gains energy as it "jumps" from a shell to another shell having a larger radius.

electron shells. The electrons orbit around the nucleus in the shells, each of which has a fixed radius. The amount of energy in an electron increases as its shell radius increases. An electron can "jump" to a higher level or "fall" to a lower level.

If a photon having precisely the right amount of energy (and therefore exactly the right wavelength) strikes an electron, the electron "jumps" from a given energy level to a higher one, as shown in Fig. 5-3. Conversely, if an electron "falls" from a higher energy level to a lower energy level, the atom emits a photon of a wavelength corresponding to the amount of energy the electron loses.

The Bohr model oversimplifies the actual behavior of electrons. In the modern view, electrons "swarm" around atomic nuclei so fast, and in such a complicated way, that it is never possible to pinpoint the location of any single electron. Instead, electron movement is defined in terms of probabilities. When we say that an electron is in a certain shell, we mean that at any point in time, the electron is just as likely to be located inside that shell as outside it.

Electron shells are assigned *principal quantum numbers*. The smallest principal quantum number is $n = 1$, which represents the smallest possible electron shell and therefore the lowest possible electron energy level. Progressively larger principal

quantum numbers are $n = 2$, $n = 3$, $n = 4$, and so on, representing shells of increasing radius. As the principal quantum number of an electron increases, so does the amount of energy it has, assuming all other factors are held constant.

THE RYDBERG-RITZ FORMULA

Changes in the energy contained by an electron within an atom are always attended by absorption or emission of EM waves. Let's consider the special case of hydrogen gas. Hydrogen atoms are the simplest in the universe, normally containing a single proton in the nucleus and a single electron in "orbit" around the nucleus.

Imagine a glass cylinder from which all the air has been pumped out. Suppose that a small amount of hydrogen gas is introduced, and then a direct-current (DC) electrical source is connected to electrodes at either end of the cylinder. This apparatus is called a *gas-discharge tube*. Such a tube can be constructed in any well-equipped high-school physics or chemistry lab. If the DC source voltage V is high enough, the gas inside the tube ionizes, and an electrical current I flows. As a result, a certain amount of power P is dissipated in the gas. The power, voltage, and current are related by the formula

$$P = VI$$

where P is expressed in *watts*, V is expressed in *volts*, and I is expressed in *amperes*. One watt (1 W) of power is the equivalent of energy dissipated at the rate of 1 *joule per second* (1 J/s).

If the DC voltage source remains active and stays connected to the electrodes, the gas in the tube absorbs energy over time. This causes the electrons in the atoms to attain higher energy levels, and therefore to reside in larger shells, than would be the case if there were no voltage source connected to the electrodes. As a result of the overall energy gain, the gas gets hot. This process cannot continue indefinitely. Therefore, as current continuously flows through the ionized hydrogen gas, electrons not only rise to higher energy levels (larger shells) but also constantly fall back to lower energy levels (smaller shells), emitting photons at various EM wavelengths.

Once an electron has fallen from a certain shell to another shell having a lower energy level, that electron is ready to absorb some energy from the DC source again, rise again, and then fall again. After a short initial "heating-up time," the amount of energy absorbed by the gas from the DC source, causing electrons to rise from smaller shells to larger ones, balances the energy radiated by the gas as electrons fall from larger shells back to smaller ones.

A hydrogen atom has several electron shells, so there are numerous different energy quantities that an electron can gain or lose as it moves from one shell to another. Energy is therefore radiated from ionized hydrogen gas in the form of photons having multiple discrete wavelengths λ as follows:

$$\lambda = [R_H (n_1^{-2} - n_2^{-2})]^{-1}$$

where λ is in meters, R_H is a constant called the *Rydberg constant*, n_1 is the principal quantum number of the smaller shell to which a particular electron descends, and n_2 is the principal quantum number of the larger shell from which that same electron has just fallen. The above equation is known as the *Rydberg-Ritz formula*, named after the physicists *Johannes Rydberg* and *Walter Ritz* who first published a paper about it in 1888. The value of the Rydberg constant, given by NIST to six significant figures, is

$$R_H = 1.09737 \times 10^7$$

This figure is sometimes rounded off to three significant figures as 1.10×10^7. It is defined in units of *per meter* (/m or m^{-1}). Some texts denote the constant as R with an infinity-symbol subscript (R_∞). Once in awhile you'll see it symbolized simply as R, but in that case you must not confuse it with the *universal gas constant*.

We've examined how electrons are related to EM energy for hydrogen gas. Other gases, as well as liquids and solids, can behave in a similar way, although their Rydberg constants are different. Every element, mixture, and compound has a unique set of wavelengths at which it can absorb or emit energy under certain conditions. With some materials, this phenomenon can be exploited to generate coherent light.

WAVE AMPLIFICATION

In a *gas laser*, the atoms behave a little differently than they do in a simple gas-discharge tube, because the gas is kept in a super-energized condition where an abnormally large number of electrons exist in high-energy shells. This state of affairs, called a *population inversion*, is the critical factor responsible for the *stimulated emission* of photons that makes a laser work.

When a photon strikes an electron in a gas laser tube, that photon is not absorbed. Instead, it continues to travel in the same direction, and with the same energy, as before. The electron, instead of rising to a higher shell, falls to a lower one, as shown in Fig. 5-4. When this transition occurs, the electron loses energy instead of gaining energy. Therefore, a new photon P_2 is emitted from the electron. It leaves the atom along with the original photon P_1.

If the original photon P_1 has exactly the right amount of energy, the new photon P_2 is emitted with the same wavelength, and in the same direction, as the original photon P_1. Photon P_1 then travels along with P_2 in such a way that the two wave disturbances exist in phase coincidence. When transitions of this sort occur in large

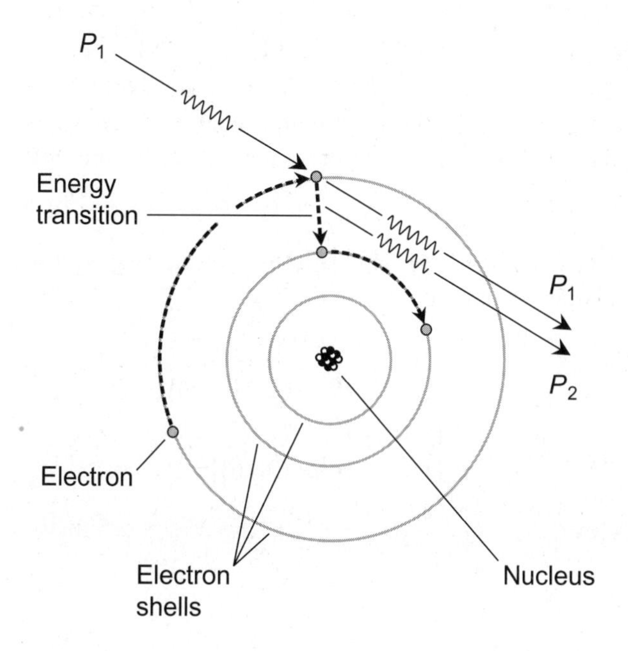

Figure 5-4 Amplification of EM energy by means of a photon interacting with an atom.

numbers, applied or incoming EM energy is amplified by the super-energized gas, a phenomenon called *wave amplification*. That's exactly what happens in a gas laser. A similar phenomenon can also occur in certain liquids and solids.

PROBLEM 5-1

Suppose that a high-voltage power supply producing 5000 volts DC is connected to a gas-discharge tube, ionizing the gas and causing a current of 0.0020 amperes to flow. How much power is dissipated in the gas?

SOLUTION 5-1

We multiply the voltage V in volts by the current I in amperes to get the power P in watts. In this case,

$$\begin{aligned} P &= VI \\ &= 5000 \times 0.0020 \\ &= 10 \text{ W} \end{aligned}$$

That's roughly the same amount of power consumed by a typical compact fluorescent lamp.

PROBLEM 5-2

Suppose an electron in a hydrogen atom falls from the shell whose principal quantum number is 4 to the shell whose principal quantum number is 2. What is the wavelength of the resulting photon? Express the answer to three significant figures.

SOLUTION 5-2

To solve this, we use the Rydberg-Ritz formula, letting $n_2 = 4$ and $n_1 = 2$. When we plug in the numbers, we obtain

$$\begin{aligned}
\lambda &= [R_H (n_1^{-2} - n_2^{-2})]^{-1} \\
&= [1.10 \times 10^7 \times (2^{-2} - 4^{-2})]^{-1} \\
&= [1.10 \times 10^7 \times (1/4 - 1/16)]^{-1} \\
&= [1.10 \times 10^7 \times (3/16)]^{-1} \\
&= (2.06 \times 10^6)^{-1} \\
&= 4.85 \times 10^{-7} \text{ m} \\
&= 485 \text{ nm}
\end{aligned}$$

Referring back to Table 1-1 in Chap. 1, we can see that this wavelength is in the blue part of the visible spectrum.

The Cavity Laser

One of the most common types of lasers, the *cavity laser*, is characterized by a *resonant cavity*, usually in the shape of a cylinder or prism, and an external source of energy. The cavity is designed so that EM waves bounce back and forth inside it, reinforcing each other in phase coincidence.

BASIC CONFIGURATION

The cavity contains, or is made of, a gas, liquid, or solid substance called the *lasing medium*. Mirrors are placed at each end of the cavity, so the EM waves reflect back and forth many times between them. One mirror reflects all of the energy that strikes it, and the other mirror is approximately 95 percent reflective. The mirrors can be flat or concave. The ends of the cavity can be perpendicular to the lengthwise cavity axis or oriented at a slant.

Figure 5-5 illustrates two common cavity laser designs. These drawings are vertically exaggerated for clarity. The waves represent the EM energy, usually IR or visible light, resonating between the reflectors. Only a few wave cycles are shown in this simplified functional drawing. In reality, thousands or millions of wave cycles occur between each reflection.

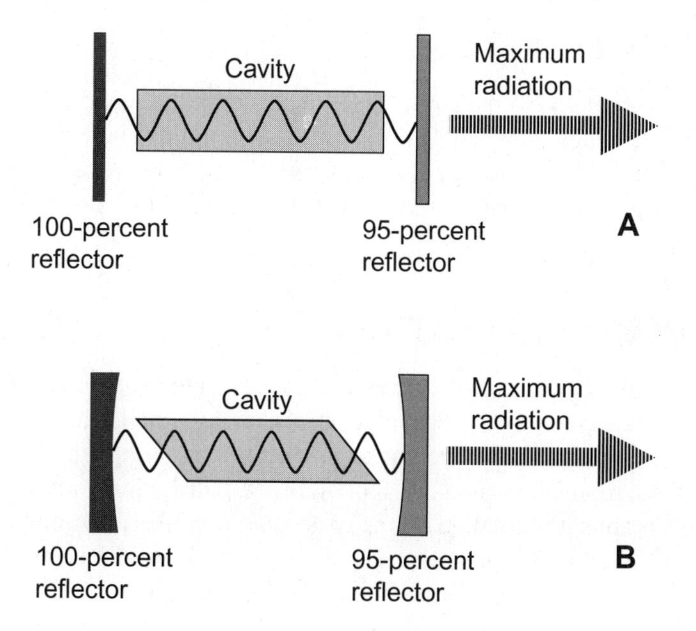

Figure 5-5 Two simple cavity lasers. At A, a flat-ended cavity and flat reflectors. At B, an angle-ended cavity and concave reflectors.

PUMPING

Energy is supplied to the lasing medium by means of a process called *pumping*, so the intensity of the beam builds to higher and higher levels as the energy is repeatedly reflected from the mirrors. The waves emerge from the partially silvered mirror in the form of coherent radiation, which concentrates the wave energy into the most intense possible beam.

BEAM RADIUS

The *beam radius* depends on the physical radius of the cavity and also on its length. Generally, small-radius cavities produce narrower beams at close range than large-radius cavities. However, large-diameter cavities tend to produce less *beam divergence*, which is a measure of the extent to which the beam radius increases with increasing distance from the source (once we get at least several cavity-lengths away). At a great distance, therefore, a large-radius laser often has a far narrower beam than a small-radius laser of the same type. Beam divergence occurs as a result of imperfections in the hardware, and also because of scattering and diffusion in the air or other medium the laser beam propagates through.

OUTPUT WAVELENGTH

The *output wavelength* from a cavity laser depends on the length of the cavity and on the atomic resonant wavelengths of the lasing medium. There are many sizes of cavity lasers that can produce an output of a given wavelength. Some cavity lasers are continuously tunable over a range of wavelengths. Most operate in the IR or visible part of the EM spectrum.

BEAM ENERGY DISTRIBUTION

The *beam energy distribution* at a given distance from the source can be defined by generating a cross-sectional graph of beam intensity versus the radius from the beam center in a plane perpendicular to the beam axis. Figures 5-6 and 5-7 show hypothetical examples for so-called spot and ring distributions, respectively. In all four of these graphs, the relative intensity is shown on the horizontal axis, and the relative distance from the beam center is shown on the vertical axis, with the intersection of the two axes representing the beam center. Figure 5-8 illustrates typical examples of spot and ring beams projected on a screen in a dark room.

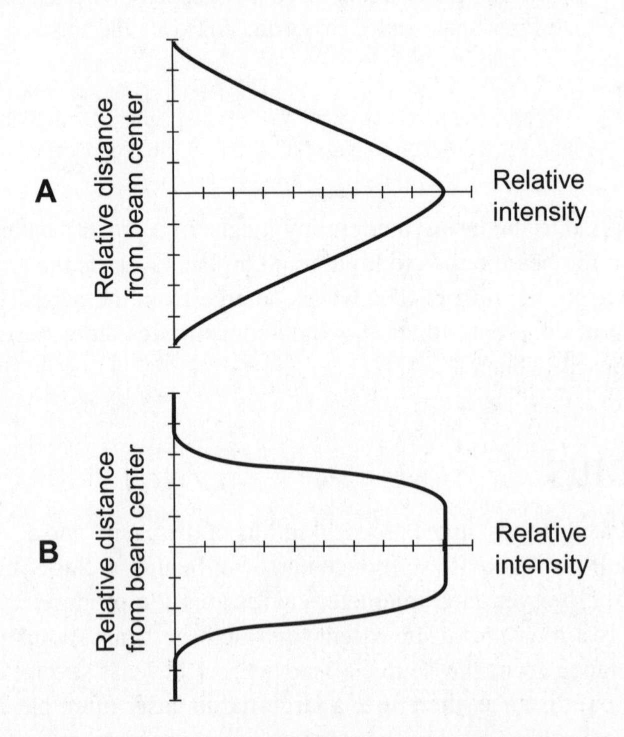

Figure 5-6 Hypothetical intensity versus radius graphs of "spot" laser-beam distributions. At A, sharp peak and gradual dropoff. At B, flat peak and abrupt dropoff.

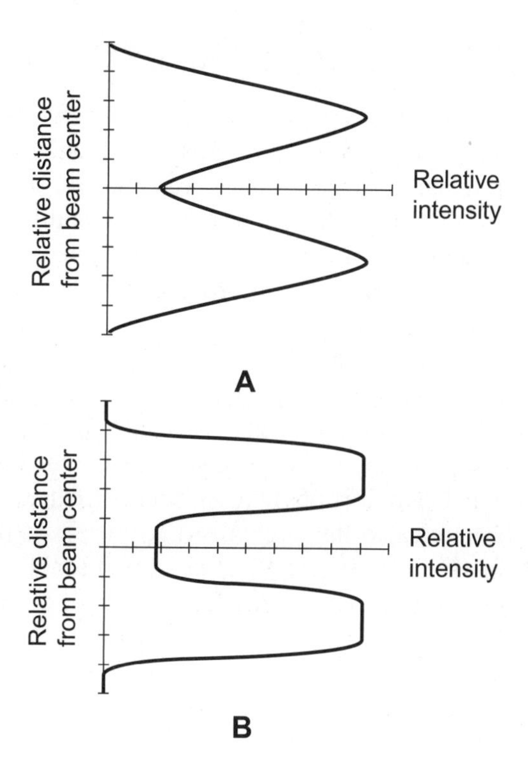

Figure 5-7 Hypothetical intensity versus radius graphs of "ring" laser-beam distributions. At A, sharp peaks and gradual dropoffs. At B, flat peaks and abrupt dropoffs.

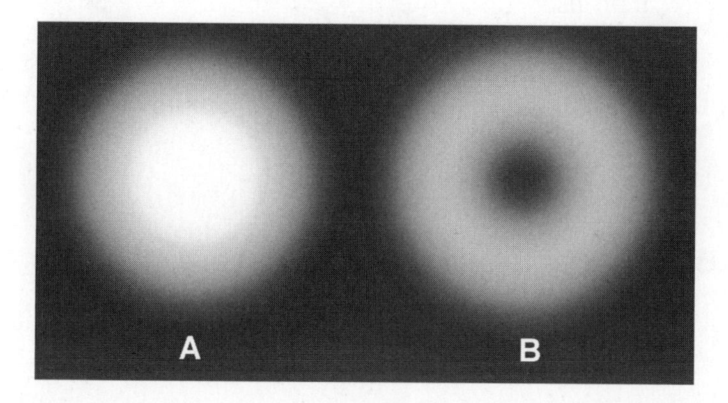

Figure 5-8 At A, a typical "spot" laser beam as it would look if projected on a screen in a dark room. At B, the projection of a "ring" beam on the same screen.

FOCUSING AND COLLIMATING THE BEAM

Because the rays from a distant light source are essentially parallel, they can be brought to a sharp focus and will cover a small area. The radius of the focused beam image is determined by the distance to the source, the size of the source, and the focal length of the lens or mirror. We are all familiar with the way sunlight can be focused to cause a piece of paper to catch fire. As the diameter of the lens or mirror increases, more light is gathered, and the focused beam becomes more intense (assuming the focal length remains constant).

When we want to focus a laser beam, the area of the lens or mirror need only be enough so that the whole beam is captured. This is, in theory, the same no matter what the distance of the laser from the lens. In practice there is some spreading of the beam, but it is not significant unless the distance from the laser is great.

With a coherent light beam, it is possible to obtain an extremely small region of light at the focus. Theoretically, this region is a geometric point regardless of the focal length of the lens or mirror (Fig. 5-9). In practice it is not quite a perfect point; if it were, it would have infinite energy density! Instead it is an extremely intense spot or ring of minuscule radius. Some laser beams can be focused with such precision that they can alter the genetic structure of a biological cell's nucleus.

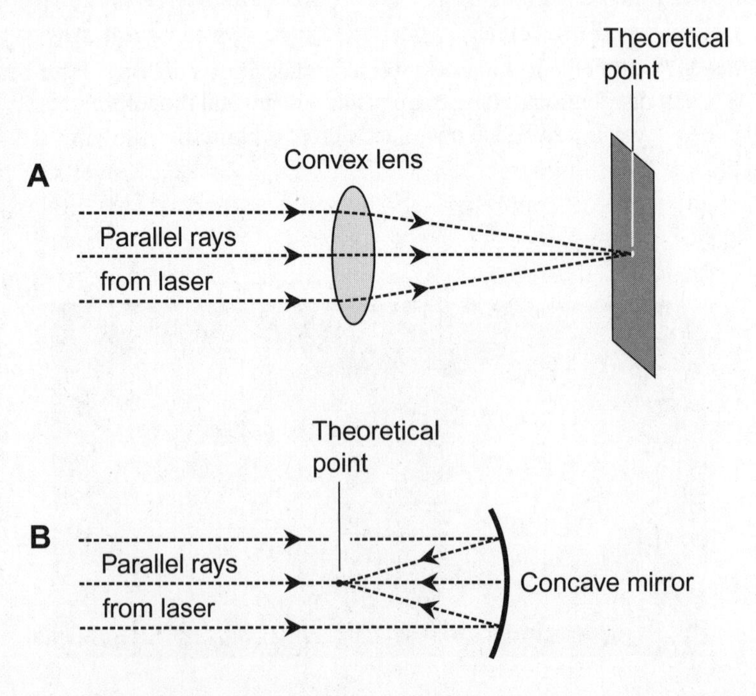

Figure 5-9 In theory, a coherent light beam can be focused to a geometric point by a convex lens (A) or by a concave mirror (B).

CONTINUOUS VERSUS PULSED

Some lasers deliver their output power at a constant rate as time passes, and therefore have equal peak and average power output. These are called *continuous-wave* (CW) *lasers*. Other lasers produce brief bursts having extreme peak power but much lower average power output. They are called *pulsed lasers*. Pulsed output can be achieved in a variety of ways.

A pulsed-laser technology known as *Q-switching* involves allowing the intensity of the light in the cavity to "build up" over a period of time while the cavity is in a state unfavorable for lasing. When the intensity reaches a certain level, the cavity conditions are adjusted, producing resonance. The result is an intense but brief "spike" of coherent radiation. The process is repeated indefinitely, producing spikes at regular intervals.

Pulsed lasing can also be obtained by supplying the cavity with bursts of energy, rather than with a continuous supply. This technique is useful with lasing media that cannot maintain CW operation. A brilliant flash lamp can be used as the pumping source in this mode.

ENERGY, EFFICIENCY, AND POWER

The *output energy* and *average output power* obtainable from a cavity laser depend on the physical size of the cavity, and on the amount of energy or power supplied by the pumping source.

The output energy is determined by measuring the total radiant energy E_{out} of the beam, in joules, over a period of time. This can be done by allowing the laser to strike a perfectly absorbing target (a so-called *black body*) that converts all the incident radiant energy to heat, and then measuring the amount of heat produced.

The efficiency *eff* of the laser can be determined by measuring the total pumping energy E_{in} over a certain period, measuring E_{out} for the same period, and then taking their ratio:

$$eff = E_{out}/E_{in}$$

The average output power $P_{avg\text{-}out}$ in watts is determined by measuring the output energy E_{out} in joules produced over certain period of time t in seconds, and then dividing the energy in joules by the number of seconds over which the energy measurement has been made:

$$P_{avg\text{-}out} = E_{out}/t$$

In terms of power, the efficiency *eff* can be determined by dividing the average output power $P_{avg\text{-}out}$ by the average input power $P_{avg\text{-}in}$:

$$eff = P_{avg\text{-}out}/P_{avg\text{-}in}$$

The peak power output $P_{\text{pk-out}}$ of a pulsed laser can be difficult to determine unless the pulses are *rectangular*: zero rise time, zero decay time, and constant amplitude for the duration of the pulse. In this situation (which is a theoretical ideal), the peak power output is equal to the average power output divided by the duty cycle D, which is the proportion of the time the laser is active:

$$P_{\text{pk-out}} = P_{\text{avg-out}}/D$$

In a pulsed laser, D is always smaller than 1 (usually much smaller), so P_{pk} is always greater than P_{avg} (usually much greater). If the peak output power and the duty cycle are known, the average power can be determined by the formula

$$P_{\text{avg-out}} = P_{\text{pk-out}}\ D$$

If the peak and average power output are both known, then the duty cycle is

$$D = P_{\text{avg-out}}/P_{\text{pk-out}}$$

PROBLEM 5-3

Suppose a laser produces rectangular pulses at a duty cycle of 2.0 percent (0.020), and having a peak power output of 1200 W. If the average input power is 30 W, what is the efficiency of this device?

SOLUTION 5-3

First, let's calculate the average output power. It's equal to the peak output power times the duty cycle:

$$\begin{aligned} P_{\text{avg-out}} &= P_{\text{pk-out}}\ D \\ &= 1200 \times 0.020 \\ &= 24\ \text{W} \end{aligned}$$

To determine the efficiency, we divide the average output power by the average input power, obtaining

$$\begin{aligned} \mathit{eff} &= P_{\text{avg-out}}/P_{\text{avg-in}} \\ &= 24/30 \\ &= 0.80 \end{aligned}$$

PROBLEM 5-4

Imagine that we modify the laser described in Problem 5-3 and its solution so its duty cycle becomes half as great, while the average input and output power do not change. What happens to the peak power output? What happens to the efficiency?

SOLUTION 5-4

The peak power output doubles to 2400 W. The new duty cycle is 0.010, but the average output power is still 24 W. Therefore

$$\begin{aligned} P_{\text{pk-out}} &= P_{\text{avg-out}}/D \\ &= 24/0.010 \\ &= 2400 \text{ W} \\ &= 2.4 \text{ kW} \end{aligned}$$

The abbreviation kW stands for *kilowatts*, or units of 10^3 W. The efficiency remains the same, because the average input and output power both stay the same.

Semiconductor Lasers

The term *semiconductor* arises from the ability of certain materials to conduct "part time." The conductivity can be controlled to produce various useful electrical, electronic, and optoelectronic effects. Common semiconductor materials include *silicon*, *germanium*, *selenium*, compounds of *cadmium*, compounds of *gallium*, compounds of *indium*, and various *metal oxides*.

THE SEMICONDUCTOR DIODE

There are two principal categories of semiconductors: *N type*, in which the electrical charge carriers are mainly electrons, and *P type*, in which the charge carriers are primarily *holes* (atoms having electron shortages). When wafers of N-type and P-type semiconductor material are placed in direct physical contact, the result is a *P-N junction*. When electrodes or wires are connected to the semiconductor wafers, we get a device called a *diode*. Figure 5-10 shows the engineer's symbol for a diode.

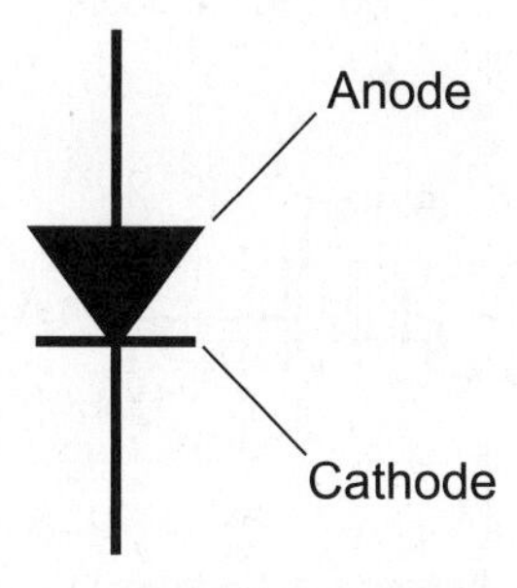

Figure 5-10 Anode and cathode symbology for a diode.

The N-type material is represented by the short, straight line in the symbol, and forms the *cathode*. The P-type material is represented by the arrow, and forms the *anode*.

BIAS

When a battery and a resistor are connected in series with a semiconductor diode as shown in Fig. 5-11A, electric current flows if the negative battery terminal is connected to the cathode and the positive terminal is connected to the anode. The resistor protects the diode from destruction by excessive current that could result from direct

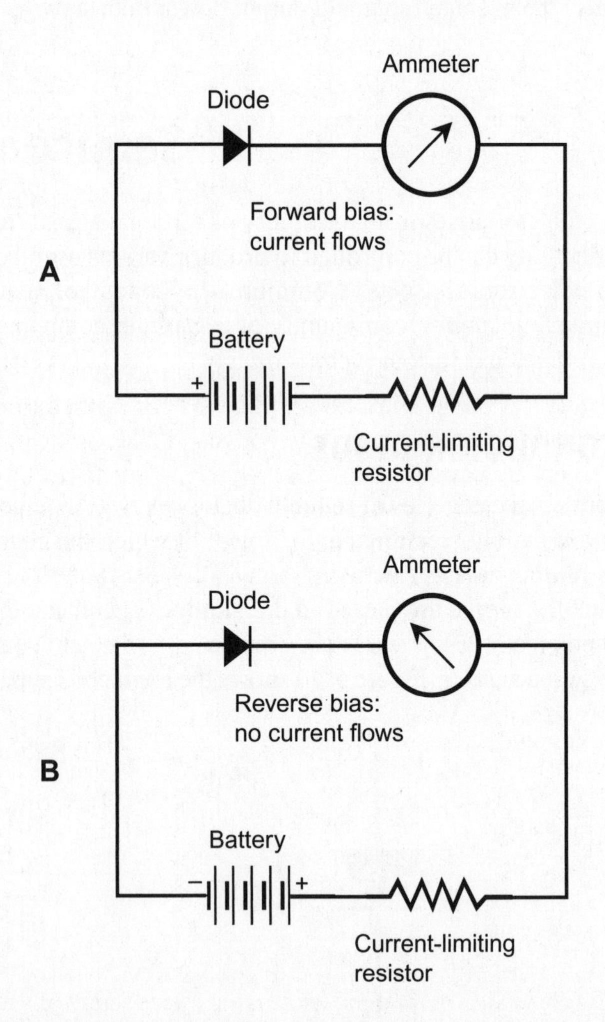

Figure 5-11 At A, forward bias of a diode results in a flow of electric current. At B, reverse bias produces essentially no current under normal conditions.

connection to the battery. This condition is called *forward bias*. If the battery is connected the other way as shown in Fig. 5-11B, current does not flow through the diode unless the battery voltage is exceptionally high. This condition is called *reverse bias*.

In a forward-biased diode, a certain minimum voltage called the *forward breakover voltage* is necessary for conduction to occur. Depending on the type of material, it varies from about 0.3 volt to 1 volt. In a germanium diode, for example, the forward breakover voltage is about 0.3 volt. In silicon it is approximately 0.6 volt. If the voltage across the junction is not at least as great as the forward breakover value, the diode will not conduct.

When a diode is reverse biased and the voltage is increased steadily, no current flows until a certain critical voltage is reached. Then the junction abruptly begins to conduct, and continues to conduct as the reverse voltage is further increased. This phenomenon is called the *avalanche effect*. The minimum voltage necessary to cause reverse conduction, the *avalanche voltage*, is considerably greater than, and is of the opposite polarity from, the forward breakover voltage.

PHOTOEMISSION

Some semiconductor diodes emit radiant energy when forward-biased so that current flows. This phenomenon, called *photoemission*, occurs as electrons in the atoms, driven to higher shells from the energy provided by the battery or other DC source, fall back down to lower energy states. The emission wavelength depends on the mixture of semiconductors in the diode. Most diodes produce emission in the visible or IR range.

The intensity of energy emission from a *light-emitting diode* (LED) or *IR-emitting diode* (IRED) depends on the forward current through the junction. As the current rises, the brightness increases, up to a certain point. If the current continues to rise past that point, no further increase in brilliance takes place. The LED or IRED is then said to be in a state of *saturation*.

A *laser diode*, also called an *injection laser*, is a special LED or IRED with a relatively large and flat P-N junction. If the forward current through the junction is below a certain threshold level, the device behaves like an ordinary LED or IRED, producing incoherent light or IR radiation. When a threshold current is reached, the emissions become coherent. Laser diodes, as well as conventional LEDs and IREDs, are designed to work properly only when forward-biased. They operate like ordinary diodes when reverse-biased, but do not normally radiate under those conditions.

GALLIUM-ARSENIDE LASER

Gallium arsenide (GaAs) is a compound used in the manufacture of various semiconductor devices, including diodes, LEDs, IREDs, *field-effect transistors* (FETs), and *integrated circuits* (ICs). Pure GaAs is an N-type semiconductor material. Because

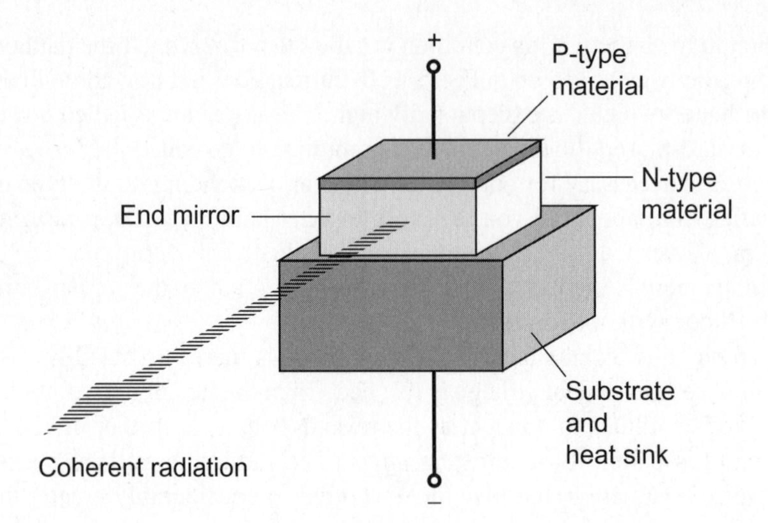

Figure 5-12 Simplified functional diagram of a laser diode. Coherent radiation is emitted in the plane of the P-N junction.

the charge carriers move fast and easily, GaAs is said to have high *carrier mobility*. Gallium arsenide can tolerate exposure to high levels of ionizing radiation (such as X rays and gamma rays), so GaAs components are ideal for use in outer space.

GaAs LEDs are used in numeric displays for consumer items such as electronic clocks. They can also be found in sophisticated systems such as radio transceivers, test instruments, calculators, and computers. GaAs IREDs, which have a primary emission wavelength of approximately 900 nm, are used in *fiberoptic communications systems*, robotic *proximity sensors*, handheld *remote-control* boxes for home appliances, and *motion detectors* for security systems.

All GaAs LEDs and IREDs can be switched rapidly on and off because of the high carrier mobility of the substance. Some such devices can be modulated at frequencies in excess of 1 GHz (10^9 Hz), making them ideal for use in broadband or high-speed optical communications systems. Figure 5-12 illustrates the construction of a typical GaAs laser diode.

HYBRID SILICON LASER

In order to meet the demand for inexpensive, mass-producible silicon-based LEDs and IREDs, the *hybrid silicon laser* (HSL) was developed. Silicon is combined in layers with certain other semiconductor compounds called *III-V materials*. These compounds, which include *indium phosphide* (InP) and GaAs, allow for energy emission when excited by either an electrical current or an external source of coherent

light. The lasing medium is surrounded by mirrors to form an enclosed medium called a *waveguide*, which produces coherent light by creating a resonant state where the waves reinforce each other.

QUANTUM CASCADE LASER

In some semiconductor lasers, electrons make "cascading" jumps among multiple atoms as shown in the simplified illustration of Fig. 5-13. This type of device, called a *quantum cascade laser* (QCL), produces more output than a conventional semiconductor laser because of a "chain reaction" of energy transitions. In Fig. 5-13, only two semiconductor layers are shown, but a typical QCL has many layers, producing a massive cascade effect. The emission wavelength, which depends on the thickness of the semiconductor layers, is in the middle-IR to far-IR range.

VERTICAL-CAVITY SURFACE-EMITTING LASER

In a *vertical-cavity surface-emitting laser* (VCSEL), a coherent visible or IR beam is emitted perpendicular to the plane of the P-N junction, is in contrast to the usual laser diode where beam emerges in the plane of the junction. Lasing takes place inside a region called the *quantum well*, which is sandwiched between two semiconductor layers that also serve as end reflectors (Fig. 5-14).

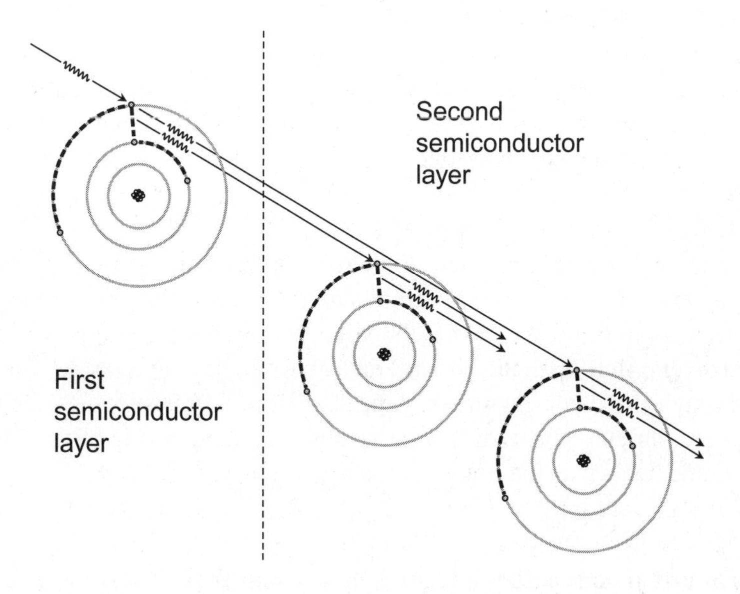

Figure 5-13 In a QCL, multiple energy transitions produce greater output than can be obtained with a conventional semiconductor laser.

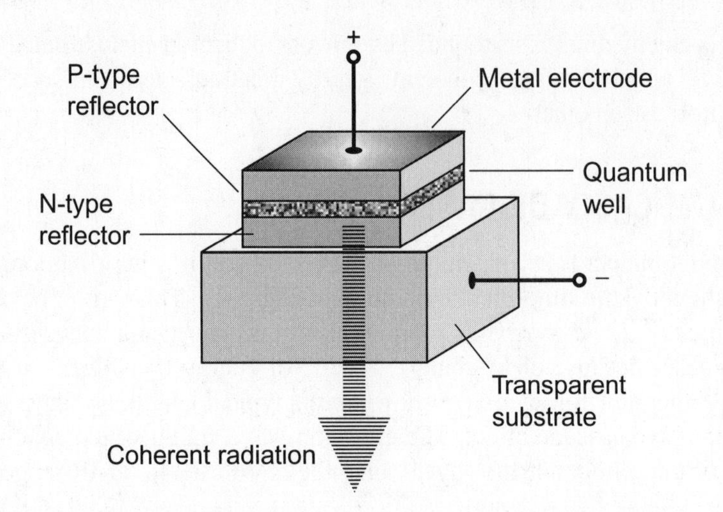

Figure 5-14 Simplified functional diagram of a VCSEL. Coherent radiation is emitted at right angles to the plane of the P-N junction.

The VCSEL design offers some advantages over conventional laser diodes. During manufacture, testing can be done at multiple stages prior to production. In operation, VCSELs usually produce less beam divergence than conventional laser diodes, because the emitting surface dimensions are larger.

Vertical-cavity devices based on GaAs operate at wavelengths from the visible red (about 650 nm) to the near-IR (about 1300 nm). Devices based on InP can function at wavelengths up to approximately 2000 nm. The wavelength can be adjusted by varying the thickness of the reflecting layers.

PROBLEM 5-5

Suppose a laser diode is connected in a circuit with a battery and a *potentiometer* (variable resistor) as shown in Fig. 5-15. The P-N junction is forward-biased. The potentiometer is initially set so that the voltage across the diode is below the forward breakover point. Then the potentiometer resistance is gradually reduced so the current through the diode increases. A current-limiting resistor protects the diode and the potentiometer, and avoids excessive power drain on the battery. How does the laser diode behave as the potentiometer resistance goes down?

SOLUTION 5-5

At first, the P-N junction does not conduct, because the voltage across it is below the forward breakover point. The ammeter registers zero. No energy is emitted. When forward breakover occurs, current begins to flow. The ammeter shows some

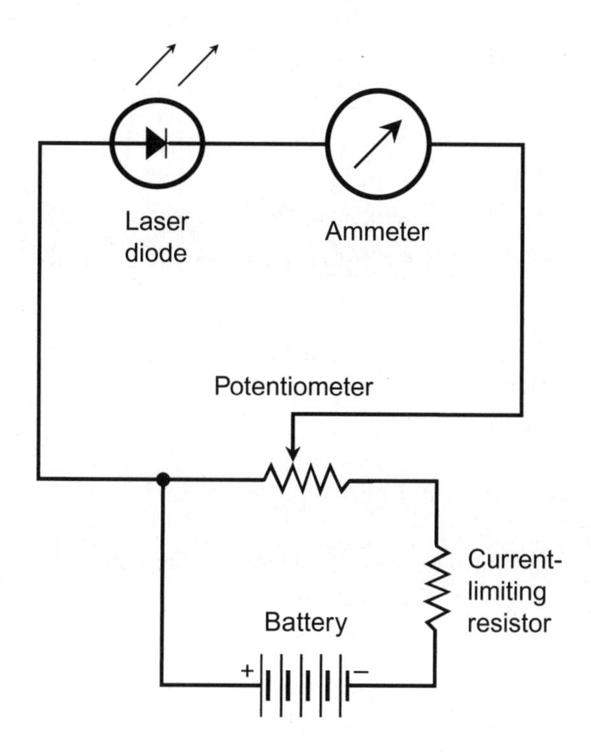

Figure 5-15 Illustration for Problem 5-5 and its solution.

current, and the diode emits incoherent light or IR. As the potentiometer resistance keeps decreasing, the ammeter registers increasing current. When the lasing point is reached, the emission becomes coherent. If the current is further increased, the ammeter reading rises to a certain value and then levels off at the saturation point. If the current rises higher still, no changes occur in the behavior of the diode, assuming the current-limiting resistor has enough resistance to prevent destruction of the P-N junction.

Solid-State Lasers

Solid-state lasers include the *ruby laser*, the *crystal laser*, and various lasers using a solid combination of *yttrium*, *aluminum*, and *garnet* (called YAG). A solid-state laser is optically pumped by a *flash tube* that surrounds the lasing medium.

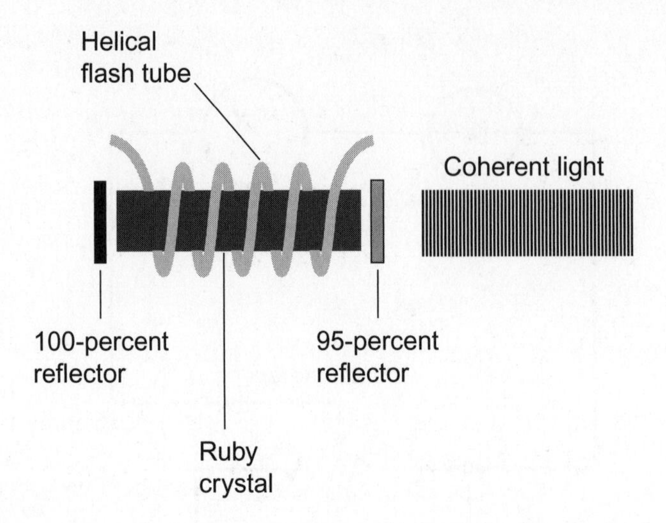

Figure 5-16 Simplified functional diagram of an optically pumped ruby laser.

RUBY LASER

The lasing medium in a ruby laser is primarily *aluminum oxide* with a trace of *chromium* added. The ruby crystal is a cylindrical piece of this material, which appears as a reddish solid with reflective surfaces at each end. One reflector is totally silvered so it is 100 percent reflective. The other is partially silvered so it's about 95 percent reflective. The laser beam emerges from the partially silvered end. Figure 5-16 is a simplified diagram of the device. The flash tube emits brilliant, brief pulses of visible light that cause energy transitions in the electrons of the ruby crystal. Coherent emission occurs in pulses at a wavelength of approximately 694 nm, which is at the red end of the visible spectrum.

With 1 W of power input to the flash tube, a typical ruby laser produces a few milliwatts (mW) of coherent light output, where 1 mW is equal to 10^{-3} W. The rest of the input power is dissipated as heat or radiated as stray light from the flash tube. The output pulses have duration on the order of 500 microseconds (500 μs) to 1 millisecond (1 ms). In some arrangements, a large ruby laser is pumped by a smaller one. The larger laser acts as an amplifier to obtain peak power output far greater than is possible with a single device.

CRYSTAL LASER

A solid-state laser having multiple compounds in the lasing medium is called a *mixed crystal laser*. A common cavity material is a mixture of gallium arsenide and *gallium antimonide*, termed *gallium-arsenide-antimonide* (GaAsSb). Other materials used in

mixed crystal lasers include *gallium-arsenide-phosphide* (GaAsP) and *aluminum-gallium-arsenide* (AlGaAs).

Solid-state lasing crystals are "grown" in a liquid solution in a manner similar to semiconductor diode and transistor crystals. The semiconductor materials are *doped* (mixed with trace amounts of impurities) in various concentrations to obtain N-type and P-type properties. The emission wavelength can be adjusted by varying the relative concentrations of the substances in the cavity medium. In a GaAsSb laser, the wavelength can be tuned continuously from roughly 900 to 1200 nm in the near-IR range.

An *insulating crystal laser* consists of a rare-earth element that is slightly doped. The resulting laser emission is usually at near-IR or visible red wavelengths. The insulating crystal laser is designed to operate at a cool temperature, so an elaborate heat-dissipation system is required. Overheating will result in deterioration of the lasing properties of the crystal. Wavelength adjustment is difficult, so the insulating crystal laser is generally used at only a single frequency.

NEODYMIUM LASER

The *neodymium-yttrium-aluminum-garnet* (or *neodymium-YAG*) laser is a specialized solid-state laser that combines liquid and solid materials. The elemental neodymium is in a liquid solution and is confined within a solid YAG crystal. Energy is delivered into the laser medium by arc lamps or other bright sources of light focused on the crystal.

The wavelength of the neodymium-YAG laser is about 1065 nm in the near-IR portion of the spectrum. The efficiency is on the order of 1 percent. High-power neodymium-YAG lasers, like insulating crystal lasers, require a method of cooling to prevent damage to the crystal. The neodymium-YAG laser can operate in the pulsed mode or in the continuous mode.

A neodymium-based laser can also take the form of an insulating crystal device. Glass can be doped with neodymium to obtain laser operation with sufficiently intense optical pumping. When maintained at a proper temperature (20°C or lower), the efficiency of a *neodymium-glass laser* is about 3 percent when glass of high optical quality is used. The peak output power can be as high as 1 *terawatt* (1 TW), which is 10^{12} W. The duty cycle is low, so the average input and output power are much less than 1 TW.

PROBLEM 5-6

Suppose that a neodymium-glass laser produces 500 gigawatts (500 GW) of peak output power, where 1 GW = 10^9 W. If the average output power is 30 W and we assume that the output pulses are rectangular, what is the duty cycle as a percentage? If the efficiency is 3.00 percent, what is the average input power?

SOLUTION 5-6

To calculate the duty cycle D, we remember the formula for D in terms of the peak power output $P_{pk\text{-}out}$ and the average power output $P_{avg\text{-}out}$:

$$D = P_{avg\text{-}out}/P_{pk\text{-}out}$$

Plugging in the values $P_{avg\text{-}out} = 30$ and $P_{pk\text{-}out} = 500 \times 10^9$ (both in watts), we get

$$\begin{aligned} D &= 30/(500 \times 10^9) \\ &= 30/(5.00 \times 10^{11}) \\ &= 6.0 \times 10^{-11} \end{aligned}$$

Expressed as a percentage, the duty cycle $D_{\%}$ is

$$\begin{aligned} D_{\%} &= (6.0 \times 10^{-11}) \times 100 \\ &= 6.0 \times 10^{-9} \text{ percent} \end{aligned}$$

The average input power can be determined by plugging in the values $eff = 0.0300$ and $P_{avg\text{-}out} = 30$ to the formula for efficiency in terms of average input and output power:

$$eff = P_{avg\text{-}out}/P_{avg\text{-}in}$$

This gives us

$$0.0300 = 30/P_{avg\text{-}in}$$

Therefore,

$$\begin{aligned} P_{avg\text{-}in} &= 30/0.0300 \\ &= 1.0 \times 10^3 \text{ W} \end{aligned}$$

That's 1.0 kilowatt (1.0 kW).

Other Noteworthy Lasers

Some of the earliest lasers employed gas and liquid media, and their basic designs have not changed for decades. Here are three examples.

HELIUM-NEON LASER

Small lasers using gas-filled tubes containing helium and neon are available through scientific hobby companies. The *helium-neon* (He-Ne) *laser* emits vivid red light. The gas is excited by an electric current, typically producing emission at 633 nm. Some designs produce IR output at 1150 nm or 3390 nm. The output power is typically in the range of 10 mW to 100 mW, and the efficiency is approximately 5 percent at the visible red wavelength.

The lasing wavelength is determined by energy transitions shared by the two gases. Helium and neon have identical transitions at some wavelengths when their concentration levels are at a certain value and in the correct proportion with respect to each other. The device is pumped by means of electron collision. Energy transitions in the electron shells of the gases occur because the applied current ionizes the gases, producing free electrons. If a high-frequency alternating current (AC) is passed through the ionized gas, the output is continuous. This offers advantages for optical communications because a continuous laser is easy to modulate with digital or analog data. A continuous-mode He-Ne laser can be modulated by many signals at the same time, each signal having a different frequency.

Figure 5-17 is a pictorial diagram of a He-Ne laser excited by *radio-frequency* (RF) energy. Some He-Ne lasers employ flat mirrors at each end of the resonant

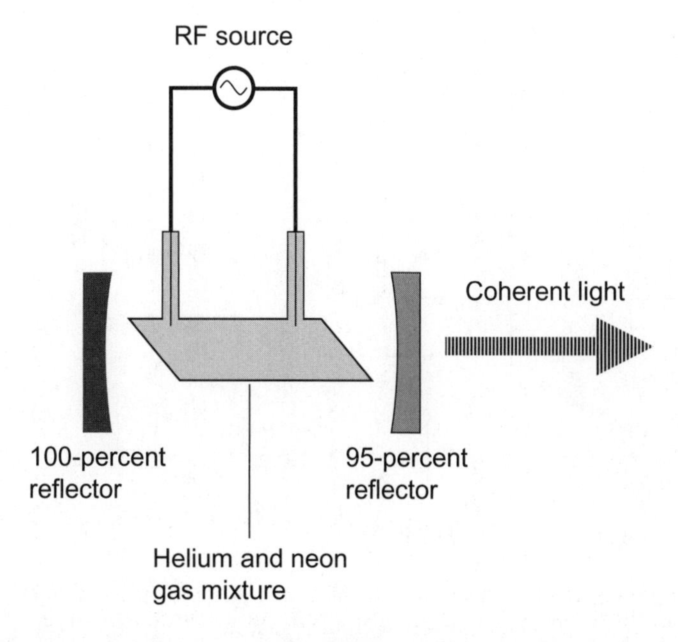

Figure 5-17 Simplified functional diagram of an RF-pumped He-Ne laser.

tube, and other designs (such as the one shown here) use concave mirrors and a tube with angled ends. The mirrors are of the *first-surface* type, meaning that the silvered surface is on the inside, facing the interior of the tube.

NITROGEN-CARBON-DIOXIDE-HELIUM LASER

A mixture of nitrogen, carbon dioxide, and helium can be used to make a laser that operates at a wavelength of about 1060 nm in the IR spectrum. Energy at this wavelength propagates with exceptionally low loss over long distances through clear air. The *nitrogen-carbon-dioxide-helium* (N-CO_2-He) *laser* can generate several kilowatts of continuous output power, and pulse peaks in the gigawatt range. The gases can be excited by direct current (DC) as shown in Fig. 5-18.

ARGON-ION LASER

The *argon-ion laser* is a gas laser operating at about 480 nm, producing blue visible light. The ionized argon is kept at low pressure. The ionization and energy input are produced by passing an electric current through the gas. The power output is low, but so is the efficiency, so a cooling system is usually necessary. The pictorial representation of the argon-ion laser is essentially the same as that for the N-CO_2-He laser (Fig. 5-18).

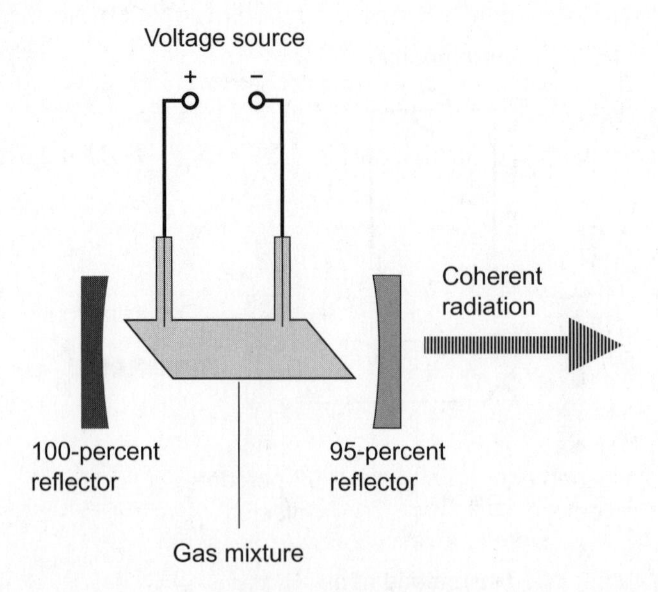

Figure 5-18 Simplified functional diagram of a gas laser excited by DC from an external voltage source.

MISCELLANEOUS GAS LASERS

Many other gases, such as mercury vapor, can be used to generate laser emissions by means of DC electric discharge. Hydrogen and xenon produce UV laser energy. Some gases, such as oxygen and chlorine, can be used in lasers excited by visible light. Atomic collisions, caused by application of energy in various forms, can result in the emission of laser energy in the IR, visible, and UV parts of the spectrum.

LIQUID LASERS

A laser cavity can be filled with an element or compound in a liquid state. A typical *liquid laser* can produce energy in the near IR, visible, or UV range. The device is pumped by an external source of visible light. Some liquid lasers produce continuous output; others produce pulsed output. In the continuous mode, the pumping source is usually a gas laser or solid-state laser. In the pulsed mode, pumping can be done by a source of incoherent light such as a flash tube. The lasing medium is a colored dye.

In some liquid lasers, the emission wavelength can be varied over a continuous range. Such a laser operates at a wavelength determined by a chemical dye dissolved in the liquid. Like their gas and solid-state counterparts, a liquid laser must be effectively cooled if high power output is required. Otherwise, the cavity enclosure may be damaged or destroyed by overheating.

PROBLEM 5-7

What is the advantage of using first-surface mirrors at the ends of the cavity in a gas or liquid laser?

SOLUTION 5-7

When IR, visible light, or UV is reflected from a first-surface mirror, the radiation encounters the reflective surface directly. In a conventional *second-surface mirror* (such as the type used on bathroom walls), the radiation must pass through the glass before and after striking the reflective surface. Because some incident radiation is inevitably reflected from an unsilvered glass surface, a second-surface mirror produces two reflected images that interfere with each other, reducing the efficiency of a laser device. First-surface mirrors are more expensive to produce than second-surface mirrors, and they are more easily damaged. However, because of their superior optical properties, first-surface mirrors are preferred in all precision optical instruments including telescopes, microscopes, cameras, and lasers.

Quiz

This is an open book quiz. You may refer to the text in this chapter. A good score is 8 correct. Answers are in the back of the book.

1. Fill in the blank to make the following sentence true: "The ratio of the peak power to the average power in a pulsed laser depends on the ________, assuming all other factors are held constant."
   (a) wavelength
   (b) energy input
   (c) energy output
   (d) duty cycle
2. Which of the following wavelengths is characteristic of the output of a He-Ne gas laser?
   (a) 1060 nm.
   (b) 900 nm.
   (c) 633 nm.
   (d) Anything in the UV range.
3. If a laser diode is reverse-biased below the avalanche voltage,
   (a) current flows through the P-N junction, and the device emits coherent radiation.
   (b) current flows through the P-N junction, and the device emits incoherent radiation.
   (c) current flows through the P-N junction, but the device does not radiate.
   (d) no current flows through the P-N junction.
4. Assuming all other factors are held constant, the output wavelength of a pulsed cavity laser depends on
   (a) the length of the cavity.
   (b) the duty cycle.
   (c) the peak energy output.
   (d) the shape of the output pulses.

5. The charge carriers in P-type semiconductor material are mainly
   (a) holes.
   (b) electrons.
   (c) protons.
   (d) nuclei.
6. When an electron moves from a certain shell in an atom to another shell having a smaller radius, the electron
   (a) gains energy.
   (b) emits a photon.
   (c) becomes coherent.
   (d) attains a shorter wavelength.
7. Suppose that the peak output power from a neodymium-glass laser is 660 GW. If the average output power is 33 W and we assume that the output pulses are rectangular, what is the duty cycle expressed as a percentage?
   (a) 2.2×10^{-11} percent.
   (b) 5.0×10^{-10} percent.
   (c) 2.0×10^{-9} percent.
   (d) None of the above.
8. If the average input power to the device described in Problem 7 is 1.0 kW, what is the efficiency expressed as a percentage?
   (a) 2.2 percent.
   (b) 3.3 percent.
   (c) 5.0 percent.
   (d) More information is needed to answer this question.
9. Coherent EM radiation occurs at
   (a) multiple frequencies, with all waves in phase coincidence.
   (b) multiple frequencies and random phases.
   (c) a single frequency, with all waves in phase coincidence.
   (d) a single frequency and random phases.

10. Which of the following is a significant advantage of a VCSEL over a conventional injection laser?
    (a) The VCSEL typically has less beam divergence than a conventional injection laser.
    (b) The VCSEL does not require semiconductor materials to function properly.
    (c) The VCSEL can work without any source of electrical power.
    (d) The VCSEL emits its energy in the plane of the P-N junction.

CHAPTER 6

Optical Data Transmission

Visible light, IR, and UV rays can be used to transmit and receive information in much the same way as can electric currents or radio waves. In this chapter, we'll examine a few of these technologies.

Optical and Image Transducers

An *optical transducer* converts visible light, IR, or UV energy to some other form or vice versa. Such devices include LEDs and IREDs, which we learned about in the last chapter. In this section, we'll examine several other forms, and also look at some devices that make use of them.

PHOTOCONDUCTIVITY

Many materials exhibit electrical conductance that varies with the intensity of illumination by IR, visible light, UV, X rays, or gamma rays. This effect is called *photoconductivity*, and substances that exhibit it are known as *photoconductive materials*. Photoconductivity occurs in many substances, but it is most pronounced in semiconductor materials such as silicon, gallium arsenide (GaAs), germanium, and some sulfides.

A sample of photoconductive material has a certain conductance when there is no illumination. As the illumination level increases, the conductance improves. There is a limit to the extent that the conductance continues to increase as the illumination becomes more intense. As the brightness of the radiation source rises beyond this limit, the conductivity of the material does not improve further. This condition is called *saturation*, and the conductance in this state is called the *saturation conductance*.

PHOTODIODE

A *photodiode* is a semiconductor diode whose conductance varies depending on the intensity of illumination that strikes its P-N junction. Most diodes having P-N junctions exhibit photoconductivity, but an ordinary diode does not work as a photodiode because its housing is opaque, and/or because the P-N junction is between two opaque pieces of semiconductor material. A photodiode is designed so that its P-N junction is easily exposed to external sources of radiation, especially at IR, visible, and UV wavelengths. The device has a relatively large P-N junction surface area to maximize the amount of radiant energy that strikes the boundary between the two types of semiconductor material.

In total darkness, a reverse-based photodiode does not conduct (unless the reverse voltage exceeds the avalanche voltage). When the P-N junction is illuminated by a source of increasing brilliance, the conductivity of the photodiode increases up to a well-defined saturation conductance. When the device is operating in the so-called *saturation region*, the conductance will not increase no matter how bright the illumination gets. This effect is illustrated graphically in Fig. 6-1, assuming that a constant reverse-bias voltage is applied to the diode. The relative level of illumination is shown on the horizontal axis, and the relative reverse current is shown on the vertical axis. The saturation region is the flat part of the curve.

A photodiode is normally connected in series with the circuit to be controlled, and is always reverse-biased. Care must be exercised to ensure that the photodiode is not forced to carry too much current, because excessive current will destroy the P-N junction. The maximum current can be limited by placing a resistor in series with the device.

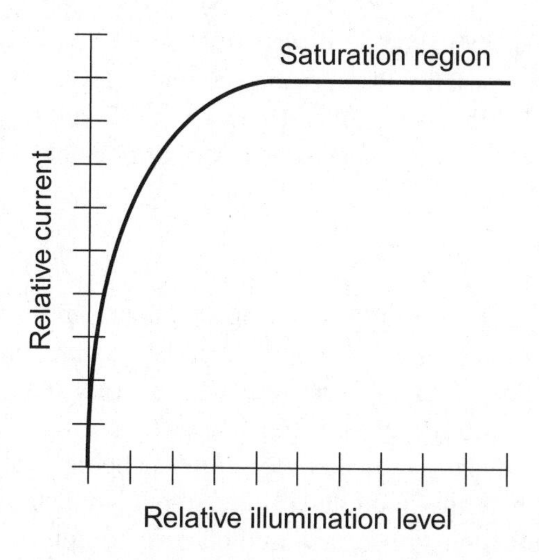

Figure 6-1 Electric current through a typical photodiode as a function of illumination intensity, assuming the applied reverse voltage remains constant.

PHOTOTRANSISTOR

A *phototransistor* is a specialized *semiconductor transistor* whose electrical conductance varies depending on the extent of illumination. As is the case with diodes, most transistors exhibit photoconductive properties. Ordinary transistors are enclosed in opaque packages to eliminate *noise* that would otherwise be caused by illumination of the semiconductor junctions. Phototransistors have transparent or translucent packages, so that light can reach the internal junctions.

A *bipolar phototransistor* is constructed in such a way that the P-N junction between the control electrode (or *base*) and the output electrode (or *collector*) can receive the maximum possible amount of illumination. The most common bipolar phototransistors are NPN devices. That means there is a layer of P-type semiconductor material sandwiched in between two layers of N-type material. The outer N-type layers form the input electrode (or *emitter*) and the collector; the central P-type layer forms the base. Most bipolar phototransistors have electrical terminals only at the emitter and collector. A bipolar phototransistor functions in much the same way as a photodiode. The effective conductance increases as the illuminance increases, but only up to the saturation point.

A *photo-field-effect transistor* (or photoFET) is another device that exhibits photoconductive properties. The construction differs somewhat from that of a bipolar phototransistor, but the results are similar in practice. A photoFET has a transparent window in its housing, and is constructed in such a way that the boundaries between the P- and N-type materials have the largest possible surface area. The main difference between photoFETs and other photoelectric devices is the fact that photoFETs have higher *impedance* (effective resistance to alternating current). Because of this increased input resistance, photoFETs draw very little electric current. This property can be used to advantage in applications where weak sources of illumination must be detected.

OPTOISOLATOR

Optical coupling is a method of transferring a signal from one electronic circuit to another by converting the signal to a modulated IR, visible, or UV beam, allowing the beam to cross a physical gap, and finally recovering the information from the beam and converting it back to its original electronic form. A component specifically designed to do this is called an *optoisolator*. The input signal is applied to an LED or IRED. The resulting modulated-light signal is intercepted by a photodiode or phototransistor that converts it back into electrical impulses. Most optoisolators contain the transmitter and receptor in an opaque outer package with a glass or plastic transmission medium between them.

Optoisolators are used when it is necessary to prevent electrical interaction between two successive *stages* (such as amplifiers) in a multiple-stage electronic circuit. Normally, changes in the input signal to an electronic amplifier cause the input characteristics to vary. This in turn affects the behavior of the preceding stage. Optical coupling avoids this problem by providing total electrical isolation. When an optoisolator is used between stages, the first stage continues to function normally even if the input to the second stage opens up or shorts out.

OPTICAL SHAFT ENCODER

An alternative to mechanical switches, which corrode or wear out with time, is a device called an *optical shaft encoder*. This consists of an LED or IRED, a pair of *photoreceptors* (usually photodiodes or phototransistors), and a rotatable disk with alternating transparent and opaque bands called a *chopping wheel*.

Figure 6-2 is a functional diagram of an optical shaft encoder. The LEDs or IREDs shine on the photodetectors, but their beams must pass through the chopping wheel. The wheel is attached to the control shaft (not shown). As the control shaft is rotated, the light beams are interrupted. The two sets of output pulses are out of phase by a certain amount when the shaft is rotated clockwise; they are out of phase in the opposite sense when the shaft is rotated counterclockwise. This difference allows for the adjustment of an electrical circuit in discrete steps.

A typical application for the optical shaft encoder is the tuning dial in a high-end digital radio receiver. If you have an audio system and its FM radio tuner has a rotatable dial and a digital display, the tuning mechanism in the receiver probably employs an optical shaft encoder.

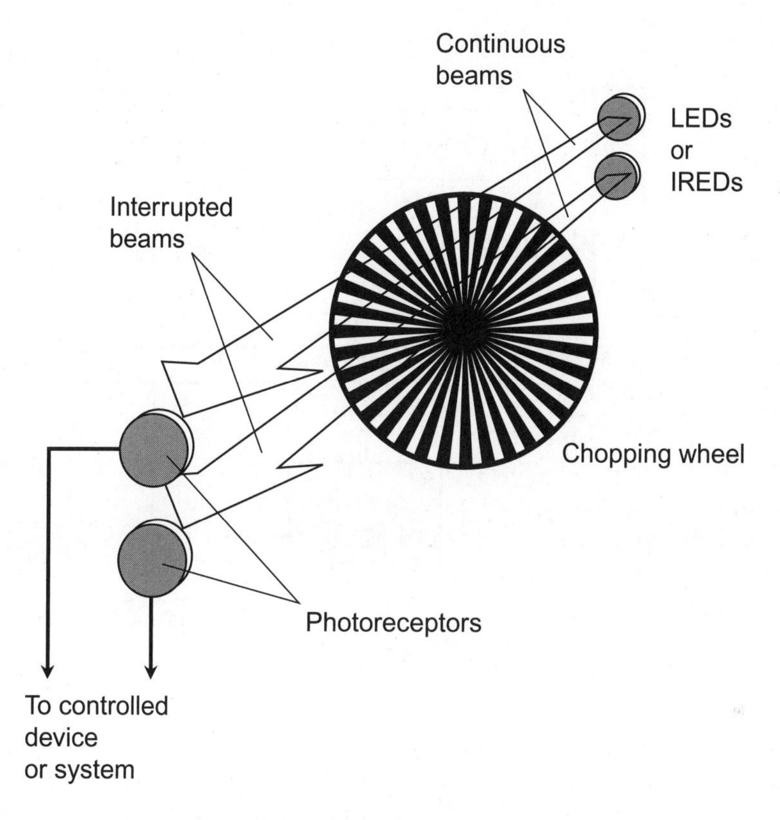

Figure 6-2 An optical shaft encoder. For clarity, the shaft is not shown. (It is attached to the center of the chopping wheel.)

ELECTRIC EYE

An *electric eye* is a sensing device that uses radiant energy beams to detect objects. Usually, the output of an electric eye controls an external machine or system. For example, an electric eye can be set up to detect or count people entering or leaving a building. Another application is the counting of items on a fast-moving assembly line; each item breaks the light beam once, and a circuit counts the number of interruptions. Elevators commonly employ electric eyes in their doors, so the doors will not close on people who hesitate on their way in or out. Automatic garage-door control systems use electric eyes to prevent accidental closure of the door on a person, animal, or other obstruction.

The typical electric eye has a visible or IR source, usually a laser diode, and a receptor such as a photodiode. Lenses ensure that the transmitted beam is narrow, and that the receptor does not respond to stray illumination. These devices are connected to an actuating circuit as shown in Fig. 6-3. When something interrupts the

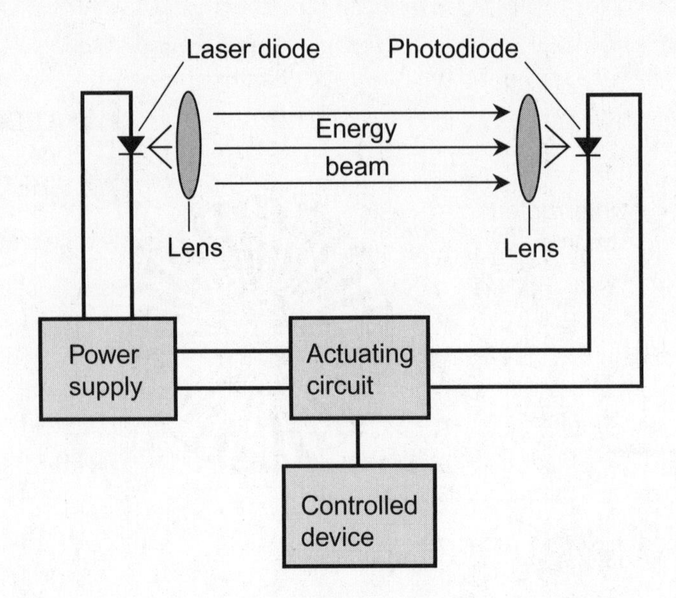

Figure 6-3 An electric eye detects objects that interrupt the path of a visible-light or IR beam.

light beam, the voltage or current passing through, or generated by, the sensor changes dramatically. It is easy for electronic circuits to detect this voltage or current change. Using amplifiers, even the smallest change can be used to control large and powerful machines. Infrared electric eyes are ideal for use in burglar alarms, because an intruder cannot see the beams.

OPTICAL SCANNER

An *optical scanner* is a machine that converts hard-copy text and graphics into digital form suitable for storage and processing in a computer. The most sophisticated optical scanners can render color images, text, photographs, and everything else needed to make a complete, accurate digital record of any document.

When printed matter is scanned, a thin laser beam moves across the page in a set of parallel lines called the *raster*, in the same fashion as the beam travels in an old-fashioned television camera. The laser beam follows the page, just as your eyes do when you read lines of text. Variations in the reflectivity of the paper causes the reflected laser light to be modulated as it scans. The modulation pattern is translated by computer software into digital code. Color scanners use three different light beams (red, green, and blue) to get three independent, overlapping monochromatic images. These are processed and combined to reproduce the original full-color image according to the RGB color model.

The image resolution of an optical scanner is usually expressed in *dots per inch* (dpi). The higher the dpi figure, the more details the scanner can "see." For reliable scanning of text at standard font sizes, an image resolution of at least 150 dpi is usually necessary, but 200 or 300 dpi is better. In the reproduction of images for casual viewing, 300 dpi is usually sufficient, but for publication-quality detail, at least 800 dpi is recommended and 1200 dpi is better.

As the image resolution increases, the image consumes more storage space in a computer. When all other factors are held constant, the stored image size is directly proportional to the square of the dpi specification. For example, doubling the dpi figure causes the digital image size (in kilobytes or megabytes) to increase by a factor of 4. For any particular image resolution and physical dimensions, a grayscale scanned digital image consumes more storage space than a black-and-white scanned image, and a color scanned image consumes more storage space than a grayscale image.

OPTICAL CHARACTER RECOGNITION

Electronic devices, working with computers, can translate printed text into digital data that can be stored, transmitted, and processed in the same way as if it had been typed on a keyboard. The technology that facilitates this, *optical character recognition* (OCR), is widely used by writers, editors, and publishers to transfer printed data to digital media. Advanced OCR software can recognize mathematical symbols and other exotic notation, as well as uppercase and lowercase letters, numbers, and punctuation marks.

Some *smart robots* incorporate OCR technology into *machine vision* systems, enabling them to read labels and signs. The technology exists, for example, to build a robot with *artificial intelligence* (AI), along with OCR, that could get in your car and drive it anywhere, reading road signs along thc way. Perhaps someday, this will be commonly done. Or you might hand a grocery list to your *personal robot* and say, "Please go get these things at the supermarket," and the robot will come back an hour later with exactly what you ordered. For a machine to read something at a distance, such as a sign, the image is observed with a video camera, rather than by reflecting a scanned laser beam off the surface. This video image is then translated by OCR software into digital data.

Whenever OCR is used in an attempt to digitize a printed document, the results must always be checked. No OCR system is perfect. Errors can creep into places, and occur in ways, that challenge the imagination!

OPTICAL COMPUTER

Optical computer technology makes use of laser beams, rather than electric currents, to perform the internal digital computations. In a conventional computer, electrical charge carriers pass from atom to atom in the conductors and semiconductors. In a wire, the charge carriers are primarily electrons. The resulting current flows through a copper wire at approximately 3×10^7 m/s, which is 10 percent of the speed of light. A current consisting mostly of holes in a semiconductor material flows somewhat more slowly than a current consisting mainly of electrons in a plain metal wire.

Visible or IR beams travel at a sizable fraction of the speed of light through a transparent glass or plastic *optical fiber*. This translates directly into higher computing speeds and greater processing power than can be obtained with conventional computers using metallic or semiconductor materials to carry electrical currents. Light beams have another advantage: the ability to pass through each other within a single medium without interaction. Because of this property, optical computers can perform millions of operations in parallel, a capability unheard of with electrical devices.

In an optical computer, *logic 1* (also called the *high state*) is usually represented by the presence of a light beam or pulse, while *logic 0* (the *low state*) is represented by the absence of a light beam or pulse. However, there are other ways to encode *binary* (two-state) logic data into light. For example, red light can represent logic 1 and blue light can represent logic 0. Or the polarization might be altered; vertical = 1 and horizontal = 0. Combinations of two beams in parallel can represent four (or 2^2) different digital states: 00, 01, 10, and 11. Three parallel beams can represent eight (or 2^3) different digital states: 000, 001, 010, 011, 100, 101, 110, and 111. This relation continues so that, for example, seven parallel beams can provide the basis for a *128-bit processor*. In a conventional electronic *integrated circuit* (IC) or *chip*, this would require seven electrical conductors in parallel. In an optical computer chip, a single optical fiber could carry seven different laser beams at different wavelengths.

PROBLEM 6-1

Suppose a scanned black-and-white line drawing consumes 1.50 megabytes (MB) of storage space at 300 dpi. How much storage space will it consume if it is scanned in the same mode at 800 dpi?

SOLUTION 6-1

The stored image size varies in proportion to the square of the number of dots per inch. If we let x represent the storage space consumed by the 800-dpi image, then

$$x/1.50 = (800/300)^2$$

When we multiply through by 1.50 and then work out the arithmetic, we obtain

$$\begin{aligned} x &= (800/300)^2 \times 1.50 \\ &= 10.7 \text{ MB} \end{aligned}$$

Modulating a Light Beam

Electromagnetic energy of any frequency can be modulated for the purpose of transmitting information. The only constraint is that the frequency of the *carrier wave* be at least several times the highest modulating frequency. *Modulated light* has recently become a significant means of conveying data, although the concept has existed for more than a century.

THE EARLIEST SCHEMES

The first communications system using modulated light was designed in the late 1700s by a Frenchman named *Claude Chappe*. He used towers on hilltops, equipped with semaphores, to portray the various letters of the alphabet. By relaying signals of this form, people could send messages rapidly and accurately over distances of thousands of kilometers when the weather was good.

Alexander Graham Bell was the next to use modulated light for communications. In 1880 he demonstrated the *photophone* (or light-beam telephone). This device transmitted voice information over sunlight reflected from a mirror that was made to vibrate with voice information. The signal was received by photosensitive devices similar to modern photocells. The system worked quite well under ideal conditions, but it proved useless in foggy or cloudy weather, it would not work at night, and the workable communications range was limited by the primitive nature of the hardware. The original photophone was put in the Smithsonian Institution in Washington, U.S.A. Then it was virtually forgotten for decades.

The photophone idea was revived in the mid-1900s when the laser was invented and demonstrated. Engineers originally believed that laser communications would be limited to line-of-sight modes, but they knew that that narrow laser light beams would be difficult to intercept or "jam." Such communications would therefore be highly secure—an asset that interested government agencies, the military, and large corporations who wanted an efficient means of conveying sensitive proprietary information from point to point.

A SIMPLE TRANSMITTER

Figure 6-4 is a block diagram of a simple modulated-light voice transmitter that can be built for a few dollars using parts available from electronics supply houses. This circuit uses a microphone, an audio amplifier, a transformer, a current-limiting resistor, an LED, and a source of direct-current (DC) power such as a battery. Complete audio amplifier modules are sold in some hobby stores for reasonable prices.

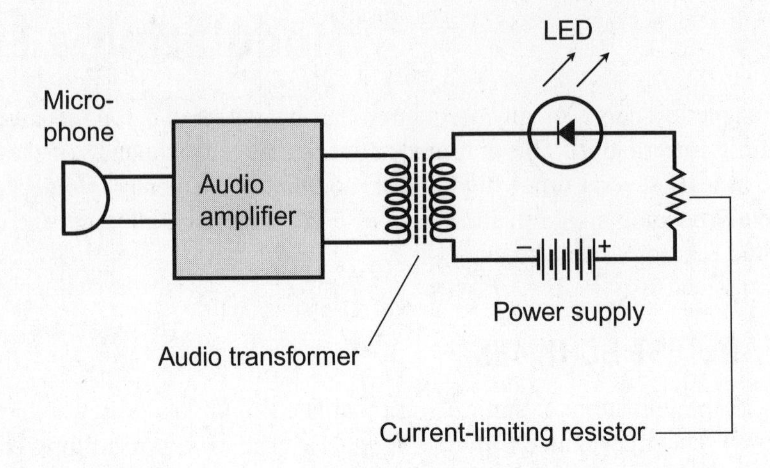

Figure 6-4 A simple modulated-light transmitter for voice data.

Alternatively, a simple audio amplifier circuit can be built using a rugged bipolar transistor or *operational amplifier* ("op amp") chip. Ideally, the module or amplifier should have a built-in gain control.

The DC from the battery or power supply provides constant illumination for the LED. The current-limiting resistor should have a resistance that is low enough to allow the LED to glow even when there is no output from the audio amplifier, but high enough to prevent damage to the LED by the power source. When someone speaks into the microphone, the amplifier produces alternating-current (AC) output at audio frequencies (AF). In the LED, the AF waveform appears superimposed on the DC from the power supply.

To operate this system, the gain of the amplifier should be initially set at minimum. Then, as the operator speaks into the microphone in a normal voice, the gain should be increased until the LED begins to flicker on voice peaks. Increasing the gain further than this point will cause distortion in the modulated-light signal.

A SIMPLE RECEIVER

Modulated-light beams can be detected using conventional photodiodes, as long as the modulating frequency is not too high (less than about 10^5 Hz or 100 kHz). Voice signals require a maximum frequency of only about 3 kHz. At audio frequencies, a circuit such as the one in Fig. 6-5 will provide sufficient output to drive a headset. The same type of audio amplifier module can be used for the receiver, as is used for the transmitter described in the previous section. For use with speakers, the output of the module can be fed to the external audio input of a high-fidelity amplifier system.

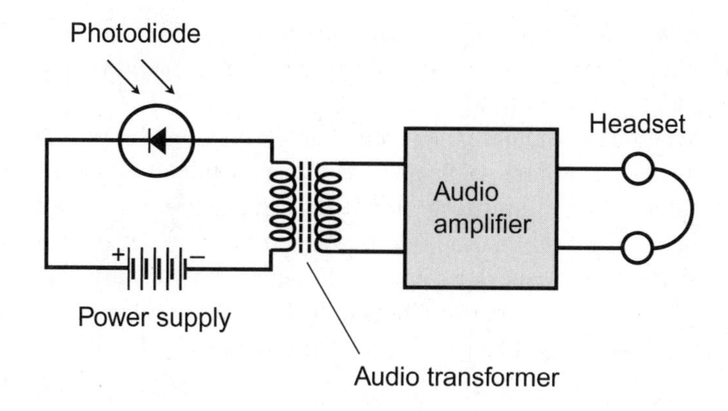

Figure 6-5 A simple modulated-light receiver for voice data.

Using the simple transmitter and receiver described here, line-of-sight modulated-light communications can be accomplished over distances up to several meters. If a small reflector, such as the type found in lanterns and flashlights, is added at the transmitter, the range can be increased to several hundred meters, especially at night. It is important that the photodiode not be exposed to a bright external source of light, because such light may saturate the photodiode and render it insensitive to the tiny fluctuations in illumination from the transmitter. It is also important that the receiver not be exposed to the light output from lamps that operate from 60-Hz utility AC power sources. All such lamps are modulated by the AC, producing severe interference to *baseband* (audio-frequency) modulated-light signals.

Using this receiver, some interesting things can be discovered. For example, the visible emissions of a cathode-ray tube (CRT), which is used in older television sets and computer monitors, produce bizarre sounds when demodulated by this circuit. It's also fun to "listen" to sunlight shining through the leaves of a tree on a windy day, or reflecting off a lake whose water is rippled by a breeze.

EXTENDING THE RANGE

The communications range of any line-of-sight modulated-light system can be increased in several ways, including *collimation* of the transmitted beam (making the rays parallel), maximizing the receiver *aperture*, and increasing the power of the transmitted signal.

A large paraboloidal or spherical reflector can be used to collimate the light beam from the transmitter LED. The larger the reflector is, the narrower the beam will be, and the farther it can propagate on a line of sight. Large reflectors are expensive; an alternative is a *Fresnel lens* of the type used in overhead projectors. These lenses are made of flat plastic, etched with circular grooves that cause the material to behave like a convex lens. Fresnel lenses are typically several centimeters square. They can be obtained from some hobby stores and catalogs. The LED is placed at the focal point of the lens.

A similar scheme can increase the receiver aperture. However, it is also important that the receiver's "field of view" be as narrow as possible to allow reception of the transmitted beam while minimizing interference from sources in other directions. One scheme is to insert the photodiode in the eyepiece holder of a telescope. The telescope's "finder" can be used to visually zero in on the transmitter. A Fresnel lens can provide a large receiver aperture, but an opaque box or other enclosure must be employed to keep stray light from reaching the photodiode.

PROBLEM 6-2

Can a modulated-light system such as the one shown in Figs. 6-4 and 6-5 be used to communicate using data instead of voice mode?

SOLUTION 6-2

Yes. A computer and modem can replace the microphone in Fig. 6-4 and the headset in Fig. 6-5. Using conventional audio amplifiers, data speeds in excess of 50 kilobits per second (kbps) can be realized, comparable to the performance of a typical dialup Internet connection. With broadband high-fidelity audio amplifiers, data speeds may reach or exceed 1 megabit per second (1 Mbps), approaching the performance of a cable or satellite Internet connection. The maximum data speed is limited by the response time of the LED in the transmitter and the photocell in the receiver.

INVERSE SQUARE LAW

Generally, the light from a point source, or from any relatively distant source, becomes less intense according to the square of the distance. If d_1 and d_2 are distances from a point source and P_1 and P_2 are the intensities (power densities) as observed from these distances, respectively, and all measurements are in the same units, then

$$P_2/P_1 = d_1^2/d_2^2$$

This principle, illustrated in Fig. 6-6, is known as the *inverse square law*. The drawing shows how the power from a point source spreads out over four times the area when the distance from a point source doubles.

In practical applications, the inverse square law applies at great distances to collimated radiation, just as it does to an uncollimated point source. At a certain distance from a collimated source, the rays effectively diverge because the illuminating object (such as an LED) at the collimator's focus is not a perfect geometric point. It therefore casts a magnified real image that gets larger as the distance is

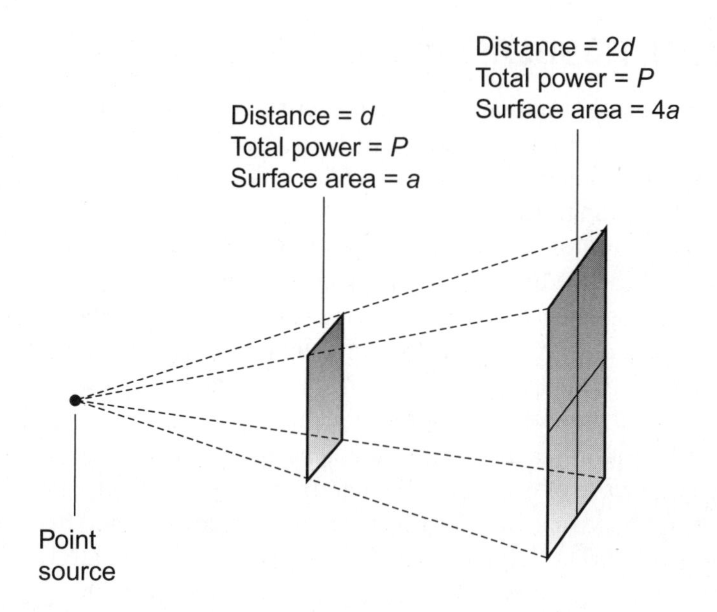

Figure 6-6 Radiated light from a point source spreads out with increasing distance according to the inverse square law.

increased, doubling in height and width as the distance is doubled. If the distance is multiplied by n, the area of the image grows by a factor of n^2 while the amount of radiant power remains the same. If a receptor is small compared to the image, increasing the distance to the source causes the intercepted power to diminish according to the same inverse square relation as would be the case if the source were a point without a collimator.

In the earth's atmosphere, some visible light is absorbed by air molecules, especially at certain wavelengths. Most of the absorption occurs at the blue and violet regions. Dust and water vapor, as well as pollutants such as particulate matter, ozone, and carbon monoxide, all increase the absorption of light in the lower atmosphere. Water droplets and particulate matter cause scattering as well, as we have learned. Absorption and scattering cause attenuation at a more rapid rate than would be the case with the inverse square law in a vacuum.

An ideal laser creates a beam of light with rays that are perfectly parallel. Therefore the intensity does not change, in theory, with increasing distance from the source. In practice, this ideal can only be approached, but the laser comes closer than any conventional radiant-energy source to this theoretical state of perfection. A sophisticated laser concentrates its energy into a beam that remains narrow for many kilometers. This is why lasers lend themselves to line-of-sight, long-distance communication.

AMPLITUDE MODULATION

In order to convey information, some characteristic of a light beam must vary. The most common way to impress information onto a beam of light is to vary its brightness. This is called *amplitude modulation* (AM). The principle resembles that used in old-fashioned radio transmission.

Binary digital amplitude modulation of a visible, IR, or UV source can be accomplished by switching the beam on and off using fast-acting shutters. This scheme is still occasionally used for signaling between ships at sea. Sunlight can be reflected from a mirror on a clear day and the reflected light rays aimed, seen, and read visually for distances limited only by the horizon. In a more sophisticated system, a light source is modulated by the audio-frequency energy from a voice amplifier, as described earlier in this chapter. The modulated-light signal varies in amplitude as if it were a radio wave of exceptionally high frequency and short wavelength.

The range of the visual wavelengths from 750 nm down to 400 nm corresponds to frequencies of 400 to 750 terahertz (THz), where 1 THz equals 10^{12} Hz. These frequencies are so high that light beams can be modulated with signals into the ultra-high-frequency (UHF) radio range up to hundreds of gigahertz (GHz), where 1 GHz equals 10^9 Hz. A signal at 500 GHz, for example, has a frequency of only 0.1 percent that of a visible-light beam at 500 THz. This is why it is possible to encode thousands of radio-frequency (RF) signals onto a single beam of light.

PROBLEM 6-3

What is the highest signal frequency that a visible, IR, or UV energy beam can efficiently carry?

SOLUTION 6-3

In communications practice, the general rule is that the modulating signal frequency must be less than 10 percent of the carrier frequency. In the case of a near-IR beam with a wavelength of 900 nm and a corresponding frequency of 333 THz, for example, the highest modulating signal frequency would be approximately 33.3 THz, which is well above the frequency of a typical microwave wireless signal.

LINE-OF-SIGHT COMPUTER COMMUNICATIONS

The complex nature of line-of-sight, high-speed, computer-to-computer data transfer requires broad-bandwidth signals that can be easily accommodated by laser systems. The laser is practically immune to interference or "jamming" unless an opaque object gets in the way of the beam.

There is some evidence to suggest that lasers of certain wavelengths propagate well for moderate distances under water. Sound waves or low-frequency radio

waves have been used for this purpose, but neither of these emission types can carry high-speed information because of their low frequencies. Acoustic waves also suffer from low propagation speed.

For line-of-sight communications through the earth's lower atmosphere, the He-Ne and neodymium-YAG lasers are effective because their emission wavelengths are not prone to scattering. These types of lasers are useful for communications between aircraft and also between oceangoing ships.

It is possible to communicate using lasers over paths not directly linked by lines of sight. Strategically located, flat-surface mirrors can be used to deflect the beams. Such reflectors need not be large, but they must be kept clean and oriented with precision.

"CLOUDBOUNCE" AND ATMOSPHERIC SCATTER

Designing a high-power modulated-light communications system and getting it to work, could be an interesting experiment for a hobbyist with a lot of extra money, a fondness for "tinkering," and a willingness to accept the possibility of failure. (I've never actually built this apparatus, and I cannot guarantee that anyone who does will meet with success.) A bright lamp such as a sodium-vapor type can replace the LED or IRED, and a high-power audio amplifier can be used in the transmitting system. If two large paraboloidal or spherical reflectors can be procured, one for the transmitter and another for the receiver, it might be possible to communicate by means of light scattering from clouds or within the atmosphere.

Figure 6-7A shows an example of how a "cloudbounce" modulated-light communication could work when conditions are partly cloudy or overcast. The obtainable range depends on the thickness of the cloud deck, its reflectivity, and its altitude. The transmitter beam is aimed at the sky so that a spot of light falls on the clouds midway between the receiver and the transmitter. The receiver reflector is aimed so the maximum amount of signal light falls on the photodetector at its focus. This technique should work in daylight as well as in darkness, as long as the amount of daylight falling on the photodetector is not so great as to cause it to go near, or into, a state of saturation.

At night, the background glow from outdoor lighting systems will cause a 60-Hz "hum" problem. This interference might be overcome by providing the transmitter lamp with a digital *phase-shift-keyed* (PSK) signal from a personal computer (PC) modulating a baseband carrier at a frequency of, say, 5 kHz. A high-power audio amplifier could be used, and its output connected to the lamp. The receiving station could employ a sensitive audio amplifier with an audio filter to get rid of the "hum." The output of this amplifier could be connected to another PC equipped with PSK demodulating software. Another way to get rid of 60-Hz "hum" interference might be to employ a separate photodetector, without a reflector, to pick up the general urban lamp glow and introduce its "hum" into the receiver input in phase opposition

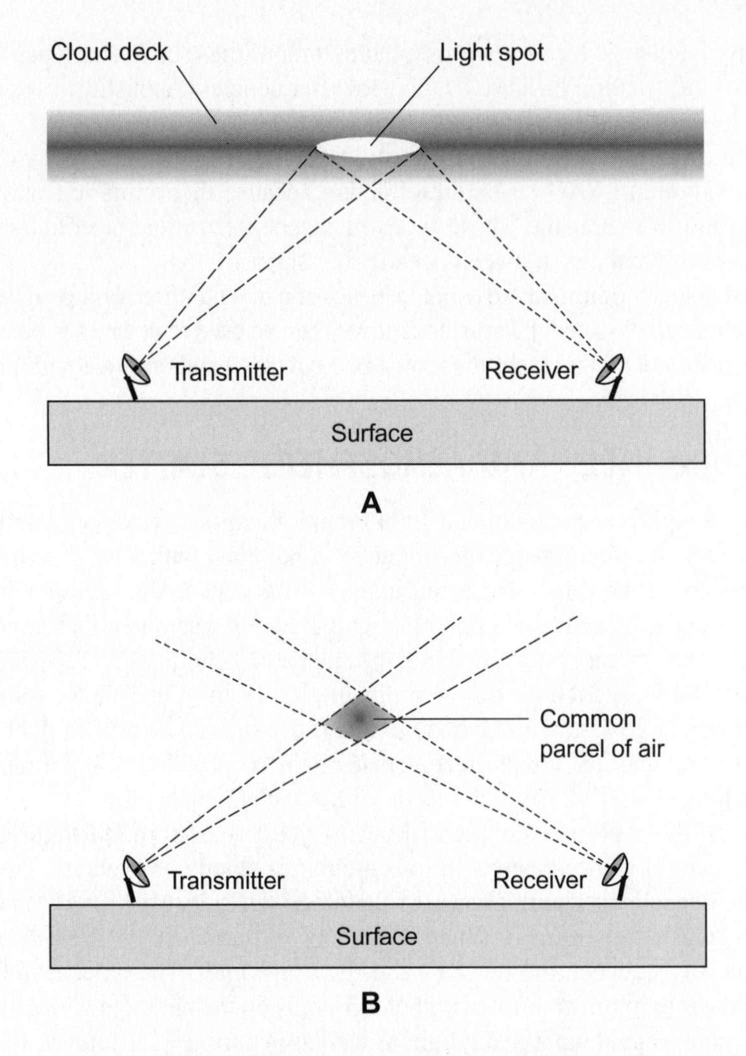

Figure 6-7 "Cloudbounce" (A) and atmospheric-scatter (B) modulated-light communications.

with, but at the same amplitude as, the "hum" from the signal photodetector. This would cancel out the "hum" but would not affect the desired signals.

Figure 6-7B illustrates how a sensitive receiver could be used to pick up the scattered light from a powerful transmitter when the sky is partly cloudy or cloudless. The receiving and transmitting reflectors would be aimed at a common parcel of air. This scheme should work best in air containing just a little dust or haze. If the air is too clear, not enough light will be scattered; if there is too much dust or haze, beam propagation will be poor.

Warning! *Anyone attempting "cloudbounce" or atmospheric-scatter experiments should check with military and civilian aviation authorities to ensure that the bright lights will not adversely affect the operation of aircraft in the vicinity. In some locations, experiments of this type will be forbidden.*

PROBLEM 6-4

How should we expect the beam from a He-Ne laser to scatter over long distances through the earth's lower atmosphere, compared with the beam from an argon-ion laser?

SOLUTION 6-4

The beam from a He-Ne laser should scatter less than that from an argon-ion laser. The output of the He-Ne laser is at 633 nm in the red part of the visible spectrum, while the argon-ion laser produces output at 480 nm in the blue range. Red light scatters less than blue light, as is evidenced by the fact that the sun often appears red at sunrise and sunset, and the sky is blue on a clear day.

Fiberoptics

In 1970, *Robert Maurer* of Corning Glass Works demonstrated the practicality of *fiberoptic communication* after an exceptional grade of glass became available from Standard Telecommunication Laboratories. Since that time, fiberoptic communications systems have gained widespread popularity throughout the world.

ADVANTAGES

Besides allowing signals to be sent at high speeds, optical fibers offer immunity from one of the oldest problems in communications: *electromagnetic interference* (EMI). A strong radio signal, thunderstorm, solar flare, or nearby high-tension power line does not affect visible light or IR rays traveling along an optical fiber. Conversely, the signals in an optical fiber do not cause EMI to external devices or systems. All of the signal energy stays within the optical fiber. The data contained in the visible light or IR rays traveling along an optical fiber is extremely difficult to intercept. The minerals from which glass fibers are made are cheap and plentiful, and can be "mined" with minimal environmental impact. Fiberoptic cables can be submerged or buried, and they do not corrode as metal conductors do. Fiberoptic cables last longer than wire cables, and they require less frequent maintenance.

LIGHT SOURCES

Two types of light source are customarily used in fiberoptic communications systems: the laser diode and the conventional LED/IRED. The laser diode produces a beam that spreads out more rapidly with distance than large lasers such as the cavity type, but beam divergence is not a problem in fiberoptics because the fiber keeps the beam confined. A collimating lens can be used at the input end of the fiber, keeping most or all of the light rays nearly parallel so that none of them escape the fiber.

A laser diode or LED/IRED emits energy when sufficient current passes through it. If the no-signal current is kept within a certain range, the instantaneous emitted energy is proportional to the instantaneous applied current and the response rate is rapid, so the beam can readily be modulated by varying the current in sync with the data to be transmitted. It is easy to obtain effective beam amplitude modulation in this way.

As the modulated beam passes along the optical fiber, the relative intensity varies in the same proportion all along the fiber, even though the absolute intensity decreases because of loss. Therefore, the signal at the receiving end of the fiber, although weaker than the signal at the transmitting end, is identical in modulation characteristics.

MULTIMODE FIBER DESIGNS

There are two basic types of optical fiber. In a *multimode fiber*, the transmission medium should have a diameter of at least 10 times the longest wavelength to be carried. For visible-light systems, that's approximately 7.5 micrometers (μm), where 1 μm equals 1000 nm, which in turn equals 10^{-6} m. In an IR system, the minimum required fiber diameter is, of course, larger than this. In a *single-mode fiber*, the transmission-medium diameter can range down to approximately 1 wavelength. The following discussions concern only multimode media.

Multimode optical fibers are made from glass or plastic to which impurities have been added. The impurities change the refractive index. A typical multimode optical fiber has a *core* surrounded by a tubular *cladding*. The cladding has a lower refractive index than the core, which constitutes the transmission medium. There are two basic multimode designs, called the *step-index* optical fiber and the *graded-index* optical fiber.

In a step-index fiber (Fig. 6-8A), the core has a uniform index of refraction and the cladding has a lower index of refraction, also uniform. The transition at the boundary is abrupt. Ray *X* enters the core parallel to the fiber axis and travels without striking the boundary unless there is a bend in the fiber. If there is a bend, ray *X* veers off center and behaves like ray *Y*, striking the boundary. Each time ray *Y* encounters the boundary, total internal reflection occurs, so ray *Y* stays within the core.

In a graded-index optical fiber (Fig. 6-8B), the core has a refractive index that is greatest along the central axis and steadily decreases outward from the center. At the boundary,

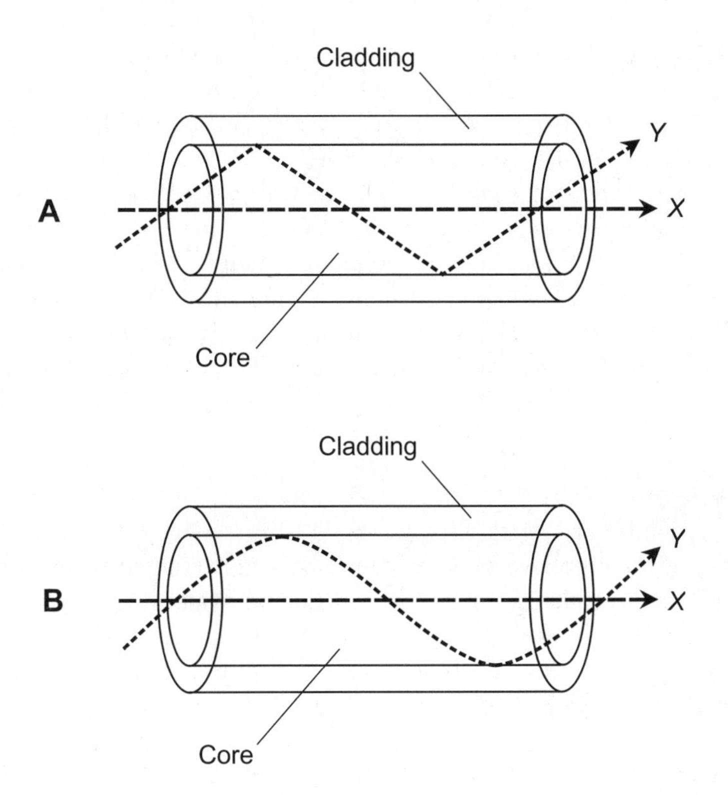

Figure 6-8 At A, a step-index optical fiber. At B, a graded-index fiber. In both designs, rays *X* and *Y* are confined to the core.

there is an abrupt drop in the refractive index. Ray *X* enters the core parallel to the fiber axis and travels without striking the boundary unless there is a bend in the fiber. If there is a bend, ray *X* veers off center and behaves like ray *Y*. As ray *Y* moves away from the center of the core, the index of refraction decreases, bending the ray back toward the center. If ray *Y* encounters an especially sharp bend in the fiber, the ray strikes the boundary between the core and the cladding, and total internal reflection occurs.

FIBER BUNDLING

Optical fibers can be bundled into cable, in the same way as wires are bundled. The individual fibers are protected from damage by layers of insulation made from materials such as polyethylene and polyurethane. The entire bundle is encased in a tough, durable outer jacket to protect it from the elements. Each fiber in the bundle can carry numerous rays of visible light or IR, each ray having a different wavelength. Every individual ray can be modulated with thousands of RF signals, each signal having a different *carrier frequency*.

Because the frequencies of visible light and IR are much higher than the frequencies of the signals used to modulate them, the attainable bandwidth in a fiberoptic cable link is vastly greater than that of any conventional cable or wireless link, allowing much higher data-transmission speed.

Another notable advantage of fiber bundles over multiple-wire bundles is the fact that because the fibers do not carry electric currents, there is no *crosstalk* (mutual signal interference) among them. Current-carrying wires, unless individually shielded, suffer from crosstalk because the AC signal in any one of the wires can "leak" into the other wires nearby. Providing an EM shield for each wire in a large bundle makes a cable expensive, bulky, and heavy. Fiberoptic cables are far cheaper, smaller, and lighter because EM shielding is not necessary.

REPEATERS

A long-distance fiberoptic system must have *repeaters* at intervals along the length of the cable, just as is the case in conventional cable systems. The repeaters consist of receivers that amplify the signals and retransmit them, usually (but not necessarily) using the same type of visible light or IR source as the transmitter.

Figure 6-9 is a block diagram of a modulated-laser repeater consisting of a demodulator that separates the signals from the faint incoming visible or IR carrier, a series of amplifier stages to boost the signal strength, and a modulated laser that transmits a visible or IR beam containing all of the same signals to the destination or to another repeater.

NONCOMMUNICATIONS APPLICATIONS

Optical fibers are superior to wire cables in various applications other than communications. Any kind of signal, such as a computer command, can be carried by optical fibers. This method of carrying signals is not subject to EMI from external sources, and does not cause EMI to external devices. Optical fibers also offer relative immunity to the effects of an *electromagnetic pulse* (EMP) of the sort that can be induced by nearby lightning strikes or nuclear explosions. The EMP immunity is an important defense consideration, because multiple denotations of high-yield devices at high altitude and geographically diverse locations could cripple the conventional communications system of a nation by overloading it with destructive electric current surges.

Fiberoptic networks generate no electric sparks, so there is no risk of explosion in volatile environments. The electric shock hazard is eliminated, because optical fibers carry no electric current. The fact that fiberoptic systems are immune to electrical noise interests automobile manufacturers. The microprocessors in modern automobile systems can be adversely affected by EMI from mobile radio transmitters.

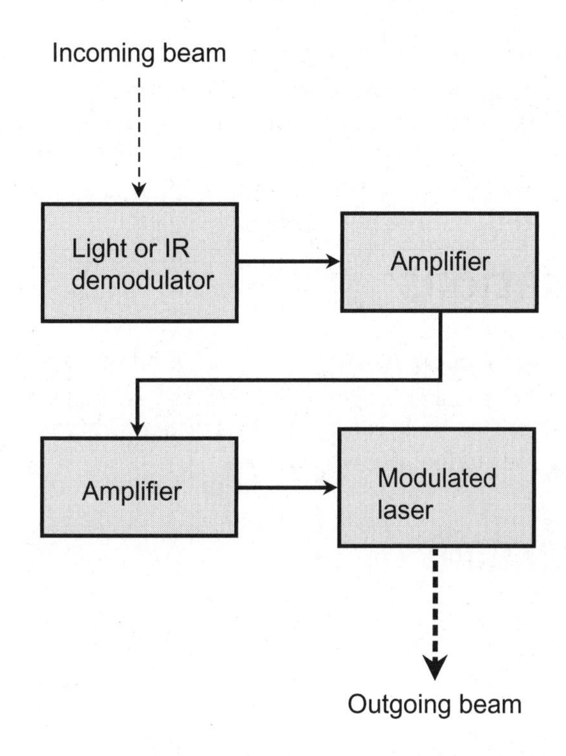

Figure 6-9 A repeater amplifies and retransmits modulated light or IR beams in a fiberoptic communications system.

The use of fiberoptic technology for transferring microcomputer commands reduces the risk of engine malfunction (or false indications of malfunction). Moreover, there is essentially no EM emission from the fiberoptic cables that might degrade the performance of a mobile telephone or radio installed in the vehicle. A fiberoptic vehicle-control system saves weight and does not corrode over time. Relatively few cables must pass through the vehicle's firewall to the dashboard, so servicing is simplified.

When a new technology is tested and developed in a system such as an automobile where malfunction carries the risk of personal injury or death, the nonessential systems are developed first, and the essential systems are developed only after the new technology has proven reliable. Fiberoptic systems will be seen first in such applications as automatic windows and door locks, and later, when the "bugs" are out, in such things as electronic steering and brake systems. A mass-produced, fully fiberoptic-controlled motor vehicle may be decades in the future as of this writing.

PROBLEM 6-5

Why do fiberoptic systems offer relative immunity from EMP effects?

SOLUTION 6-5

Optical fibers do not conduct electric currents. Therefore, powerful external electric or magnetic fields such as those in an EMP have no effect on the signals in an optical fiber.

Robotics Applications

Many robotic devices employ *machine vision* technology, also called *vision systems*. In machine vision, a charge-coupled device (CCD) gathers incoming images. The digital output from the CCD can be clarified by *digital signal processing* (DSP). The resulting data goes to the *robot controller*.

KEY SPECIFICATIONS

Two important specifications in any vision system are the *optical sensitivity* and the *optical resolution*. Sensitivity is the ability of a machine to see in dim visible light, or to detect weak impulses at invisible wavelengths such as IR or UV. Resolution is a quantitative measure of the extent to which a vision system can differentiate between objects at a distance. Sensitivity and resolution are interactive. When all other factors are held constant, improving the sensitivity requires a sacrifice in resolution, and improving the resolution requires a sacrifice in the sensitivity.

Processing full-motion video image data, and getting all the meaning from it, is a significant challenge for engineers who design machine vision systems for robots. The variables in a video image are much like those in a human voice. In order to extract the full (and applicable) meaning of an image, machine vision technology must be at least as sophisticated as high-level *speech recognition*. Fortunately, in many robot applications, it is not necessary for the robot to "comprehend" what's happening, but only to identify defined patterns or events.

Robots can be made to "see" in an environment that is dark and cold, and that radiates too little energy to be detected at any EM wavelength. In these situations, the robot provides its own illumination. This can be a simple lamp, laser, IR device, or UV device.

BAR CODING

Bar coding is a method of labeling objects. Bar-code labels or tags are extensively used in retail stores for pricing and identifying merchandise. A bar-code tag has a characteristic appearance, with parallel lines of varying width and spacing.

A laser-equipped device scans the tag, retrieving the identifying data. Bar-code tags are one method by which objects can be labeled so that a robot can identify them. This greatly simplifies the recognition process. For example, every item in a tool set can be tagged using bar-code stickers with a unique code for each tool. When a robot's controller tells the machine that it needs a certain tool, the robot can seek out the appropriate tag and carry out the movements according to the program subroutine for that tool. Even if the tool gets misplaced, as long as it is within the robot's *work envelope* or range of motion, it can easily be found.

OPTICAL TRIANGULATION

Robots can navigate in various ways. One effective technology, *optical triangulation*, is a visible-light, IR, or UV variant of the scheme that ship and aircraft captains have used ever since radio communication systems were first deployed in the early 1900s.

In optical triangulation, the robot has a direction indicator such as a compass. It also has a laser scanner that rotates in a horizontal plane. There must be at least two targets at known, but different, locations in the work environment. The targets reflect the laser beam back to the robot. The robot has an omnidirectional sensor that detects the returning beams. An onboard microcomputer gathers the data from the sensors and the direction indicator and processes the directional data to calculate the robot's exact position in the work environment from moment to moment.

In some optical triangulation systems, the directional sensor is replaced by a third target. In this type of system, there are three incoming laser beams. The robot controller can determine its position according to the three angles between these beams. If the three targets and the robot are not all located in the same geometric plane, the system can allow for navigation in three dimensions (3D).

For optical triangulation to work, it is important that the laser beams not be blocked. Some work environments contain numerous obstructions, such as stacked boxes, which interfere with laser beams and make optical triangulation impractical. If a magnetic compass is used to obtain a reference direction (magnetic north or south) for the system, the compass must not be "fooled" by stray magnetism, and the earth's magnetic field must not be obstructed or distorted by metallic walls or ceilings.

The principle of optical triangulation, using a direction sensor and two reflective targets, is shown in Fig. 6-10. The laser beams (dashed lines) arrive from different directions, depending on where the robot is located with respect to the targets. The targets are *tricorner reflectors* that send all light rays back along the path from which they arrive.

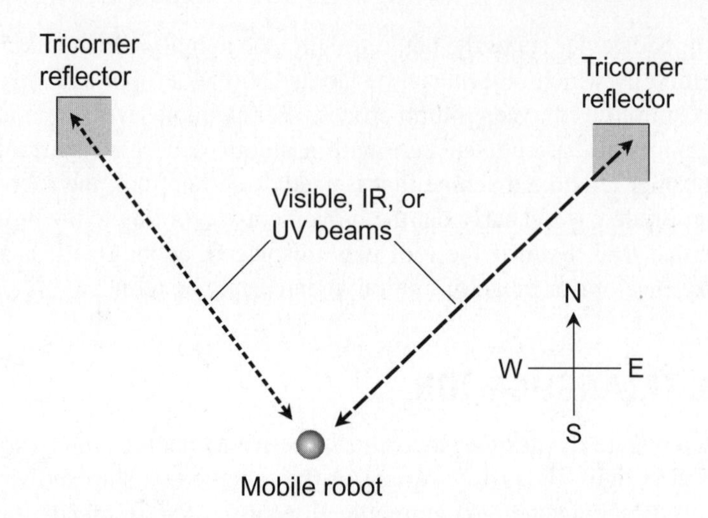

Figure 6-10 Optical triangulation can help a mobile robot navigate in its work environment.

LADAR

Ladar is an acronym that stands for *laser detection and ranging*. It is also known as *laser radar* or *lidar* (short for *light detection and ranging*). A ladar system uses a beam of visible, IR, or UV energy, rather than radio waves (as in *radar*) or acoustic waves (as in *sonar*) to perform *ranging* (distance measurement) and *range plotting* (distance-based mapping) of the environment. The device works by measuring the time it takes for a laser beam to travel to a target point, reflect from it, and then propagate back to the point of transmission.

The principal asset of ladar over other ranging methods is the fact that the laser beam is extremely narrow. This provides superior direction resolution compared with radar and sonar schemes, whose beams cannot be focused with such precision. However, ladar cannot work well through fog or precipitation, as can radar. Certain types of objects, such as mirrors oriented at a slant, do not return ladar energy and produce no echoes.

A high-level ladar system scans both ways, horizontally and vertically, thereby creating a 3D *computer map* of the environment. Less sophisticated ladar devices work in a single plane, usually horizontal, to create a two-dimensional (2D) computer map at a specific level above a floor or flat ground.

EDGE DETECTION

Edge detection is the ability of a robotic-vision system to locate boundaries. It also refers to a robot's knowledge of what to do with respect to those boundaries.

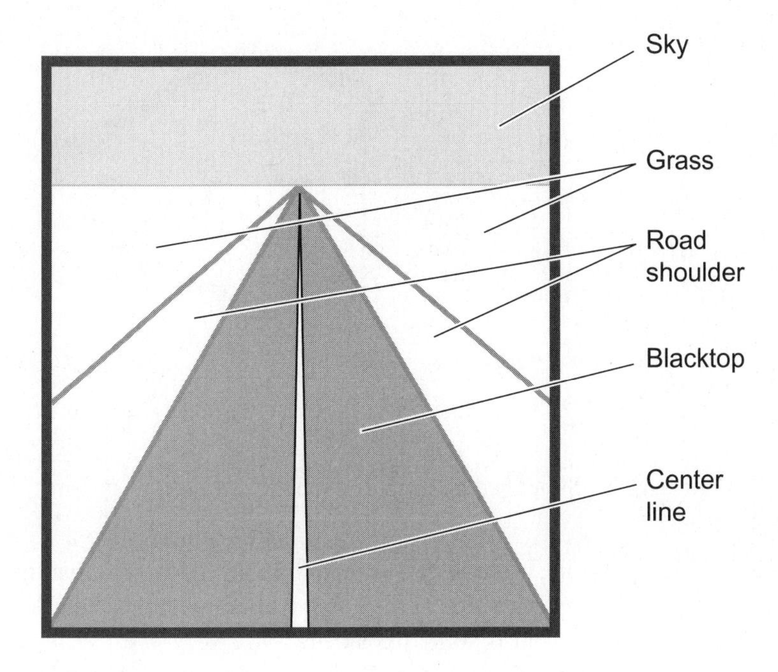

Figure 6-11 Edge detection can help a robot vehicle stay on the road.

A future robotic motor vehicle might employ edge detection to define the edges of a road and use the data to keep itself from veering off into the surrounding environment. Figure 6-11 shows what the machine-vision system "sees" after the image data has been processed by the robot controller. The robot's propulsion system must ensure that the vehicle stays a certain distance from the right-hand edge of the pavement so that it does not cross over into the lane of oncoming traffic (as it appears to be doing in this case!). The vehicle must be able to discern the difference between pavement and other surfaces such as gravel, grass, sand, or snow. Beacons, placed along the roadway at frequency intervals, can be used for this purpose, but that scheme requires the installation of the guidance system beforehand, limiting the robot vehicle to roads equipped with such navigation aids.

A personal robot equipped with edge detection can see certain contours in its work environment. This ability can keep the machine from running into obstructions or falling down stairs.

EYE-IN-HAND SYSTEM

A camera can be placed in a robotic "hand" to help the *gripper* mechanism locate and grasp objects of interest. The camera must be equipped for work at close range, from approximately 1 m down to a few millimeters (mm). The positioning error

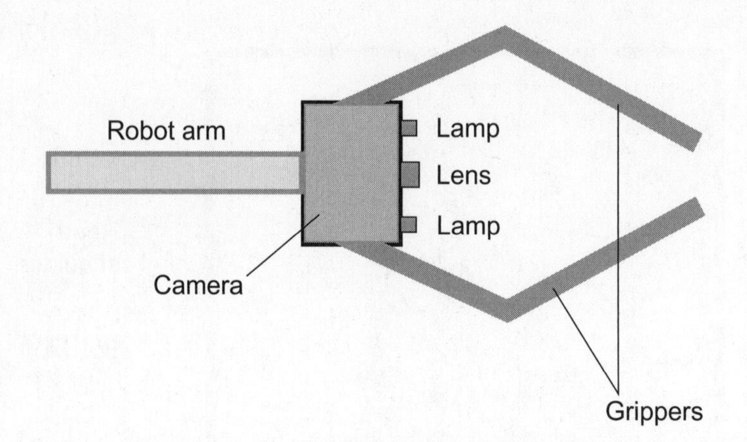

Figure 6-12 Simplified diagram of a robotic eye-in-hand system.

must be as small as possible, preferably less than 0.5 mm. To be sure that the camera gets a good image, a lamp is included in the gripper along with the camera as shown in Fig. 6-12. This arrangement is sometimes called an *eye-in-hand system.*

An effective eye-in-hand system is capable of *object recognition* for positive identification of targets in complex work environments. Size, shape, color, light reflectivity, light transmittivity, and texture are visible characteristics typical of everyday objects. The eye-in-hand system uses a *servomechanism* to regulate the movements of the robotic arm and gripper. The robot is equipped with, or has access to, a controller that processes the data from the camera and sends instructions to the motors that control movement.

FLYING EYEBALL

The *flying eyeball* is a simple form of *submarine robot* that can resolve detail underwater, and can also move around under its own power. It cannot manipulate anything because it has no robotic "arms" or *end effectors* ("hands," tools, or grippers). Flying eyeballs have been used in scientific and military applications. They are also of interest to well-funded treasure hunters!

A cable, containing the robot in a launcher housing, is dropped from a boat. When the launcher gets to the desired depth, it lets out the robot, which is connected to the launcher by a tether as shown in Fig. 6-13. The tether and the drop cable convey data back to the boat. The robot contains a video camera and one or more lamps to illuminate the underwater environment. It also has a set of thrusters (jets or propellers) that let it move around according to control commands sent through the cable and tether. Human operators on board the boat can watch the images from the video camera and guide the robot around as it examines objects on the sea floor.

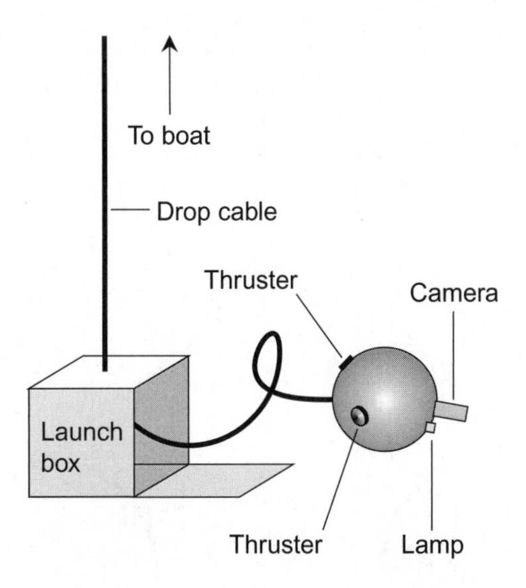

Figure 6-13 A flying eyeball is an underwater mobile robot equipped with machine vision.

In some cases, the tether can be eliminated and visible or IR beams used to convey data from the robot to the launcher. This provision gives the robot enhanced freedom of movement and eliminates concerns that the tether might get tangled up in something. However, the use of such data links limits the range to line-of-sight situations. Wireless RF data links can also be used, but EM fields at conventional RF wavelengths do not propagate well underwater except for short distances.

EPIPOLAR NAVIGATION

Epipolar navigation is a machine-vision technology that incorporates AI, allowing a robotic vehicle to determine its position and navigate a course through 3D space. Epipolar navigation works by evaluating the way a projected or viewed image of the external environment evolves from a moving reference point. The human eye and brain can do this without conscious effort. Getting a machine to do it is one of the most challenging tasks facing robotics engineers.

To illustrate epipolar navigation, imagine a robotic aircraft (also called a *drone*) flying over a small island as shown in Fig. 6-14. The robot controller has a computer map that shows the location, size, shape, and topography of the island. The drone can navigate by observing the island and scrutinizing the shape and angular size of the island's image.

As the drone travels along its assigned course through 3D space, the island seems to move underneath it. The controller sees an image of the island that constantly

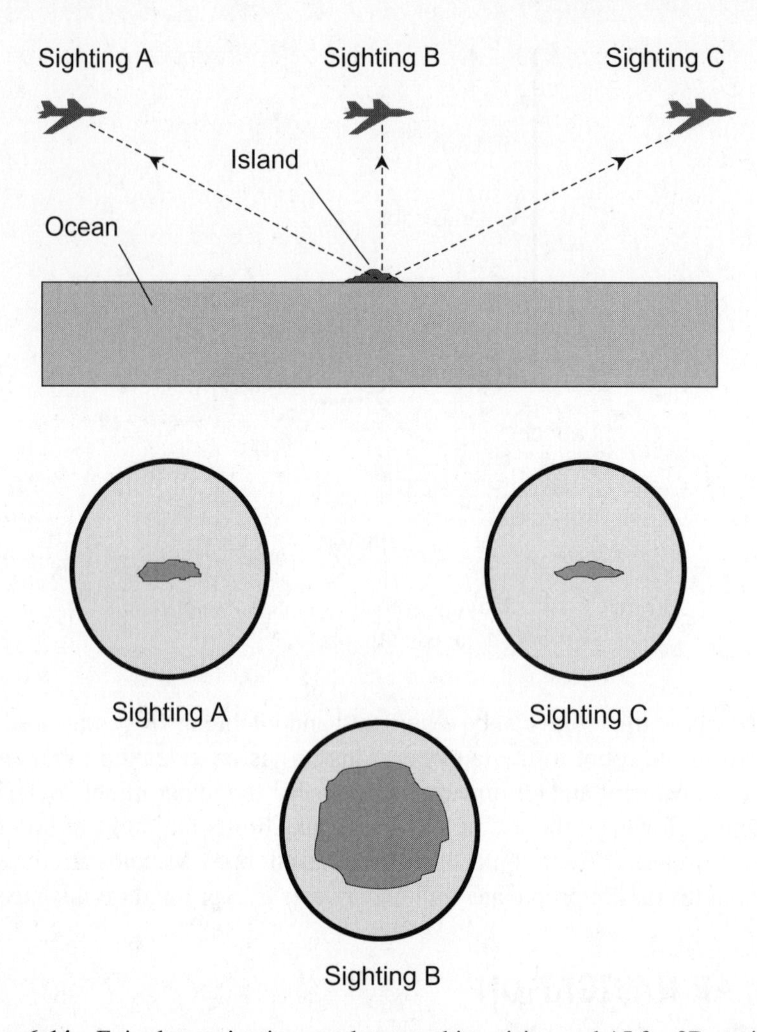

Figure 6-14 Epipolar navigation employs machine vision and AI for 3D navigation.

changes in shape and angular diameter. The controller constantly compares the shape and size of the image it sees with the actual shape and size of the island that it "knows" from its stored computer map data. From this information, the computer can precisely determine the drone's

- speed relative to the surface.
- direction relative to the surface.
- geographic latitude.
- geographic longitude.
- altitude.

There exists a *mathematical bijection* (one-to-one correspondence) between the set of all points in 3D space within sight of the island, and the angular size and shape of the island's image as seen from any particular point. The computer can therefore match, from moment to moment, the image it sees with a specific point in space.

In theory, epipolar navigation can function on any scale and at any speed. The technology might someday be used in vehicles traveling at relativistic speeds through interplanetary or interstellar space.

TEXTURE SENSING

Texture sensing is the ability of a computer or robot to ascertain the smoothness or roughness of a surface. Primitive texture sensing can be done with a laser and several light-sensitive sensors.

Figure 6-15 shows how a laser can be used to tell the difference between a shiny surface (A) and a rough or matte surface (B). A shiny surface, such as the polished hood of a car, returns visible light according to the law of reflection, which states that the angle of reflection equals the angle of incidence. But a matte or rough surface, such as a carpeted floor, scatters the light. The shiny surface reflects the beam

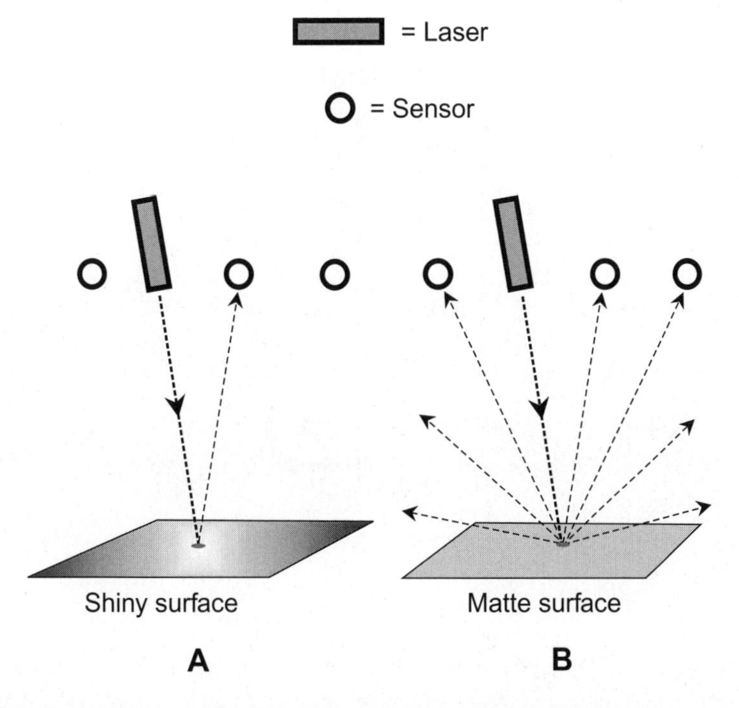

Figure 6-15 Lasers and photodetectors can be used for visual texture sensing.

back almost entirely to the single sensor in the path of the beam whose reflection angle equals its incidence angle. The matte surface reflects the beam back to all of the sensors.

The visible-light texture sensing scheme can only determine that a surface is either shiny or not shiny. A piece of drawing paper reflects the light in much the same way as a carpeted floor. The measurement of relative roughness, or of the extent to which a grain is coarse or fine, requires more sophisticated sensing technology.

PROBLEM 6-6

How can a robot perform visual range plotting?

SOLUTION 6-6

Range plotting is a process in which a graph is generated depicting the distance (range) to objects, as a function of the direction in 2D or 3D. In 2D, range plotting involves mapping the distances to various objects as a function of their direction in a defined plane. One such method is shown in Fig. 6-16. The robot is at the center of

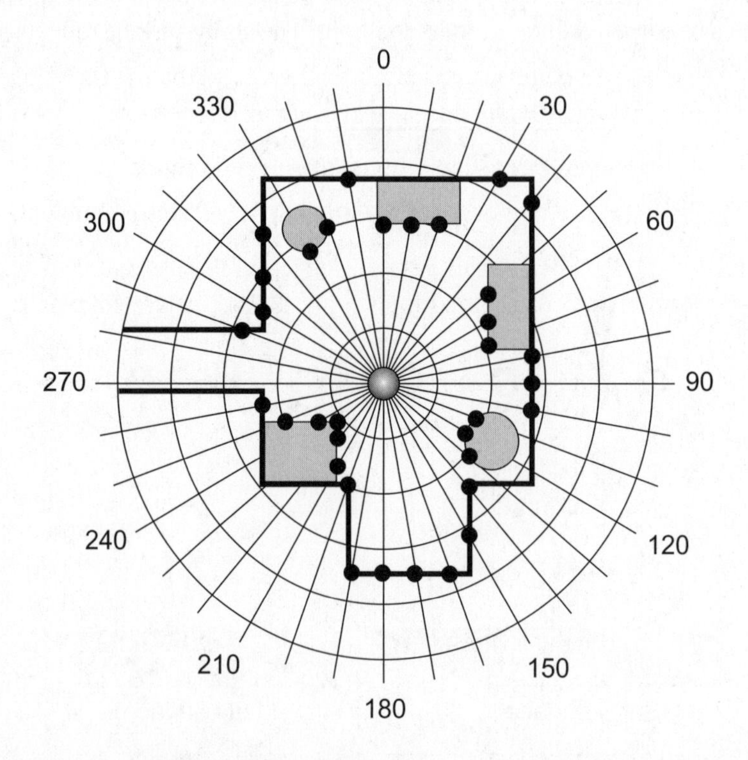

Figure 6-16 A 2D range plot with directional resolution of 10°. Azimuth angles (shown by peripheral numerals) are measured horizontally in degrees clockwise from true north or magnetic north.

the plot, in a room containing three desks (rectangles) and two floor lamps (circles). The range is measured every 10° of azimuth around a complete circle, resulting in the set of points shown. A better plot would be obtained if the range were plotted every 5°, every 2°, or even every degree or less. But no matter how fine the directional resolution might be, a 2D range plot can show things in only one plane, such as the floor level or some horizontal plane above the floor. For 3D range plotting, *spherical coordinates* are employed, in which the distance is measured for a large number of directions at all possible orientations. A 3D range plot in a room such as that depicted in Fig. 6-16 would show ceiling fixtures, objects on the floor, objects on top of the desks, and other details not visible with a 2D range plot.

Quiz

This is an open book quiz. You may refer to the text in this chapter. A good score is 8 correct. Answers are in the back of the book.

1. It is reasonable to suppose that for a drone flying at high altitude, an onboard optical epipolar navigation system might be frustrated by
   (a) a flat island with an irregular shoreline.
   (b) a mountainous island with an irregular shoreline.
   (c) EMI caused by onboard radio communications equipment.
   (d) low overcast.
2. A "cloudbounce" modulated-light communications system would likely require
   (a) permission from aviation authorities before deployment.
   (b) a high-power audio amplifier and a brilliant light source at the transmitting end.
   (c) a sensitive amplifier at the receiving end.
   (d) All of the above.
3. When a photodiode is properly biased and illuminated so that the current is below the saturation current, an increase in the illumination level will
   (a) cause the current through the device to decrease.
   (b) cause no change in the current through the device.
   (c) cause the current through the device to increase.
   (d) cause the P-N junction to emit visible, IR, or UV radiation of its own.

4. An optical triangulation system for a mobile robot might not function well in
   (a) a warehouse where large items are stacked in random locations.
   (b) a dark, open space.
   (c) a factory where there are numerous sources of EMI.
   (d) any of the above work environments.
5. Suppose each of the individual fibers inside an optical microprocessor chip carry independent binary logic signals at wavelengths of 1000, 900, 800, 700, and 600 nm. This would form the basis for a
   (a) 5-bit computer.
   (b) 25-bit computer.
   (c) 32-bit computer.
   (d) 125-bit computer.
6. The range of a line-of-sight, voice-modulated communications system using an LED at the transmitter and a photodiode at the receiver can be extended by the use of
   (a) a Fresnel lens at the transmitter to collimate the beam, and another Fresnel lens at the receiver to increase the aperture.
   (b) modulating signals having frequencies at least 1/10 that of the highest carrier-wave frequency.
   (c) a 60-Hz AC power source in place of the battery or DC power supply in the receiver to ensure that the incoming signal is adequately modulated.
   (d) an optical computer chip that can process multiple beams in parallel to maximize the amount of power delivered.
7. Fiberoptic communications systems are relatively immune to the effects of EMI and EMP because the fibers
   (a) have excellent electrical conductivity.
   (b) behave as efficient semiconductors.
   (c) short-circuit destructive currents to ground.
   (d) do not conduct electricity.

8. In an optical triangulation system, effective targets can be provided by
   - (a) black or dark spheres.
   - (b) flat mirrors.
   - (c) tricorner reflectors.
   - (d) matte surfaces.
9. If all other factors are held constant, a black-and-white optical scan of a photograph would be expected to require
   - (a) the same computer storage space as a grayscale or color scan.
   - (b) less computer storage space than a grayscale or color scan.
   - (c) more computer storage space than a grayscale or color scan.
   - (d) more computer storage space than a grayscale scan, but less than a color scan.
10. The minimum core diameter of a multimode optical fiber designed to carry IR signals at a wavelength of 1500 nm is approximately
    - (a) 1500 μm.
    - (b) 150 μm.
    - (c) 15 μm.
    - (d) 1.5 μm.

CHAPTER 7

Optics in the Field

Optical devices can be used to measure distance, shape, and time. In industry, lasers can cut, weld, drill, solder, assist in component manufacture, and enhance quality assurance. Focused or coherent light beams can produce images for entertainment purposes.

Ranging and Alignment

A laser beam follows a straight-line path unless refracted or reflected by intervening media. Lasers can therefore render any geometric object that can be physically drawn with drafting tools.

SURVEYING AND LEVELING

Lasers are useful in surveying where it is necessary to define straight lines in three-dimensional space. A laser facilitates the precise determination of any point that lies along a straight line that has been set up by sighting between two specific end points. Figure 7-1 shows an example in which the facades of buildings are aligned parallel to the edge of a straight roadway.

Lasers can be employed for determining where the foundation for a building should be laid, where a highway should go, where the property boundaries for a certain piece of real estate are, and other normal surveying applications, with less time and effort than is required by traditional surveying methods. The laser also provides measurement accuracy impossible to achieve using other methods.

Reference points in aerial photographs can be located using laser devices for measuring distances and angles. This technique is useful in making and reading maps of surface features and topography with accuracy to within a few millimeters. Computers can be used to store map data for large areas in minute detail.

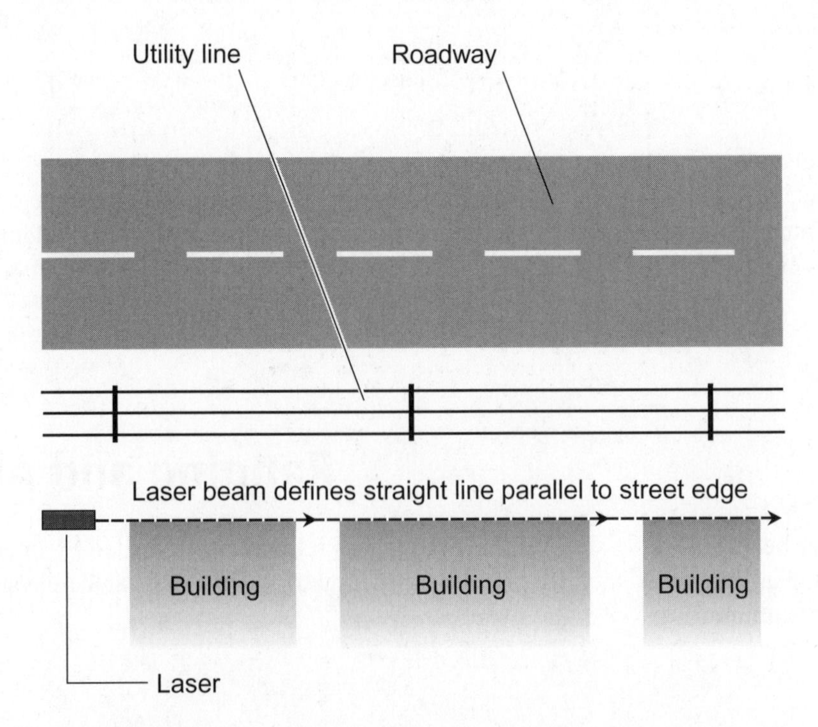

Figure 7-1 A laser can be used to align objects such as the foundations and walls of buildings.

MEASURING MEDIUM AND LARGE DISTANCES

Visible light or IR can be used to measure spans too large for a "tape measure." A common device of this type employs a pulsed light source, a mirror, and a timing device. This method does not necessarily require a laser, but the parallel rays of light from a laser travel great distances with very little attenuation, facilitating the measurement of extreme distances that cannot be accurately determined in any other way. The classic example is the measurement of the distance between the earth and the moon, approximately 400,000 km (252,000 mi). A ray of light takes approximately 2.7 s to make a round trip from the earth to the moon and back.

Figure 7-2 illustrates the basic apparatus for measuring long distances with a laser. The pulse from the laser passes through a beam splitter, and a small percentage of the output is collected by a photodiode or phototransistor. The electrical pulse is displayed on an oscilloscope or fed to an electronic interval timer. The greater part of the beam travels to the object in question, where a mirror or tricorner reflector is placed so that the beam returns exactly along the path by which it came. The orientation of the tricorner reflector is not critical; it will return any beam of light regardless of its arrival direction. The beam arrives back at the apparatus where the beam splitter delivers it to the detection device. The distance is then determined by either of the equations

$$d_{\text{km}} = (1.49896 \times 10^5)\, t$$

or

$$d_{\text{mi}} = (9.3141 \times 10^4)\, t$$

where d_{km} is the one-way distance in kilometers, d_{mi} is the one-way distance in *statute miles* (units of 5280 ft), and t is the round-trip time in seconds for any given pulse.

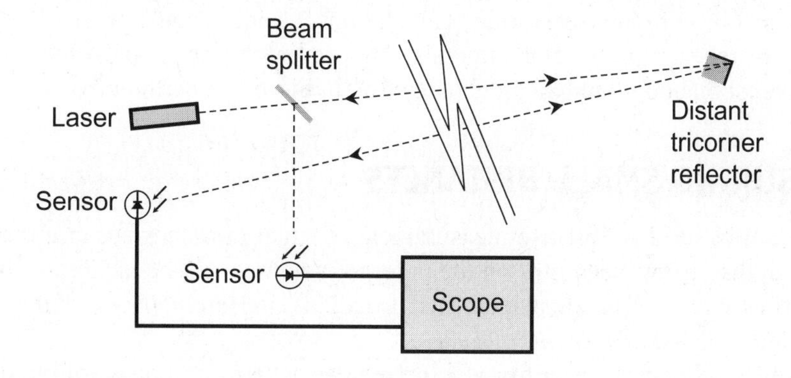

Figure 7-2 A pulsed laser, a beam splitter, a pair of sensors, an oscilloscope, and a distant reflector can be used to measure long distances with precision.

Using an apparatus similar to that shown in Fig. 7-2, the distance between specific points on the earth and the moon is measurable with such accuracy that changes can be detected from moment to moment as the earth rotates on its axis and the moon revolves around the earth.

PRECISION TERRESTRIAL DISTANCE MEASUREMENT

Lasers can be used to measure distances between mountain peaks on the earth's surface. These distances vary because the land is not absolutely rigid. A mountain range may be rising by 1 cm per year because of the movement of a continent, for example. This sort of change could not be detected accurately until the advent of laser devices for distance measurement. While a change of 1 cm a year may not seem like much, it amounts to 1 km in 1000 centuries—and in geological terms, 1000 centuries is not a long time.

Over longer periods, whole ranges of mountains buckle up from the crust and erode into the sea. Tectonic plates slide over the earth's mantle. These movements are "jerky," attended by violent events such as volcanoes and earthquakes. By observing patterns of movement of the earth's crust in a particular place, and by comparing these patterns with the occurrence of historical upheavals, it becomes possible to make medium- and long-range predictions for major geological events.

The *Laser Geodynamic Satellite*, launched in 1976, was among the first devices that could provide exact measurements of terrestrial distances. The satellite was covered with tricorner reflectors. Laser pulses were directed at the satellite, and the round-trip transit time was thereby measured from a specific point on the surface. The precise position of the satellite in the sky (compass direction and elevation above the horizon), as seen from the transmitting point, was also determined. This test was duplicated at several locations around the world. From the resulting data, the relative distance between any two points could be determined to within a few centimeters by computer-performed calculations. Changes could be detected as the tests were repeated at regular intervals. This revealed minute movements of the earth's crust that could indicate the danger of an imminent earthquake.

MEASURING SMALL DISTANCES

Lasers can be used for precise measurement of small distances because they emit energy at discrete wavelengths on the order of nanometers. The technique used for measuring microscopic displacements is called *interferometry*, and the device employed is called a *laser interferometer*.

Waves have the property of producing regular patterns as a result of phase addition and cancellation. Perhaps you have seen this effect in waves on a body of water, are familiar with the directional properties of radio antennas, or have worked

with optical diffraction gratings. The particular pattern produced depends on the wavelength of the light and the separation between sources or reflected beams. Because visible light has small wavelengths, tiny displacements produce interference patterns, as we have seen. The number of bright and dark spots, rings, or bands in an interference pattern indicates how far apart the sources of light are.

Waves that have equal amplitude and equal wavelength may add together in phase or completely cancel each other. If they arrive in some intermediate phase relationship, then the amplitude of the result is somewhere in between that of complete addition and complete cancellation. Phase differences can be measured in fractions of a wavelength, in degrees of phase (where 0° represents phase coincidence and 180° represents phase opposition), or in radians of phase (where 0 represents phase coincidence and π represents phase opposition). The output of a laser is such that all the waves are of the same wavelength and are in the same phase. Therefore, the laser is especially good for producing interference patterns. If a laser beam travels over two different paths to the same target, and if the paths differ in length, wave interference occurs at the target. Usually, this interference takes the form of a pattern of alternating bright and dark bands.

The laser interferometer consists of a laser light source, two mirrors, a beam splitter, and a target. The beam splitter transmits half of the laser energy and reflects the other half. Each of the beams is aimed at one of the mirrors, and this mirror reflects the beams to the target where an interference pattern is produced (Fig. 7-3). One

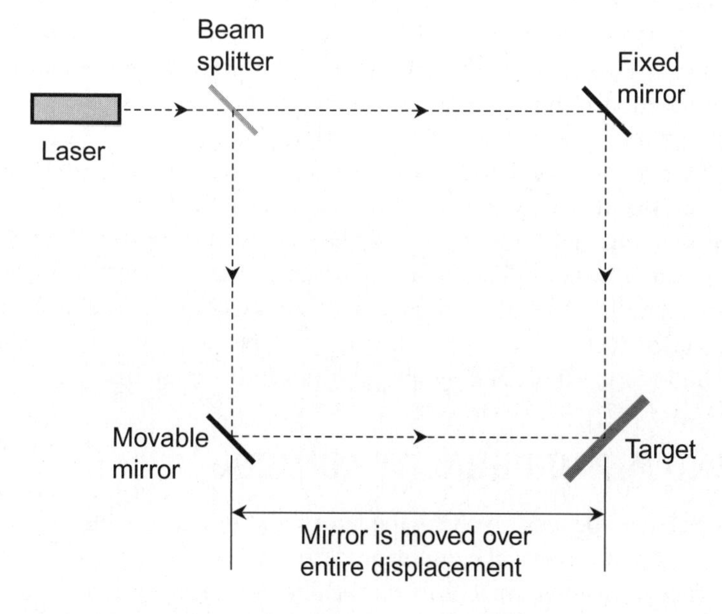

Figure 7-3 A laser interferometer can be used to measure small distances in industrial applications.

mirror is moved while the other is fixed. The movable mirror is displaced slowly over the distance to be measured. As this is done, the light at the target gets alternately brighter and darker. Each cycle from light to light or dark to dark indicates that the path difference for the two beams has changed by one wavelength. Based on the wavelength of the light used and the particular geometry of the interferometer arrangement, the displacement of the movable mirror can be determined with precision.

Interferometers are commonly employed to ensure that the geometric dimensions of sensitive apparatus stay the same over time. Temperature changes, mechanical vibration or blows, and normal use can all cause the alignment of equipment to become inexact. Interferometers show clearly when a dimension is not correct, because the appearance of the pattern changes dramatically if movement occurs over a fraction of a single wavelength of the laser light. This is crucial in a machine shop, for example, where interferometers are set up at the beginning of the day or at periodic intervals, and the intensity of the spot or nature of the pattern is frequently observed. The interferometer output can be fed to computerized robots that automatically compensate for changes in dimensions and bring the equipment back into proper alignment before significant deviation takes place.

SMALL-SCALE SURFACE TOPOGRAPHY

The laser interferometer can be employed to accurately map the topography (that is, the "flatness" or "bumpiness") of the surface of an object on a small scale. Tiny irregularities show up as changes in the interference pattern. A perfectly flat surface produces fringes that are equally spaced and fairly wide. A surface that is not flat produces irregularly shaped fringes that vary in their spacing or appear as concentric circles. Figure 7-4 illustrates the basic principle.

The surface to be tested is placed underneath a pane of glass that has been checked to ensure that it is perfectly flat for all practical purposes. A spread-out laser beam is shone through the pane of glass at the surface to be tested. Some of the beam reflects from the glass, but most is transmitted to the surface and reflected back up through the glass. Both reflected beams meet at a movable sensor. A flat surface is indicated by widely spaced fringes or peaks having a regular contour. Bumps or hollows are indicated by irregularities in the contour.

LENS AND MIRROR QUALITY CONTROL

Interferometers can be used to check the lenses and mirrors in telescopes, cameras, microscopes, and other optical equipment. *Astigmatism*, or the tendency of a convex lens or concave mirror to have a different focal length for light in one plane of polarization as compared with another plane, is easily detected. A standard sample, known to be essentially perfect, is compared with the specimen under test. The two are

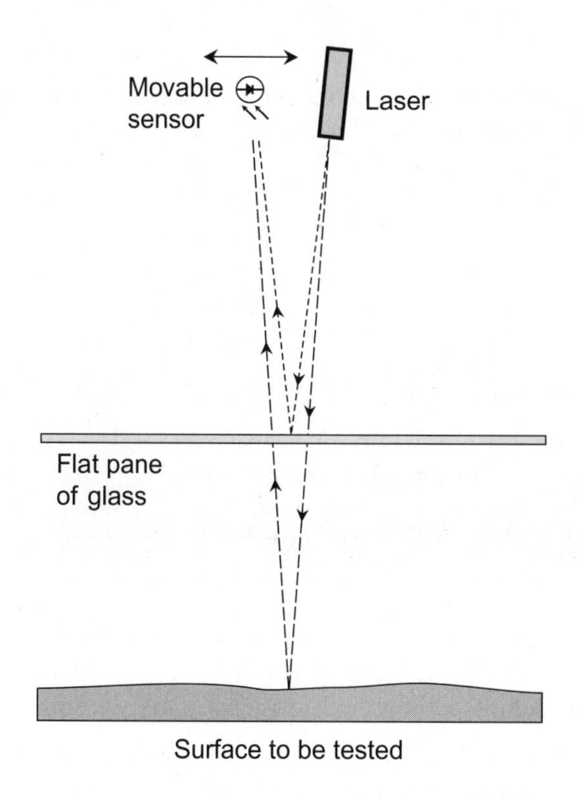

Figure 7-4 Apparatus for evaluating the topography of a surface on a small scale.

placed close together; the light transmitted and reflected interferes in the same way as in the example of Fig. 7-4. Deviations from perfection are seen as irregularities in the pattern of interference. Even microscopic deviations can be detected and astigmatism practically eliminated in mirror-grinding and lens-grinding processes.

Lasers can be employed for making sure that lens edges are smooth and of the correct shape. Rough lens edges can be detected in two ways. One method uses a small beam directed at a spot on the lens edge; rough spots cause irregular reflection of the light. The other scheme employs two laser beams that intersect at the point where they strike the lens edge; irregularities in the interference pattern indicate rough spots. The same techniques can be used to check the edges of other hardware items such as cutting blades.

PROBLEM 7-1

Suppose a laser apparatus similar to that shown in Fig. 7-2 is used to measure the distance between two broadcast towers located on hilltops within a direct line of sight. The time delay is measured and found to be 25.353 μs. How far apart are the towers to the nearest meter?

SOLUTION 7-1

If we let t represent the time delay in seconds, we can use the formula for the separation distance d_{km} in kilometers, which is

$$d_{km} = (1.49896 \times 10^5)\, t$$

In this case,

$$\begin{aligned} t &= 25.353\ \mu s \\ &= 2.5353 \times 10^{-5}\ s \end{aligned}$$

The distance in kilometers, rounded to four significant figures, is

$$\begin{aligned} d_{km} &= 1.49896 \times 10^5 \times 2.5353 \times 10^{-5} \\ &= 3.800 \end{aligned}$$

This is 3.800 km or 3800 m, accurate to the nearest meter.

PROBLEM 7-2

Suppose the captain of a Navy vessel wants to maintain a separation distance (range) of exactly 2000 ft between the center of her ship and the center of a nearby friendly destroyer. Both vessels are equipped with systems of the type shown in Fig. 7-2. What time delay, to the nearest nanosecond, will ensure that this range is maintained?

SOLUTION 7-2

We use the formula for the distance in statute miles as a function of delay time. A statute mile is 5280 ft. (A ship's captain would more likely think in terms of *nautical miles*, which are units measuring approximately 6076 ft, but as long as we use the correct formula, our calculations will work out properly.) The desired separation distance in statute miles is

$$d_{mi} = 2000/5280 = 0.3788$$

We know that

$$d_{mi} = (9.3141 \times 10^4)\, t$$

Plugging in 0.3788 for d_{mi}, we get

$$0.3788 = (9.3141 \times 10^4)\, t$$

Therefore

$$\begin{aligned} t &= 0.3788/(9.3141 \times 10^4) \\ &= 4.067 \times 10^{-6}\ s \\ &= 4067\ ns \end{aligned}$$

Industrial Applications

In manufacturing and mass production, the best type of laser for a given task depends primarily on the amount of power required for the application. Drilling, cutting, and welding demand high output power. Interferometers and counting devices can function with low-power lasers. Medium-power lasers are used for writing serial numbers on production units, for small-scale soldering, and for intra-factory optical communications.

ASSEMBLY LINES

Items on an assembly line can be counted by breaking a laser beam with the objects as they pass along. A running tally of items can proceed at speeds of up to several hundred per second. Laser counting systems have been used in the newspaper-publishing industry to accurately count the number of pages printed each day. The cost of laser-based counting systems has declined over the years, making them affordable and practical for an increasing number of medium-sized and large enterprises. (Perhaps needless to say, the Internet is gradually rendering the hard-copy newspaper-publishing business obsolete!)

Suppose that all of the objects on a certain conveyor belt are meant to turn out identical. This state of affairs can quickly be verified by shining laser beams at each object and recording the scattered light. To ensure uniformity of height in a batch of bottles, for example, a laser is shone over the tops of the bottles as they go by. Another laser is shone below the level of the bottle tops. In the ideal case, the first laser beam should never be broken, while the second beam should always be broken. Any deviation from this pattern indicates a defective bottle, which can be located and removed by a robot.

Another application is the manufacture of hypodermic needles. Samples are compared with a standard needle known to be free of flaws. The light from a laser is directed at the needle tip, and the tip scatters the light in a certain way. The needles are inspected at high speed by sensors connected to computers. Needles can be sorted quickly and efficiently in this way. Robots are programmed to pick out and discard defectives as they pass by on an assembly line.

EXAMINING SOLDER JOINTS

The laser provides a way to actively test the behavior of solder joints under thermal stress. The laser heats the material to be tested, and the cooling curve is plotted and analyzed using a computer. Laser pulses heat the solder joint, but not enough to liquefy the solder. In this way, the joint can be heated without damaging it. The behavior of cooling is monitored by a computer connected to an IR transducer. The computer outputs a graphical curve that can be compared with standard thermal curves for normal joints and for joints with specific kinds of defects. If there is a problem with a particular solder joint, such as its being "cold" (not enough heat was applied when the joint was made), the problem can be found and corrected before a defective module or device is shipped out.

EXAMINING CLOTH

In the textile industry, fabrics were once inspected manually. The work was tedious, time-consuming, and expensive. Lasers provide a way to check cloth for defects using far fewer human hours at a fraction of the cost. The laser energy is split into three beams. As a band of cloth moves by, the laser beams scan laterally across the cloth at several thousand meters per second. If a flaw is found, ink can be squirted on the defective area from a high-speed nozzle to mark its location. The laser makes it possible to determine the nature of flaws as well as their presence and location. All data is recorded by a computer for future reference. Laser scanners equipped with optical sensors can determine if a particular defect can be repaired without cutting out and replacing a section of the cloth.

DRILLING HOLES

Lasers can be used to drill tiny holes with great precision. The beam is focused using a convex lens to a sharp, tiny point, concentrating all the energy into an area of microscopic dimensions. The principle is similar to that used in a solar oven, but on a smaller scale. One of the main advantages to using lasers for hole drilling is the fact that the tool does not touch the material being drilled. There is no drill bit to break, slip, clean, or replace. Brittle, fragile, or hard materials can be drilled through without damage to the drill or the material being drilled.

Soft materials such as rubber, which have a tendency to wander out of place when mechanical drills are used, are handled readily by a laser drill, because the laser does not involve physical contact. There is no friction, so extraneous and destructive heating of surrounding material is kept to a minimum. Lasers of the N-CO_2-He type are widely used for drilling into rubber and flexible plastic. The smallest holes can be made at the rate of more than 10 per second in actual production, using lasers of about 150 W output power. Lasers can also be used to drill holes in capsules for the timed release of certain pharmaceuticals. The holes must be precisely sized to ensure that the medication is absorbed at the proper rate. Lasers provide a predictable hole size, fast production, and low failure rate, which is crucial when drugs are involved.

In the manufacture of wire, a hole is drilled in a quartz or diamond sample, and the wire is then pulled through the hole. A single hole in a diamond sample requires a few minutes to produce with a high-power laser. This compares with hours or days using mechanical drills. Lasers are also used for drilling and cutting ceramic materials, which are widely used in electronics because of their dielectric properties and their physical ruggedness. The problem with mechanical tools when used with ceramics is not a matter of hardness, but of fragility; ceramics break easily. Because a laser beam induces no mechanical stress, the risk of a mishap is minimal when lasers are used instead of mechanical drills for ceramics work.

Lasers can drill holes in circuit boards for electronic devices and computers. The glass-epoxy circuit board lends itself especially well to laser drilling, because such materials are fragile. In mass production, laser devices are programmed and operated by robots to ensure that the holes are drilled in the correct places with the correct diameters. Lasers are also used to "score" circuit boards so they can be broken apart without shattering or other damage. Tiny pits are drilled along the line where the material is to be broken.

LIMITATIONS OF LASERS FOR DRILLING

Laser drilling has many advantages, chiefly speed, predictability, and repeatability. But there are some significant limitations to the technology. Metals are difficult to drill using lasers, because they tend to reflect most of the visible light or IR energy that strikes them. Glass, quartz, and diamond are difficult to drill because these substances tend to transmit most of the laser light. These problems force the use of high-power lasers to ensure that enough power is absorbed to accomplish the desired task. For example, a metal alloy that reflects 95 percent of the laser power, and which melts when it absorbs 5 W over a circular region of a specific diameter for 30 s, requires a 100-W laser to produce a hole of that diameter in 30 s.

A pulsed drilling laser and support equipment can cost in the tens of thousands of dollars. Personnel must be trained to use a laser drill. The required amount of overall average output power needed to drill a hole in a certain length of time increases according to the cross-sectional area, which is proportional to the square of the diameter of the hole. For example, the average laser output power required to drill a 5-mm hole through a sample in 15 s is 25 times the average output power needed for a laser of the same wavelength to drill a 1-mm hole through the same sample in 15 s.

Material melted or vaporized by a laser does not disappear, but recondenses, often in the hole. Condensation also can occur on anything near the hole, whether it be the substance being drilled, or something else. The material around the hole may be partially melted and then resolidify unevenly, resulting in a "ragged" hole.

Perhaps the most significant problem in laser drilling is the fact that, because the laser beam must be focused by means of a lens, the hole resembles a cone more than

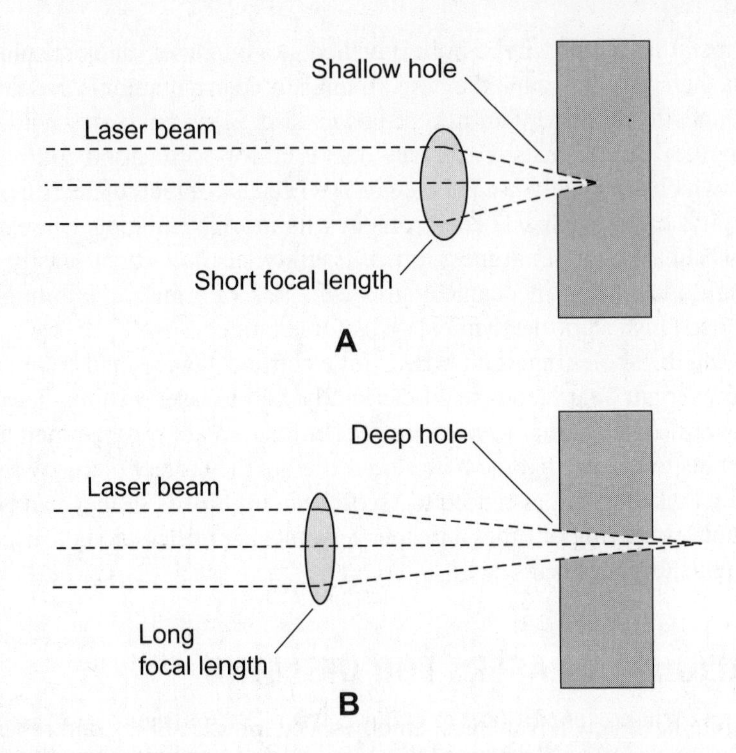

Figure 7-5 A lens with a short focal length produces a conical hole with shallow depth and sharply slanted sides, as shown at A. If the lens has a longer focal length, the hole is deeper and the sides less sharply slanted, as shown at B.

the long, thin cylinder that would represent an ideal hole. This problem can be made less severe by using a lens with a long focal length, as shown in Fig. 7-5. The shorter focal length, shown at A, results in a shallow, conical hole with a large apex angle. The longer focal length, at B, reduces the apex angle and results in a deeper, more nearly cylindrical hole.

CUTTING

Lasers cut light-absorbing and IR-absorbing materials by melting or vaporizing the material. Vapor is blown away by a gas jet so that it will not cool and resolidify in or near the cut. Materials that reflect or transmit most incident laser energy are burned away by a high-speed jet of oxygen that encourages oxidation of the material and drives away debris. The laser does not actually do the cutting in these cases, but heats the material to a temperature sufficient to allow combustion when the oxygen is supplied. The process is similar to that of fanning hot coals into flame.

A laser can be used to cut almost anything that will absorb at least some of the incident radiant energy at visible or IR wavelengths. Such materials include wood, paper, plastics, rubber, cloth, and certain metals. In some scenarios, the laser offers no improvement over traditional saws or blades for cutting. For wood cutting, the laser doesn't work very well with thick samples and boards, although it can sometimes do a reasonable job of cutting thin veneers. The heat from the laser chars thick planks near the cutting edge, a problem that occurs less often with traditional metal blades.

Lasers are used to cut flexible plastics and rubber that is deformed easily by mechanical tools. Lasers can seal as well as cut certain kinds of plastics if the correct temperature and rate of cutting are used. Lasers can cut patterns from pieces of cloth rapidly and accurately. The main difficulty, as with laser wood cutting, is the risk of burning or charring the material. Some fabrics, such as synthetics, tend to melt when exposed to the concentrated energy from a focused laser. If the cutting is done rapidly and air-blowing or suction devices are used to speed cooling and remove smoke, this problem can be avoided in some cases.

MARKING

Lasers are well suited for marking or etching items. A thin laser beam can produce an indelible pattern, or inscribe a serial number or identification code on an object. Extremely hard materials, such as quartz, jewels, and some metals, lend themselves to this process because mechanical inscription is difficult. The laser inscription process is quicker than traditional methods, and is less expensive when done by automated robots in large quantities. A pulsed laser leaves a string of tiny dots that form recognizable characters or are arranged in a pattern encoded to represent data. Continuous lasers can be used for etching or engraving.

WIRE AND CABLE STRIPPING

Lasers have been used for stripping insulation or enamel from wires or cables. Coaxial cable lends itself especially well to stripping by this means. This type of cable has a wire center conductor surrounded by a cylindrical wire-mesh braid. Insulation (the *dielectric*) separates the center conductor from the braid, and an insulating jacket surrounds the braid. Any technician who has worked with coaxial cable knows that it is difficult to strip without damaging the braid. Lasers make short work of it, and can speed the process of connector installation in mass production without damaging the cable. The plastic or polyethylene insulating substances absorb the laser energy readily, while the metal conductors, usually copper, do not. This process can be particularly useful when small coaxial cables are installed in compact electronic devices.

WELDING

In large-scale welding applications, the traditional blowtorch and arc are more practical than the laser. Super-high-power lasers require bulky power supplies, which are difficult to transport. Another problem is the sheer cost of building and maintaining a high-power laser.

In medium-scale laser-based welding, the power requirements are more modest (on the order of a few kilowatts). The metal need only be melted, not removed or vaporized. Laser-based welding is most easily done with substances that have fairly low melting points. The technology is ideal for use in cramped situations or in a vacuum.

In small-scale welding, lasers are practical. Pulsed lasers are more commonly used than continuous lasers. Either type will work, but high energy is easily obtained and regulated with the pulsed laser. The amount of energy produced by a pulsed laser over a given length depends on three factors as shown in Fig. 7-6, assuming the ideal case where the device produces rectangular pulses:

1. The pulse period X
2. The peak output power Y
3. The pulse duration (or width) Z

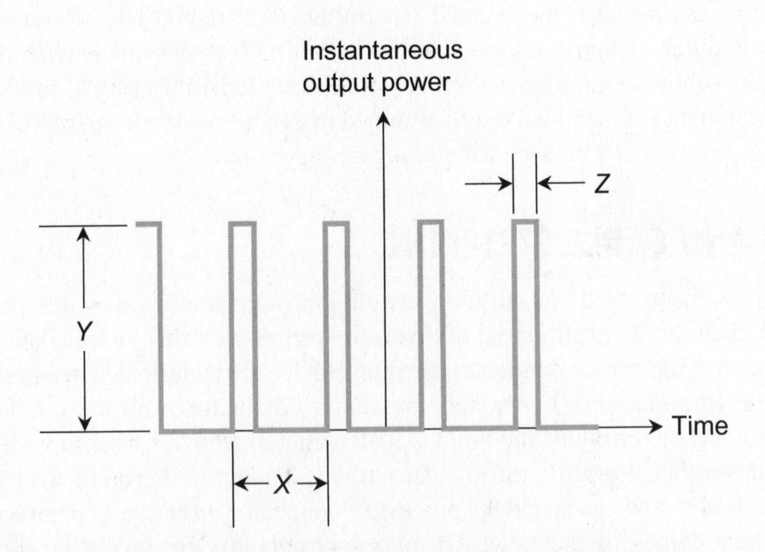

Figure 7-6 The energy output that a pulsed laser produces in a fixed length of time depends on the pulse period (X), the peak pulse output power (Y), and the pulse duration (Z).

The output energy E produced over a given length of time varies as follows:

- If X and Y are held constant, then E is directly proportional to Z.
- If X and Z are held constant, then E is directly proportional to Y.
- If Y and Z are held constant, then E is inversely proportional to X.

Welding that is performed on an extremely small scale is called *microwelding*. If the welded material is provided with a heat sink, tiny samples of metal can be made to melt without vaporizing. Sometimes inert gases such as argon are used in the welding process to prevent oxidation or combustion of the metal. The laser can be focused on a microscopic point, making it feasible to work with integrated circuits and microchips. When the scale is reduced to the point that the welds involve only a few molecules or atoms at a time, the operation can be called *nanowelding*.

The technicians and engineers who work in aerospace and deep-space programs are especially interested in laser welding of connections in electronic circuits. The environment often involves extremely low air pressure or a vacuum, so blowtorches will not function unless oxygen tanks are carried along. In the manufacture of semiconductor transistors, dissimilar metals or alloys must often be welded at junctions of microscopic size. The laser does this with a minimum number of accidents, improving the quality and reducing the rejection rate in mass production.

SOLDERING

Soldering is a low-temperature form of welding using specialized alloys that adhere readily to some metals, especially copper. Soldering of printed-circuit boards can be tedious and difficult when done manually, but laser-equipped robots can carry out this task in mass production. Small, pulsed N-CO_2-He lasers are ideally suited for soldering. Laser soldering is faster and more precise than older methods, and can apply exactly the same amount of energy to every joint. The amount of applied energy can be programmed into a computer. A laser can remove wire insulation at the same time that it heats a joint for soldering. Faulty joints are essentially eliminated in the production process.

HEAT TREATING

Lasers are used to treat metals in the manufacture of wire. Soft-drawn wire is easy to work with because it holds its shape when bent, but has relatively poor tensile strength and will stretch to some extent, a poor characteristic for use in structural supports or utility lines. Hard-drawn copper wire is better suited for these applications, but it is difficult to work with; when supplied on spools, it tends to coil back up if allowed to go slack. Hard-drawn wire is also difficult to splice, and tends to be more brittle than soft-drawn wire. Nevertheless, hard-drawn wire is preferred for use in applications where tensile strength is paramount.

In order to obtain a certain hardness, wire is heat treated or mixed with various other metals to form alloys. Heat treating often changes the way the atoms in the substance are arranged. A vivid example in nature is provided by the element carbon. When the atoms are arranged in a certain way, we have *graphite*, used for marking and lubrication. When the atoms are arranged another way, the same substance becomes coal or carbon powder. When arranged still another way, the atoms form diamond.

Metals can be hardened by heating or by application of tremendous pressure. Generally, heat is easier to apply than extreme pressure and is therefore used more often, especially in the annealing process. In order to enhance the absorption of the laser energy, the metal to be treated is coated with a substance that readily absorbs energy at the wavelength of the laser. The depth of the heat treatment depends on the amount of energy absorbed by a specific area of the surface over a given period of time.

PROBLEM 7-3

Which type of laser is most practical for small-scale welding? How much average power output do such lasers have? Can lasers be used to weld dissimilar metals to each other?

SOLUTION 7-3

For use in small-scale welding, a laser must be capable of producing several kilowatts of continuous output power. The N-CO_2-He laser is a good choice for such applications. Lasers can heat metals more quickly, and in more localized fashion, than conventional welding equipment. For this reason, lasers can sometimes produce durable welds between dissimilar metals such as copper and steel. With conventional welding torches, the difference in thermal conductivity between dissimilar metals usually makes welding impractical.

PROBLEM 7-4

In the situation shown by the pulsed-laser output graph (Fig. 7-6), what will happen to the energy output over a fixed length of time if

- the pulse period X doubles while all other factors remain constant?
- the peak output power Y doubles while all other factors remain constant?
- the pulse duration Z doubles while all other factors remain constant?

SOLUTION 7-4

The pulse period X is the time interval between the leading edges (beginnings) of successive pulses. If the pulse period doubles while the peak output power and pulse duration remain the same, the duty cycle is halved, so the energy output is also

halved. If the peak output power *Y* doubles while the pulse period and the duration remain constant, the energy output doubles. If the pulse duration *Z* doubles while the period and peak output power remain constant, the energy output doubles.

Other Applications of Lasers

Engineers, inventors, and artists are constantly testing lasers in new situations. Here are a few of the more interesting "everyday" examples of how lasers and other narrow-beam optical devices have been implemented since they became widely available in the 1960s and 1970s.

GUIDING SAWS

In the lumber industry, laser beams are used to show workers where a saw is going to cut a log, making it easy to guide logs through cutting machines. The laser beam "draws" the line as it reflects from a mirror. The beam from the laser strikes the mirror, which rotates rapidly so that the beam describes a plane in space corresponding to the plane of the cutting saw. The beam scans over the entire range many times per second, so that the light appears continuous (Fig. 7-7).

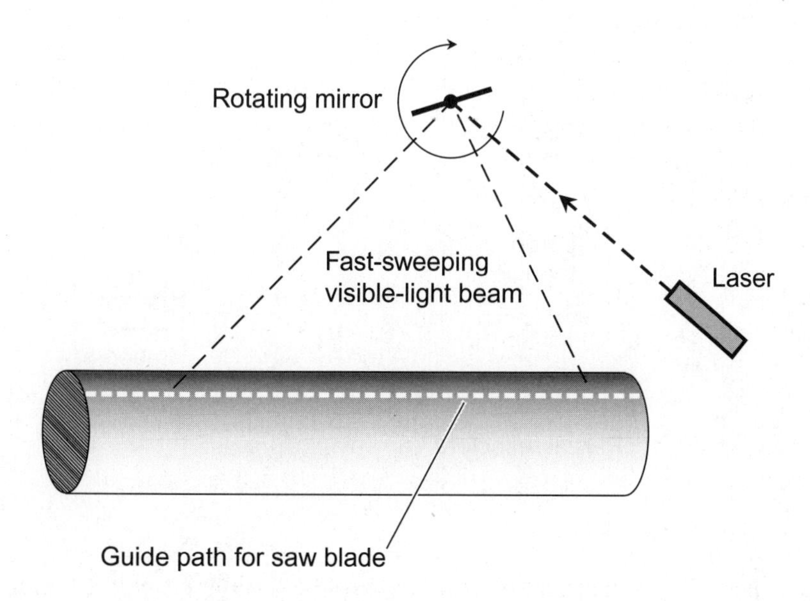

Figure 7-7 A laser beam, reflected from a rotating mirror, can "draw" a line on a log to highlight the path for a saw blade to follow.

Laser scanners can be used to precisely measure the dimensions of logs. The information is fed to a computer, which determines the best way to cut the logs in order to get the greatest quantity of usable lumber from them. This technique is especially useful in the cutting apparatus for making plywood veneers. Laser scanning of logs increases the amount of veneer obtainable by several percent over older methods. In the traditional method, a log was rotated around a lengthwise axis, but the straight line described between the chosen endpoints was not necessarily the axis that would yield the most veneer. Most logs are not perfect cylinders, so the optimum axis is not necessarily the one connecting the centers of the log ends. Laser scanning and computers help to determine the optimum axis to define the geometric cylinder within the log that contains the greatest volume of wood.

THE LASER GYROSCOPE

Lasers are useful for enhancing the accuracy of navigational systems that employ *gyroscopes*. One type of laser gyroscope consists of a resonator containing three mirrors, two of them totally reflective and one partially reflective, arranged at the vertices of a triangle. The mirrors are aligned perfectly so that the light beams propagate along the sides of the triangle. Two lasers are used. One of the beams travels clockwise around the triangle, and the other beam travels counterclockwise around the same triangle (Fig. 7-8). The entire assembly is affixed to a rigid frame. As long as the apparatus does not accelerate or rotate, the two laser beams arrive at the sensor in a certain phase.

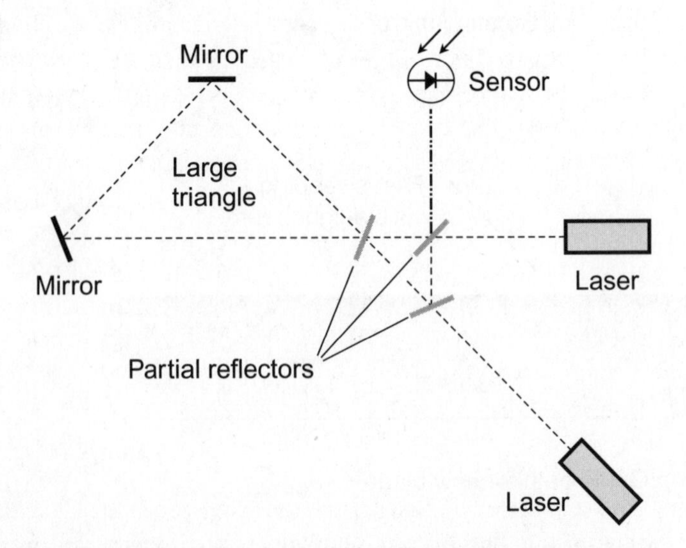

Figure 7-8 Laser-based system for determining rotational motion or acceleration.

If the structure accelerates or rotates, the mirrors move slightly with respect to the light beams, which always describe straight lines through space. As the two beams go around the circuit, one beam is forced to travel farther than the other by a distance amounting to a fraction of a wavelength. The relative path-length change shows up as a phase difference in the coherent light beams from the lasers. The phase change produces wave interference that alters the sensor output, which is constantly monitored. If the apparatus is large enough and the phase-detecting system is sensitive enough, even a tiny amount of acceleration or rotation shows up as a measurable change in the sensor output.

When three laser gyroscopes are combined, one along each of the three axes in Euclidean space, complex three-dimensional acceleration and rotational motion can be analyzed and recorded. Based on this information, an onboard computer can plot the orientation and course of an aircraft, spacecraft, or missile over a considerable period of time, even when no external points of reference are available.

POLLUTION ANALYSIS

Pollution and contaminants in gases and liquids can be detected and identified by means of a technology known as *laser spectroscopy*. The equipment and processes resemble those used by astronomers to determine the composition of interstellar and intergalactic matter. Various materials absorb energy at specific wavelengths while allowing energy at other wavelengths to pass through unaffected. The *absorption spectrum*, or pattern of visible and IR wavelengths blocked by the presence of a certain substance, can be used to identify that substance.

In a laser-based pollution analyzer, the output wavelength of a tunable laser is made to vary continuously over a certain range, so that the wavelengths of the absorption lines can be determined and the presence of specific chemicals identified. The existence of ions can also be detected. The data thereby obtained allows scientists to deduce the behavior of substances under different conditions. Changes in temperature, and exposure to ionizing radiation such as solar UV, gives rise to increased concentrations of some substances.

A familiar example of air pollution is the accumulation of *ozone gas* in the lower atmosphere in and near large population centers, particularly in the summer. Ozone, which consists of three atoms of oxygen grouped together (O_3) instead of the usual two (O_2), results from solar UV action on automobile exhaust. Ozone irritates the lungs in some persons, and can exacerbate chronic diseases such as asthma and emphysema. The chemical has an aroma resembling that of chlorine bleach, which many people find objectionable.

Ozone presents a cruel irony. This gas is largely opaque at the shortest UV wavelengths that are harmful to biological organisms. In the upper atmosphere, the presence of ozone gas is essential, because it keeps excessive short-wave solar UV radiation from reaching the surface. Human overpopulation gives rise to localized *ozone pollution* near the surface, while at the same time creating a general danger to life on the planet by producing *ozone depletion* in the stratosphere.

Sulfur dioxide and carbon monoxide are also produced by human activities. These gases accumulate in the lower atmosphere in and near cities, and downwind of industrial plants and refineries. Sulfur dioxide is a by-product of the combustion of low-grade coal, which is used in some electric power plants. Carbon monoxide is present in car exhaust, and is also produced by combustion of carbon-containing solids, liquids, and gases such as wood, oil, and methane. The laser spectroscope can determine the relative concentrations of pollutants such as these, which helps local officials decide when to issue "pollution alerts," and also to reassure the public when emission concentrations are not high enough to present a problem.

Another method by which lasers can detect and measure the extent of air pollution is by determining the extent of beam scattering caused by particulate matter. A laser can produce a thin, straight beam of light that does not diverge or lose intensity over moderate distances. Dust and other particles in the air cause scattering and also give rise to attenuation with distance. These phenomena can be measured and can provide clues as to the quality of the air in a given sample, indicating contaminant concentrations down to a few parts per billion (where a billion is equal to 10^9). Under laboratory conditions, individual, isolated atoms have been detected and identified using such equipment.

LIGHT SHOWS

Artists and musicians have been putting on laser-based optical exhibitions for decades. Three-dimensional images are formed in midair by techniques similar to holography, which we'll learn about in Chap. 8. "Laser pictures" are "drawn" in the air by tracing the outlines of geometric figures, the laser beam moving rapidly as it is reflected from mirrors that rotate in two dimensions (Fig. 7-9). Small mirrors can be mechanically moved much faster than would be possible if it were necessary to rotate the whole laser apparatus. Krypton-based lasers are especially effective because they can generate visible-light energy at multiple discrete wavelengths. These components can be geometrically split into colored beams by prisms or diffraction gratings.

A projected, moving laser produces a picture because the human eye sees a spot linger for about 1/30 of a second, even if it is flashed for just a few microseconds. If the spot moves in such a way that it retraces the same pattern at a rate of one complete cycle every 1/30 of a second or faster, then the human eye/brain sees the image as a complete image rather than as a fast-moving spot. Intense beams are necessary to produce complex images bright enough to be easily visible; the scanning process and the sheer size of the images "dilute" the energy in the beams. Because of this "brightness dilution effect," there is a practical limit to the complexity of the images that can be formed in this way. It can be quite expensive to produce large, detailed fast-moving, three-dimensional images using the laser scanning technique.

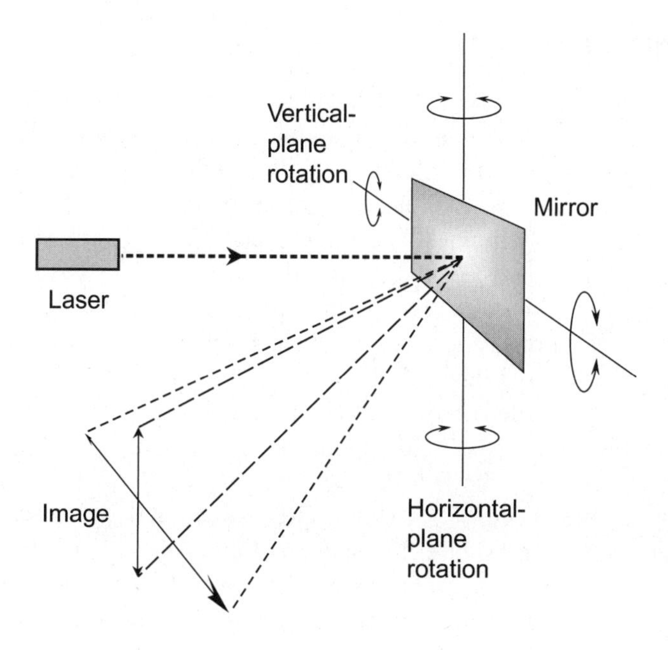

Figure 7-9 Animated images can be produced by laser beams reflected from mirrors rotating independently on perpendicular axes.

A planetarium provides an ideal environment for a laser show. The darkened dome produces a visual backdrop that seems to have infinite depth. With skillful manipulation of the laser beams, the apparent sizes of the projected images can be vast, while the actual scale of the hardware is modest. Another method of producing dramatic images by means of the laser is the *giant hologram*. A hologram is a three-dimensional rendition of an image, rendered by the interference of light beams from different lasers. With the proper diffusion in the air, large moving three-dimensional images are produced, such as illusory space ships hovering and darting around over an outdoor stadium at night.

Laser light shows are sometimes called *kinetic art* because the images change position, size, and shape. When the movements of the lasers are made to follow the sound patterns in a musical accompaniment, kinetic art can have a strong psychological impact. Some people claim that it can induce a trance-like state or mind-clearing effect. A few people have seriously suggested that modulating the lasers in kinetic art with subliminal messages could be used for "mind control."

Whenever lasers are used in public places, special care must be exercised to ensure that no one in the audience can look directly into the beams. The lasers are not strong enough to pose a fire hazard or to burn exposed skin, but they are intense enough to cause permanent eye damage if observed directly. The risk can be minimized by ensuring that the lasers are always aimed up and away from the audience. In the United States, kinetic art shows are governed by strict regulations to ensure the safety of the viewing audience.

THE CD AND THE DVD

A *compact disc* (CD) is a plastic disc having a diameter of 12 cm (approximately 4.7 in) on which data can be recorded in digital form. Any kind of data can be digitized: text, sound, images, multimedia (audio/video programs), and computer software. A *digital versatile disc* (DVD) is a CD especially designed to store full-motion audio/video data, usually movies. When a CD or DVD is manufactured, the input data is subjected to a process called *analog-to-digital (A/D) conversion*. This process changes the continuously variable audio-frequency (AF) waves into binary high and low states, or bits. These bits are then "burned" into the surface of the disc in the form of microscopic flaws or pits.

Digitized audio, recorded on the surface of a CD, is practically devoid of the hiss and crackle that was the bane of older media such as tape or vinyl disc. This exceptional clarity results from the fact that the information on the CD is binary: a bit (binary digit) is either 1 (high) or 0 (low). The distinction between these two states is more clear-cut than the subtle fluctuations of an analog signal. Likewise, digitized video images are clearer, more detailed, have better contrast, and have superior color quality compared with the images rendered on older analog media.

A CD or DVD player recovers the data from the disc without physically touching the surface. Laser beams scan the disc; the energy is scattered by the pits but is reflected, mirror-like, from the unpitted regions. Figure 7-10 shows what happens at a single reflective layer, typical of the original CD technology. Recent refinements have produced CDs and DVDs with multiple layers to increase the amount

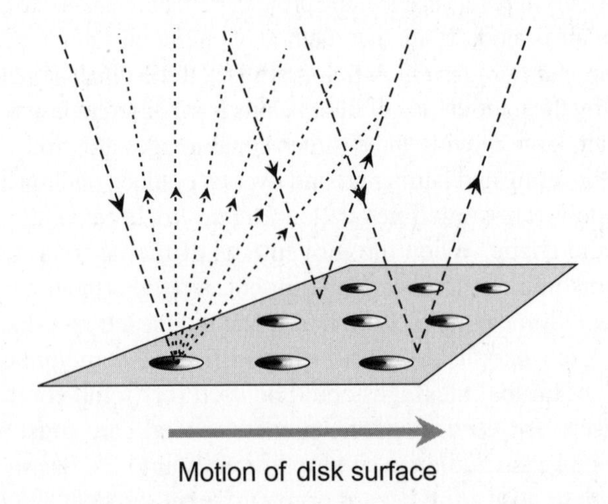

Figure 7-10 On the surface of a conventional CD, incident laser beams are reflected mirror-like from unpitted regions of the surface, but are scattered when they strike the pits.

of data that can be held on a single disc. As the lasers scan the surface, the reflected rays are digitally modulated, picked up by sensors, and converted into electrical currents. *Digital signal processing* (DSP) can drastically reduce noise that has been introduced by environmental factors beyond people's control, such as microscopic particles on the disc, or random electronic noise in circuit hardware. Then the data is subjected to *digital-to-analog* (D/A) *conversion* to restore it to its original form. Finally, the signal is amplified and sent to the output devices such as speakers, headsets, video displays, or computers.

Quiz

This is an open book quiz. You may refer to the text in this chapter. A good score is 8 correct. Answers are in the back of the book.

1. Glass can be difficult to cut with a laser because
   (a) most of the light or IR energy is reflected by the first surface of the glass.
   (b) most of the light or IR energy is transmitted through the glass.
   (c) the index of refraction differs dramatically from that of the surrounding air.
   (d) the melting point of glass is too low to allow for clean cutting.
2. When a laser beam is rotated rapidly in a geometric plane and shines on a surface, we see a bright line or curve instead of a moving spot. Why?
   (a) The human eye and brain see bright spots linger for a fraction of a second, so if the scanning takes place fast enough, we see a repeated trace as a continuous line or curve.
   (b) The energy from a fast-spinning laser is actually spread out in a continuous flat plane rather than a well-defined beam; this plane-shaped energy "zone" intersects a surface as a continuous line or curve.
   (c) The emitted energy becomes incoherent rather than coherent, so the beam tends to spread out.
   (d) The emitted wavelength no longer occurs at a single wavelength, but instead, takes place over a wide band of wavelengths.

3. A laser gyroscope works by detecting
   (a) changes in the relative wavelengths of laser beams.
   (b) changes in the phase relationship between laser beams.
   (c) changes in the output bandwidths of laser beams.
   (d) changes in the output power levels of laser beams.
4. Suppose a laser apparatus similar to that shown in Fig. 7-2 is used to measure the distance between cabins on two mountain peaks. The time delay is measured and found to be precisely 100 μs. How far apart are the cabins to the nearest hundredth of a kilometer?
   (a) 0.15 km
   (b) 1.50 km
   (c) 14.99 km
   (d) 149.99 km
5. A laser beam can be used to detect particulate pollution in the atmosphere because the presence of the particles
   (a) makes the output of the laser appear more nearly concentrated at a single wavelength.
   (b) absorbs energy from the laser beam, thereby increasing the amount of power required to properly operate the laser device itself.
   (c) makes the laser beam appear narrower, so the spot or ring projected on a distant screen has a smaller diameter than it would if the air were not polluted.
   (d) increases the amount of energy lost between a laser and a distant detector compared with the situation in unpolluted air.
6. A laser interferometer would most likely be used to determine
   (a) the distance between the earth and the moon.
   (b) the irregularities in the surface of a piece of plastic.
   (c) the intensity of the light from the sun.
   (d) the extent to which a laser beam diverges with distance.

7. Lasers are better suited than conventional methods for soldering in mass production because
   (a) robot-controlled lasers can work faster, and with greater precision, than humans wielding soldering irons.
   (b) lasers can be used to solder connections between metals that do not adhere to solder under other circumstances.
   (c) lasers can be controlled by robots, while other soldering machines cannot.
   (d) All of the above.
8. In laser-based drilling applications, the use of a lens with a long focal length to focus the beam is
   (a) undesirable, because it dilutes the laser power.
   (b) desirable, because it produces a focused energy spot of tiny diameter.
   (c) undesirable, because such lenses are easily damaged by the laser.
   (d) desirable, because such a lens allows for a deep hole.
9. In surveying applications, a ray of laser light follows a straight line between any two points in space, provided that the laser beam is not reflected from any mirrors, and also provided that
   (a) the beam divergence is zero.
   (b) the energy is confined to a single wavelength.
   (c) the phase difference between the two points is either 0° or 180°.
   (d) the index of refraction is constant between the two points.
10. Localized irregularities in a surface can be detected by observing
    (a) changes in the energy contents of reflected laser beams.
    (b) changes in the relative wavelengths of reflected laser beams.
    (c) changes in the bandwidths of reflected laser beams.
    (d) changes in interference patterns produced by reflected laser beams.

CHAPTER 8

Exotic Optics

Optical devices have applications limited only by our imaginations and the availability of physical and financial resources. In this chapter, we'll look at a few exotic optical technologies that have been developed or contemplated.

Holography

A conventional photograph can reproduce *perspective*, in which distant objects look relatively smaller than, and may be eclipsed by, nearer objects. The true recording and reproduction of 3D perspective effects require special hardware and processes to generate modified photographs called *holograms*.

PARALLAX AND ECLIPSING

In effect, a hologram is an encoded 2D rendition of a 3D scene. The method of image encoding differs from that in an ordinary photograph. All of the information needed to provide a realistic 3D image is contained in the 2D film surface. It is only necessary that the proper "decoder" be employed to recover the original image.

Stereoscopic photographs existed before holograms were conceived. A stereoscopic photograph can be produced by placing two conventional photographs, taken from slightly different directions, in a special viewing scope in which the left eye views one photograph and the right eye views the other. This gives the observer the impression of parallax. Our eyes and brains perceive depth largely because of parallax effects, as shown in Fig. 8-1A. The vividness of the effect depends on the distance between the left and right viewpoints. Such dual-photograph combinations can also give the impression of eclipsing (Fig. 8-1B).

In a real-life 3D scene, the extent of parallax changes as the observer moves toward or away from observed objects. In addition, distant objects that are eclipsed by nearer objects come into view gradually as the observer moves laterally or vertically. The vantage point in a stereoscopic photograph is fixed. The pictures can be put into motion by running two simultaneous movies, but the viewer cannot willfully change the point of view. A hologram changes all that, introducing a new psychological variable into the situation: The viewer can change the way the scene appears *at will*.

THE FIRST IDEA

Holography was originally conceived as an improvement in the *electron microscope*, a machine used for viewing extremely small objects at high magnification. The electron-beam focusing systems in the first electron microscopes were so poor that the developers decided to bypass them altogether. Engineers reasoned that if there were some way to accurately record the direction and intensity of light from an object, then all the information needed to reconstruct the image would be recorded on the film. The only remaining challenge would be to extract that information for viewing.

A *grayscale* image as rendered from a single vantage point consists of a 2D field of view, in which the intensity of light from any given direction is represented by a point of variable brightness. Engineers thought that an electron-microscope image containing all of the necessary information would look like a meaningless jumble if viewed straightaway, but if the information could be retrieved using special apparatus, the result would be superior to the images produced by electron microscopes then in existence. Optical means might be used to "correct" the electron-produced image, gleaning the field-of-view intensity and direction information from the film. This notion gave rise to an entirely new method of exposing and viewing photographs.

When the light from an object or scene strikes a photographic plate without passing through a lens, the resulting image is plain gray, without apparent detail.

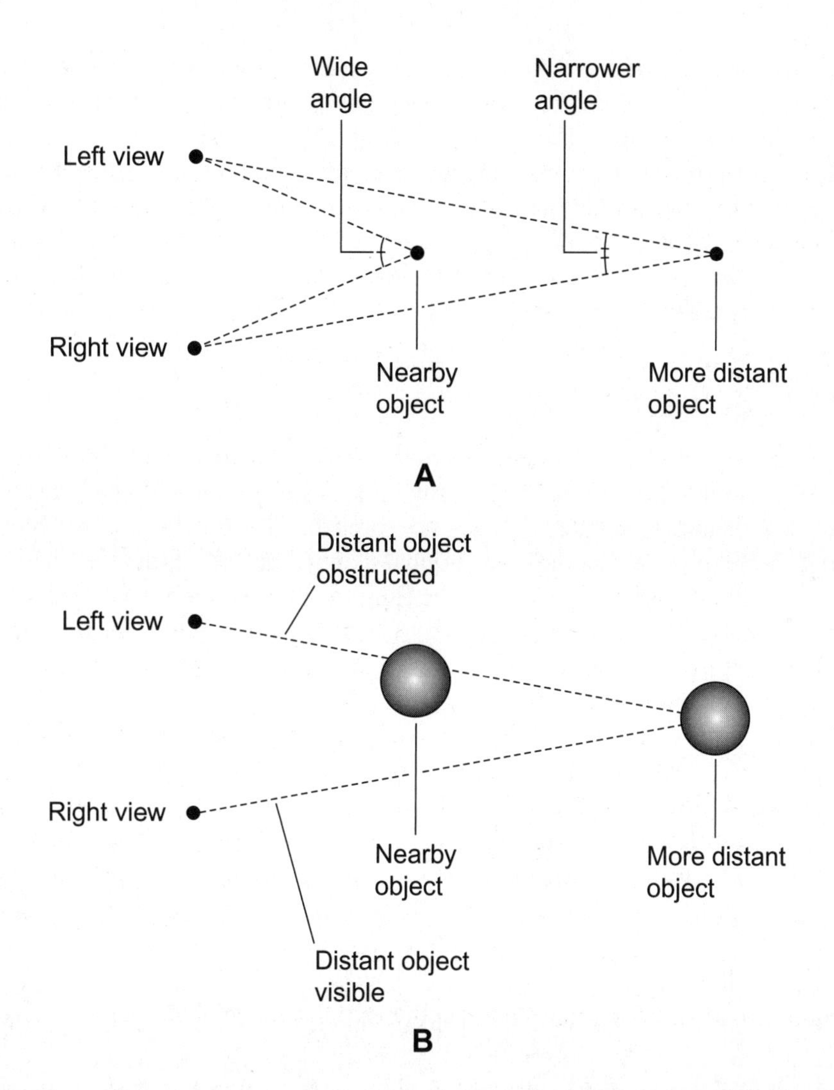

Figure 8-1 At A, parallax results in different relative images as seen with the left eye as compared with the right eye, allowing for depth perception. At B, eclipsing occurs when a distant object is blocked to a different extent as seen from either eye.

No real image results when a wavefront strikes a photographic plate; the exposure is essentially the same everywhere. However, if a reference laser beam, emanating from a point source and having coherent wavefronts, interferes with this incoming light, the image can be developed into a photograph. When this photograph is observed with the help of a reference laser oriented in a manner similar to the one used in the developing process, it appears as if the object or scene is being observed under the original conditions. Eclipsing effects are reproduced, so that objects hidden from view when seen from one viewpoint emerge when seen from another viewpoint.

This discovery was made by accident, as many important discoveries are made. The original engineers likely did not anticipate the impact that their investigations would eventually have. In fact, for a decade and a half, this discovery was passed over by the scientific establishment because the lasers necessary to make high-quality holograms were not yet available.

ENTER THE LASER

The physicists *Juris Upatnieks* and *Emmett Leith* were among the first to use lasers to make and view holograms in the United States. The interference pattern produced by the laser-generated images appeared on the film as diffuse blobs or smudges when observed in the conventional manner, but by duplicating the illumination conditions for the exposure of the film, the 3D rendition was obtained. The discovery was made independently by a Russian scientist, *Yuri Denisiuk*. He used his techniques to reproduce art so it could be enjoyed by people not able to physically visit art galleries.

The first time you see a good hologram, you might be surprised by the high image quality. Color holograms often seem to have more vivid colors than conventional color photographs or even real-life scenes. Holograms look different when viewed from different directions because they contain all the information necessary to produce accurate images as viewed from multiple points on a flat surface in space.

PROBLEM 8-1

Most binoculars provide more vivid depth perception than unaided eyes. Why?

SOLUTION 8-1

The lenses in most binoculars are set farther apart than are human eyes. The increased separation is achieved by using mirrors (in cheap binoculars) or prisms (in more expensive binoculars), as shown in Fig. 8-2. The vividness of depth perception increases in direct proportion to the extent of the parallax and eclipsing variation at a specific distance from an observed object. Variations in parallax and eclipsing are, in turn, directly proportional to the length of the *baseline* between the points of view. The increased baseline length, not the magnification, is responsible for the improved depth perception afforded by binoculars as compared with unaided eyes. Some "miniature binoculars" do not contain mirrors or prisms to increase the separation between the left and right points of view. Such devices do not offer increased depth perception compared with unaided eyes.

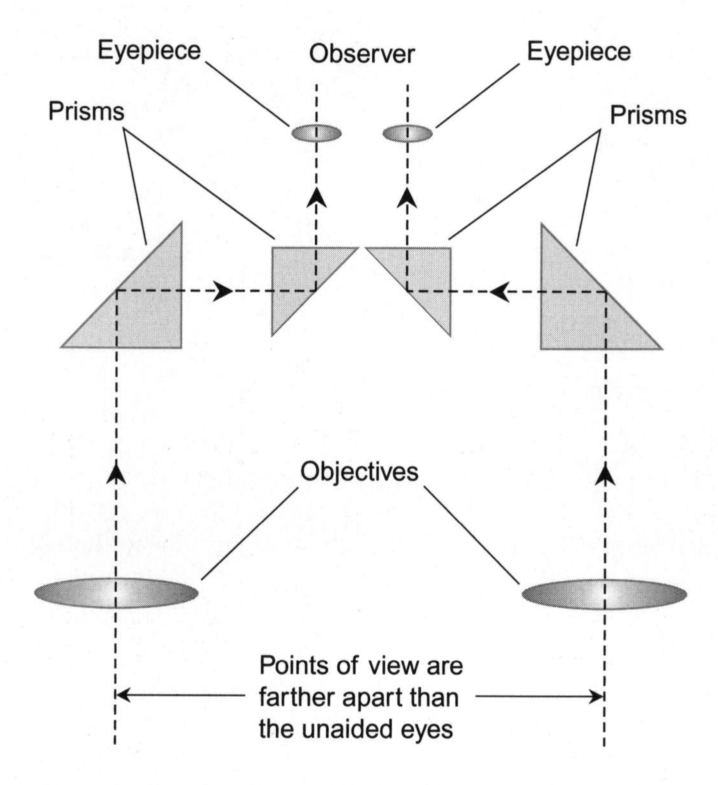

Figure 8-2 Functional diagram of a pair of binoculars. The objective lenses are set more widely apart than human eyes, increasing the baseline to provide more vivid depth perception.

How Holograms Are Made

A scene can be faithfully reproduced by recording the images as seen from many different points of view. This need not (and in fact cannot) include every possible vantage point in 3D space, but as the number of points of view increases, the image information becomes increasingly complete.

A "FLAT" 3D SCENE

A 3D scene is the combination of 2D images obtained by looking at that scene from numerous points on a plane near the subject (Fig. 8-3). This process can transfer some aspects of a 3D scene to a 2D medium because there is a one-to-one correspondence between all the points on a flat film surface and all the points on the

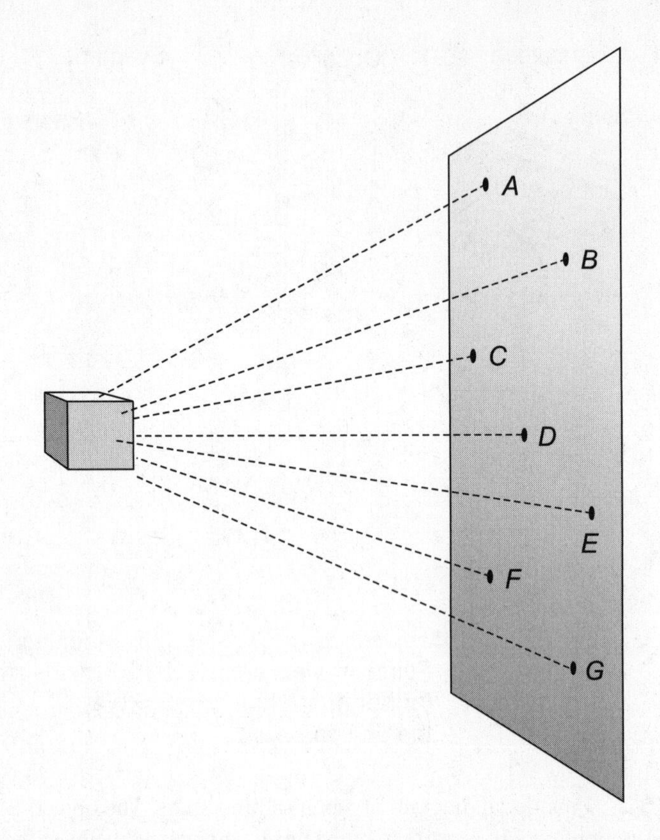

Figure 8-3 A complex object or scene presents a unique image from every viewpoint on a plane region in space. The viewpoints (such as *A* through *G* shown here) are selected to represent an optimal variety of viewpoints.

plane region. Figure 8-3 shows seven points on a plane region near an object; a real hologram takes advantage of thousands or millions of such points, with the images put together so that the "eye/brain team" can make 3D sense out of it. The result is a vivid, although limited, portrayal of the exterior of a 3D object or scene as the observer moves around with respect to the hologram.

Eclipsing effects can be directly reproduced by the holographic process, but depth is rendered only indirectly, because the photographic viewpoints are all on a single plane in space. When moving closer to the hologram or farther away from it, an observer sees a pair of images (one in each eye) whose relative viewing angle varies with the distance from the hologram. The illusion of depth is thereby created, but depth perception does not occur in quite the same way as it does in real life. A more sophisticated technique known as *volumetric holography* makes true depth reproduction possible, but that's beyond the scope of this discussion.

GATHERING THE IMAGE DATA

A photograph is a compilation of information, the simplest being coordinate position and brightness. This was how old-fashioned "black-and-white" television worked. A spot, produced by an electron beam, scanned a plane surface in a precise pattern. The brightness of the spot varied as it moved along the lines of the screen, reproducing a completely new image 30 times per second. The television signal was a complex function of amplitude versus time that contained all the information needed to produce a motion picture.

The image data in a hologram is encoded in a way that makes it unrecognizable when viewed as an ordinary photograph. We would not be entertained by watching a television program on a laboratory oscilloscope or spectrum analyzer, even though all the essential video information would be there. If we are to see a meaningful image, the information must be presented to us in the correct format. When we look at a hologram as we would view an ordinary photograph, we see nothing more than a pattern of interference fringes. A light source, coming from a single point, must be shone at the film to reproduce the holographic image.

A hologram contains redundant information, making it fundamentally different from an ordinary photograph. The holographic "code" is exposed on the film in such a manner that the hologram can be damaged or cut, and still contain essentially all of the information of the original. Some of the image resolution (detail) is lost, but the same information remains. A hologram contains more meaningful 3D information than a conventional 2D photograph taken from a single vantage point, even though the hologram might have lower resolution per view. Figure 8-4 shows an example. With a conventional photograph, only one of the views in this drawing could be obtained. With a hologram, they could all be contained in the film.

As previously mentioned, holograms were originally developed with the electron microscope in mind. Today, holography can be used with electron microscopes or conventional optical microscopes to get 3D images. Numerous and diverse

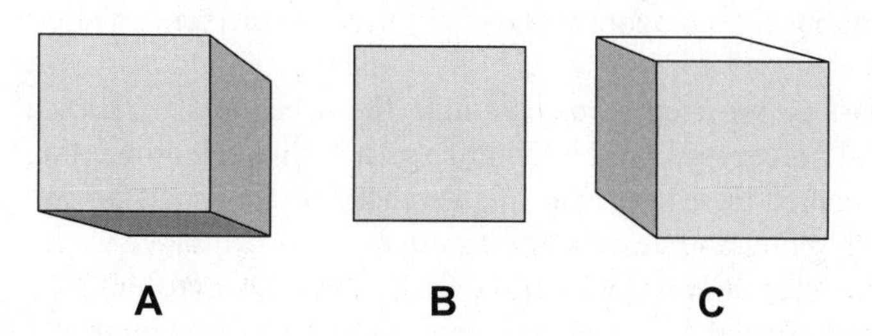

Figure 8-4 A cube as seen from below right (A), directly face-on (B), and from above left (C). Note the differences in perspective, and also the differences in shading among the cube faces produced by background illumination.

applications of this technique exist. *Holographic microscopy* has been employed to determine the velocities of tiny particles ejected by burning fuel or thrown off by bullets impacting a hard surface. Enhanced renditions of bacteria and other microorganisms have been obtained by photographing them using holographic methods.

WAVE INTERFERENCE

Interference patterns in visible light can be recorded on old-fashioned photographic emulsions to produce a hologram. The quality of the hologram is directly proportional to the sharpness of the interference patterns that can be produced by a given light source. The quality of the hologram also depends on the image resolution that the film can provide.

In a conventional photograph, a lens focuses incoherent light waves on the photographic film, producing the image. To make a hologram, no lens is involved, but coherent light is used to illuminate the scene. Parallax information is recorded by the phase differences of the light wavefronts as they strike the film at various distances from a given point on the scene. The hologram is not a record of a real image in the classical-optics sense, but a record of interference patterns.

The photographic emulsion of hologram film is similar to that of ordinary film, but the "grain" is "finer." "Fine-grain" film can record more interference lines per millimeter than "coarse-grain" film, increasing the effective viewing angle because interference fringes get closer together as the angle from the source increases. This effect is illustrated in Fig. 8-5A for a spherical set of wavefronts intersecting a flat film surface. The resulting pattern is a "bull's eye" with one light or dark center spot and fringes that get closer together with increasing radius (Fig. 8-5B).

HARDWARE

When holograms are created, the apparatus must be physically stable and vibration-free because of the microscopic dimensions of visible-light wavelengths. Motion or vibration among the components will upset the interference pattern, ruining the exposure.

Sheer mass is one method of obtaining the necessary mechanical stability for making good holograms. Objects with large mass have high inertia, so they resist rapid back-and-forth motion that might otherwise be produced by minor disturbances. Cushioning and absorbent construction are two other ways to ensure that the holographic apparatus does not move or vibrate. Air cushioning works well. A table can be constructed from heavy metal such as steel and mounted on pneumatic supports. The heavy metal provides inertia, and the pneumatic supports provide the cushioning. A basement is a better place to set up holographic equipment than an upper floor, because upper floors usually vibrate more. Traffic on nearby streets can

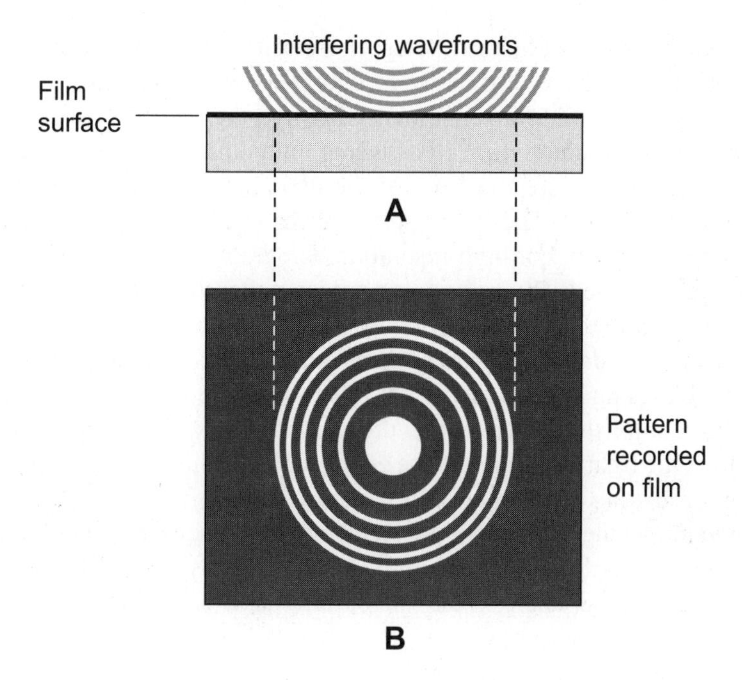

Figure 8-5 Interfering wavefronts of light strike the flat surface of a photographic film, producing an interference pattern resembling a "bull's eye." At A, the dark curves represent constructive interference, which produces bright rings on the film as shown at B.

be a problem, but the work can be done in the middle of the night or the equipment can be set up in a rural area. Common plastic "bubble wrap" can be an excellent cushion, especially the variety with bubbles having large diameter.

The choice of a laser for making a hologram depends on available funds, the wavelength desired, and the power needed. Some high-quality lasers, notably the He-Ne type, can be obtained at prices affordable to most hobbyists. "Laser pointers" are also affordable. Specialized semiconductor lasers, designed especially for holography, are available at higher cost. The light from any laser is concentrated enough to make it dangerous if viewed directly, so precautions must be taken to protect the eyes. Moderate-power lasers are better than low-power ones only insofar as the exposure time is reduced.

The production of high-quality color holograms requires the use of multiple lasers that emit coherent light at various wavelengths throughout the visible spectrum. The most common wavelengths are those that produce the red, green, and blue primary colors. Radiant light having these three colors can be combined to make full, true-color holograms. Continuous-output or pulsed lasers can be employed. The pulsed lasers are better suited to setups where there may be some vibration; pulsed lasers act like strobe lights, freezing the scene if the light is intense enough so that the entire exposure can be completed with one pulse.

Most hologram-making arrangements involve the use of one or more first-surface mirrors. A common mirror is silvered on the back surface of the glass and will not work well because the front of the glass also reflects some of the light, creating a double image. The holographic mirror is silvered on the front surface, preventing this problem. For high-resolution holography, concave collimating mirrors are sometimes used. The collimating mirror makes the wavefronts from a point source emerge parallel and flat, as if the source were at an infinite distance.

Lenses can be used to spread laser beams out, so that they illuminate larger areas than is possible using a simple laser with its thin beam. Either convex or concave lenses will accomplish the beam-spreading function (Fig. 8-6). The proper choice of lens allows the light to illuminate the entire subject, while retaining the necessary coherence for making sharp interference patterns on the film.

A hologram-making apparatus may include at least one beam splitter that creates two beams from a single laser. The simplest beam splitter is a partially reflecting, first-surface mirror set at an angle to the approaching laser beam. Some of the beam is

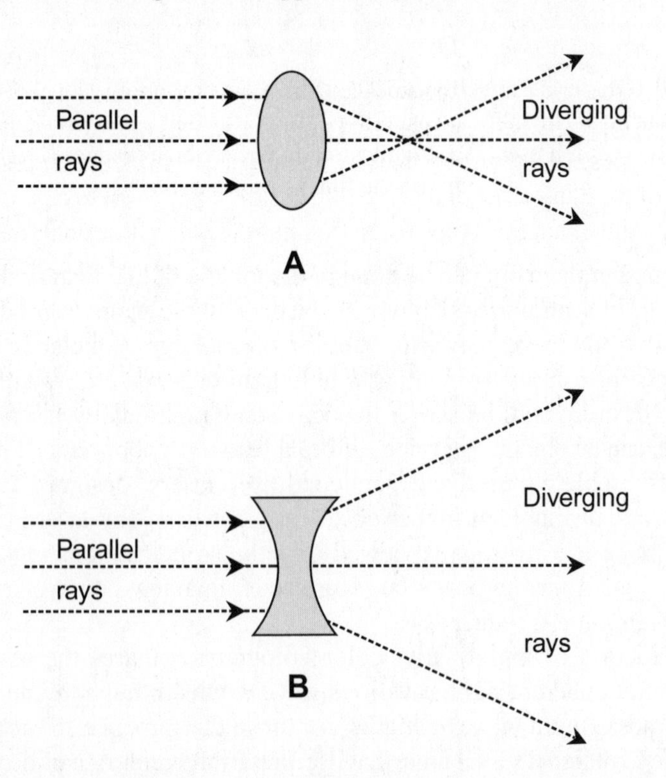

Figure 8-6 A convex lens (A) or a concave lens (B) can be used to spread out a narrow laser beam while retaining wavefront coherence.

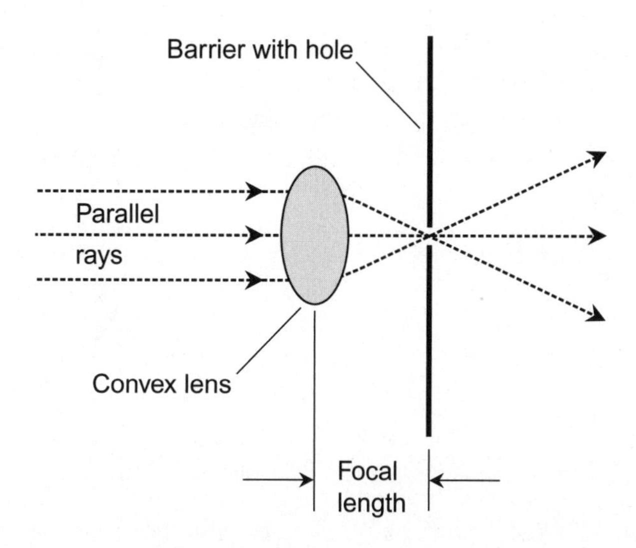

Figure 8-7 A method of eliminating stray light in a holographic apparatus. The hole is at the focal point of the lens.

reflected and some passes through the glass. A beam splitter can be silvered so that it splits the energy equally between the two beams, or it can be so made that the reflected beam contains more or less energy than the transmitted beam. In either case, the sum of the energies of the transmitted and reflected beams is equal to the energy of the incident beam minus a tiny amount lost by absorption in the glass. The ratio of transmission to reflection is specified in terms of percentages totaling 100, such as 80:20 or 50:50. Some beam splitters can be adjusted so that the ratio can be set to a desired value.

When a hologram is exposed in an environment that is not completely dark except for the light sources used for making the hologram, all stray light must be eliminated. One method of accomplishing this is to enclose the entire hologram-making apparatus in an opaque box, painted flat black on the inside, and containing a single tiny hole. The laser beam is focused to a point by a convex lens, and then passes through the hole as shown in Fig. 8-7. The hole must be precisely punched; its edge must be fine and regular to minimize optical interference effects.

EXPOSURE AND VIEWING

Figure 8-8 is a functional diagram of a simple laser-based holographic exposure apparatus. A lens spreads out the light so that it covers the entire object. This arrangement can be refined by adding a beam splitter to obtain the reference and object beams separately, as illustrated in Fig 8-9. The beam splitter allows light beams to reach the film over greatly differing paths, so that the beams arrive at significantly different

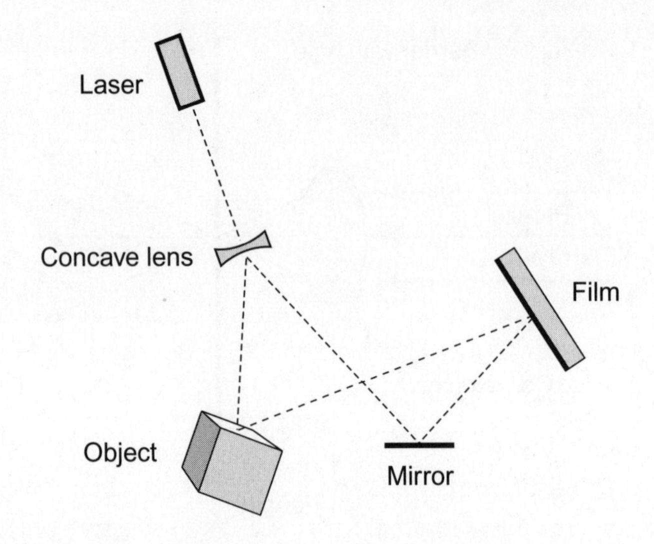

Figure 8-8 A simple arrangement for exposing a hologram.

angles on the film surface, producing a fine interference pattern and therefore high detail. Convex mirrors serve the dual purpose of spreading and reflecting the light.

In an ideal holographic exposure setup, the object and reference beams traverse distances that are nearly equal on their way from the laser to the film. This geometry minimizes the difference in the amounts of wavelength change that take place over distances.

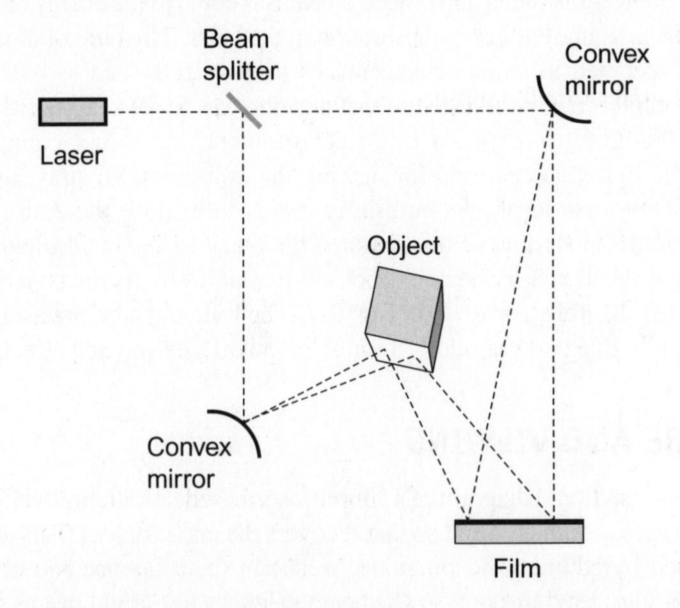

Figure 8-9 A transmission hologram-making arrangement.

If the difference in the path lengths is too great, the interference pattern will not be clear enough, and the quality of the hologram will suffer. The greatest allowable difference in path lengths for the object beam and reference beam is called the *coherence length*. Modern holographic apparatus has a coherence length in the order of 1 m.

Holograms can be categorized as *transmission type* or *reflection type*. In a transmission hologram, the object beam and reference beam strike the film from the same side. In a reflection hologram, the object beam strikes the film from one side and the reference beam strikes the film from the other side. Ordinary photographic film is opaque on one side, so a special type of film is required for the reflection hologram. The reflection hologram is an improvement over the transmission type for several reasons, the most significant being that a reflection hologram can be viewed using ordinary white light such as daylight. The reflection hologram often provides a greater range of views than the transmission type because the object is more evenly and completely illuminated during exposure.

To view a hologram, the setup arrangement is duplicated, except that the observer takes the place of the object. In either case, when the viewer changes position with respect to the illuminating light and the film, the image moves in position as if it were a real object. Parts of the image that cannot be seen from a given viewing angle may become visible from another viewing angle. The viewing angle, not the actual position of the image components on the film, determine what the observer sees. This is why a hologram can be cut in half and still retain all of the position information of the original hologram, although some detail is lost because there are fewer interference fringes in a cut piece of a hologram than in the original.

PROBLEM 8-2

Why are holograms often imprinted on items such as driver licenses, passports, and credit cards?

SOLUTION 8-2

Holograms are practically impossible to duplicate. Conventional photocopiers, cameras, or digital scanners do not reproduce the 3D effects. A fake document is therefore easy to identify as such, if the person inspecting the document is willing to take the time to see whether or not the hologram on the document "works" properly.

Photometry and Spectrometry

The intensity and spectral distribution of IR, visible, and UV radiation can be measured and analyzed by *photometers* and *spectrometers*. These devices have widespread applications, particularly in astronomy and chemistry.

PHOTOMETER

Sometimes it is necessary to measure the brightness of a celestial object, rather than merely making a good guess or a subjective comparison. A *photometer* is an instrument that does this. An astronomical photometer is a sophisticated version of the so-called light meter used by photographers to determine camera exposure time. The device consists of a photosensor placed at the focal point of a telescope. An electronic amplifier circuit increases the sensitivity of the photosensor. The output of the amplifier can be connected to a device that plots the light intensity as a function of time, or to a computer to analyze and store the intensity data.

Many celestial objects, such as variable stars and pulsars, change in visual magnitude with time. Variable stars fluctuate slowly, but some pulsars "blink" so fast that they look like ordinary stars until a graph is plotted using a photometer capable of resolving into brief intervals of time. Astronomical photometers can be made sensitive in the IR and UV ranges, as well as in the visible spectrum.

SPECTROMETER

A *spectrometer*, also known as a *spectroscope*, is a device intended for analyzing visible light at all its constituent wavelengths. Some spectrometers also work at IR or UV wavelengths. The spectra of interstellar gas and dust clouds usually contain dark *absorption lines* at certain wavelengths. Other clouds are dark except at specific *emission wavelengths* that manifest themselves as bright lines in a spectrum. The patterns of lines allow astronomers to determine the amounts of various chemicals that make up these clouds, after corrections are made for the absorption effects of the earth's atmosphere.

Figure 8-10A is a functional diagram of a simple spectrometer. The light-sensitive surface can be photographic film or a matrix of optoelectronic sensors. The maximum image resolution obtainable by a spectrometer of this type is limited by the "grain" of the film or the number of pixels per centimeter in the sensor. Higher resolution can be achieved by a scheme such as that diagrammed in Fig. 8-10B. The rotating prism causes the spectrum to sweep across the objective of a viewing scope. A light sensor connected to the scope measures the intensity of the rays. The angle of the prism at any given moment in time is fed to a computer along with the sensor output. This arrangement produces a rendition called a *spectrograph*.

LASER SPECTROMETER

The human body receives and processes chemicals from air, water, and food throughout a lifetime. The chemical processes can be evaluated by various means to determine whether or not the body is working properly. Some substances are normally excreted in certain amounts; a significant change may indicate trouble. A technique

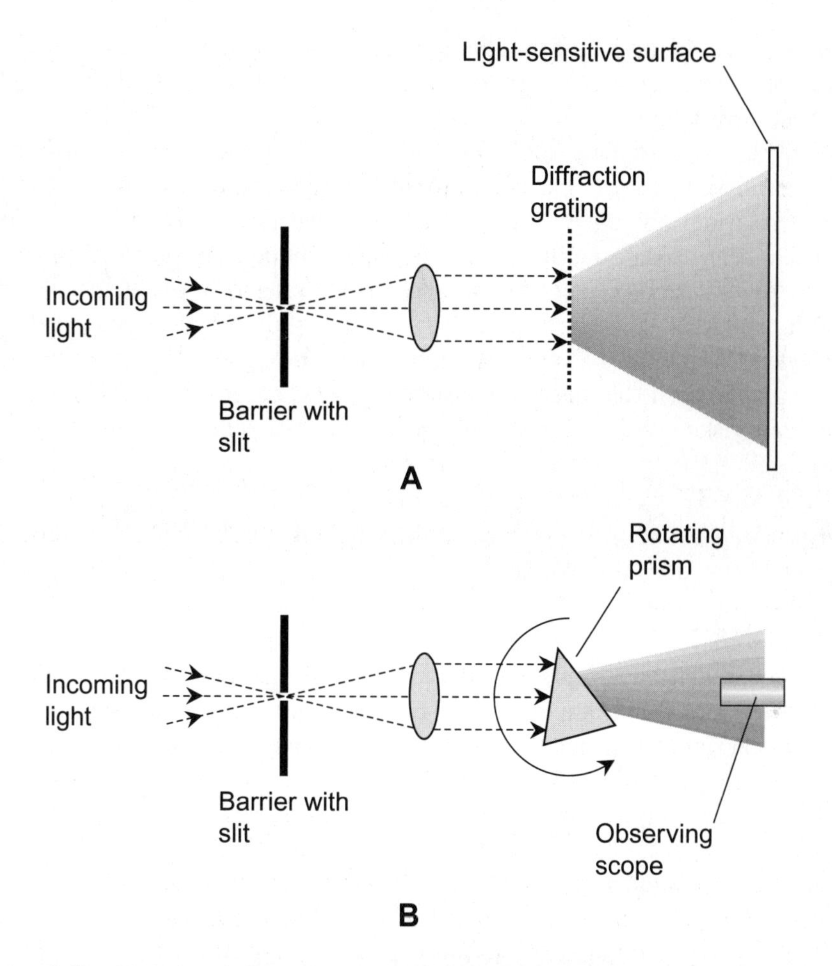

Figure 8-10 Functional diagrams of a simple spectroscope (A) and a high-resolution spectrometer (B).

has been developed for spectrometry based on the degree of IR absorption, resulting from the presence of certain chemicals in the breath, saliva, blood, urine, and feces.

The *laser spectrometer* works according to the same principle as the spectrometer used by astronomers to determine the composition of interstellar gases. A particular gas has characteristic wavelengths at which it attenuates EM radiation much more than at adjacent wavelengths. This happens because, at certain frequencies, photons striking the atoms cause electrons to move to higher-energy orbital shells. When there are several isotopes of a certain gas present together, the absorption wavelengths are slightly different for each isotope. Using conventional spectrometry, the difference is difficult or impossible to see. But with a tunable diode laser, which emits energy at discrete, precisely known, and controllable wavelengths, the difference becomes clear. The output of the laser passes through a sample of the substance to be analyzed, and the relative intensity of the transmitted energy is plotted as a function of the wavelength.

The IR laser spectrometer has applications in pollution analysis and also in diagnostic medicine. The device can determine the concentration of isotopes that are readily converted into noxious gases; a device designed and implemented by General Motors has been used primarily for determining the amounts of such isotopes in automobile exhaust. Medical applications include the diagnosis of fat malabsorption, intestinal problems, liver disease, kidney disease, pulmonary disease, and blood-electrolyte imbalances. For example, diabetes might be diagnosed by analyzing exhaled air from a subject who has been administered a tagged sugar sample. All of these tests involve zero risk of radiation exposure, and can be carried in a short time.

PROBLEM 8-3

What is an *emission spectrum*? What does it look like? What is an *absorption spectrum*? What does it look like?

SOLUTION 8-3

An emission spectrum appears as a dark band with bright lines, indicating the wavelengths at which photons are emitted by energized gas or plasma. Figure 8-11A is a hypothetical example. An absorption spectrum appears as a relatively light band with

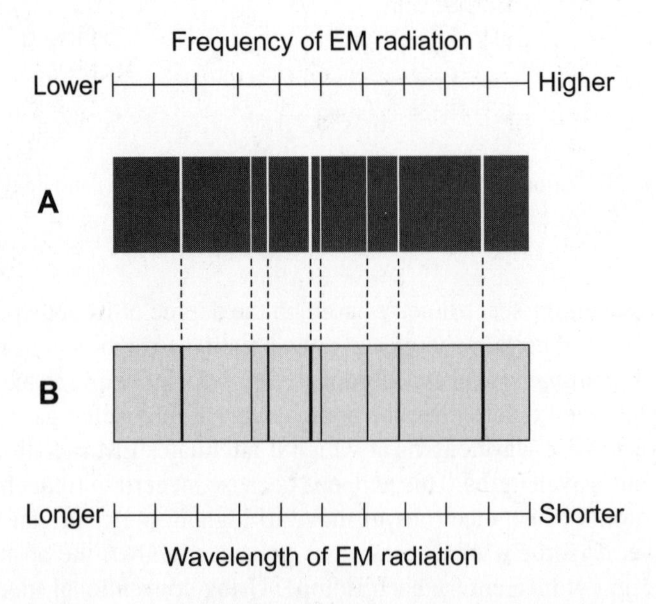

Figure 8-11 At A, a simplified hypothetical emission spectrum. At B, the absorption spectrum produced by the same substances. Most spectra are more complicated than these examples.

dark lines, indicating the wavelengths at which photons are absorbed as they pass through a cool gas cloud (Fig. 8-11B). Both of these drawings are greatly simplified for clarity; most real-life spectra have hundreds or thousands of discernible lines.

Astronomy beyond the Visible

Stars, nebulae, galaxies, quasars, and some planets radiate EM energy at all wavelengths. Astronomers have developed devices that can observe and analyze radio waves, IR, UV, X rays, gamma rays, and cosmic particles coming from these objects.

MEASURING TEMPERATURE

The EM radiation from a star can be considered to occur in a form called *blackbody radiation*. A *blackbody* is a perfect absorber and radiator of energy at all wavelengths. Any object having a temperature above *absolute zero* (approximately –273°C or –459°F) has a pattern of wavelength emissions that depends directly on the temperature. For any EM-radiating object, the emission strength is maximum at a certain defined wavelength, and decreases at longer and shorter wavelengths. If the radiation intensity is graphed as a function of wavelength or frequency, the result is a curve that resembles a statistical distribution, as shown in Fig. 8-12.

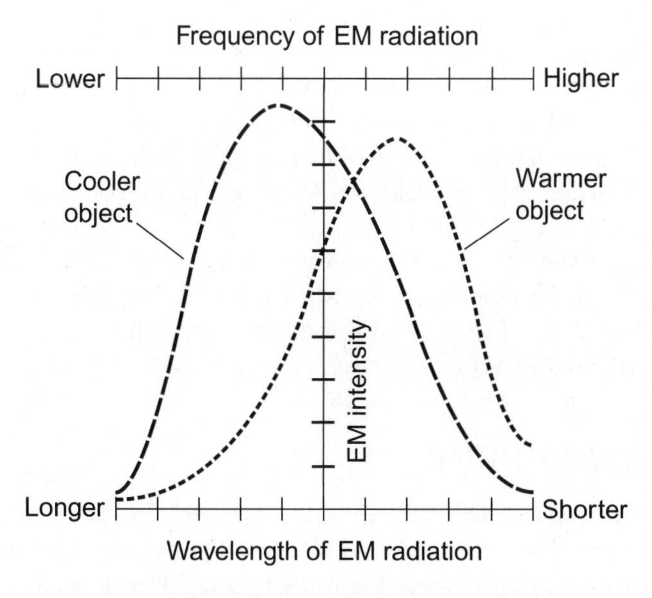

Figure 8-12 The graph of EM intensity radiated from a blackbody, as a function of frequency or wavelength, has a characteristic shape.

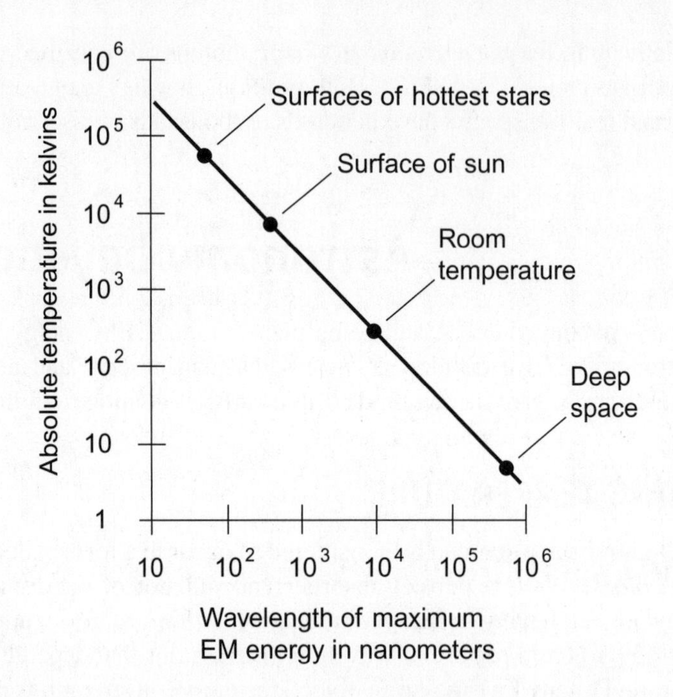

Figure 8-13 Absolute temperature as a function of the maximum-intensity wavelength for a radiating blackbody.

When astronomers talk about the temperature of a celestial object, they usually refer to the *spectral temperature*, which is determined by examining the EM radiation intensity from the object at various wavelengths. By observing an object at many different wavelengths, the point of maximum emission can be determined. Sometimes the maximum-emission wavelength can be inferred, even if observations are not directly made at that wavelength. Points are plotted on an intensity-versus-wavelength graph, and then a computer translates the set of points into a smooth curve (a process called *curve fitting*). From the maximum-emission wavelength, the temperature of the object can be estimated, based on the hypothesis that it behaves as a blackbody. Figure 8-13 is a rough logarithmic graph of the function employed by scientists for this purpose.

INFRARED ASTRONOMY

So-called *IR astronomy* has helped scientists to discover stars that radiate most of their energy at wavelengths longer than the visible. Visually, such stars appear red and dim. But, like a glowing ember in a wood stove, they are powerful sources of IR. These

stars have relatively low surface temperatures as compared with other stars. The peak wavelength at which an object radiates is a direct function of the temperature. "Cool" stars produce radiation at predominantly longer wavelengths than "hot" stars.

Infrared astronomy is important in the study of evolving stars and star systems. As an interstellar nebula contracts, it begins to heat up, and its peak radiation wavelength decreases. A cool, diffuse cloud radiates most of its energy in the RF part of the EM spectrum. Hot stars radiate largely in the UV and X-ray regions. For awhile between the initial contraction of the nebula and the birth of the star, the peak emission wavelength passes through the IR.

The observation of IR is also important in the analysis of dying stars. As a star runs out of nuclear fuel, it goes through a period of instability, and finally enters a stage during which its peak radiation wavelength steadily increases. On its way to oblivion, the star emits most of its energy in the IR for a certain period of time.

ULTRAVIOLET ASTRONOMY

As the wavelength decreases beyond the visible violet, the earth's atmosphere becomes highly absorptive at a wavelength of about 290 nm. At still shorter wavelengths, the atmosphere is essentially opaque. That's fortunate for life on our planet, because it protects the environment against excessive short-wave UV radiation from the sun. Surface-based observatories can see into outer space at wavelengths slightly shorter than that of the visible violet, but when the wavelength gets down to 290 nm, nothing can be seen. At the shortest UV wavelengths (*hard UV* and *extreme UV*), as in the case of the far IR, it is necessary to place observation apparatus above the atmosphere.

Glass is virtually opaque to UV, so cameras with glass lenses cannot be used to take photographs in this part of the spectrum. Instead, a pinhole-type device is used, limiting the amount of energy that passes into the detector. While a camera lens has a diameter of several centimeters, a pinhole is less than a millimeter across. This constraint does not present a problem for photographing the sun or the moon, but for other celestial objects, a pinhole does not allow enough energy into the device to make photographs practical.

For analysis of faint celestial objects in UV, a specialized wide-bandwidth spectrophotometer can be used (Fig. 8-14). In the long-wavelength or *soft-UV* range, a photoelectric cell can serve as a sensor. In the hard- and extreme-UV spectra, radiation counters are sometimes used, similar to the apparatus employed for the detection and measurement of X rays and gamma rays.

The sun's UV radiation has been investigated using equipment aboard rockets and satellites. The UV surface of the sun is somewhat above the visual surface. This tells us that, as the altitude above the *photosphere* (visible surface of the sun) increases,

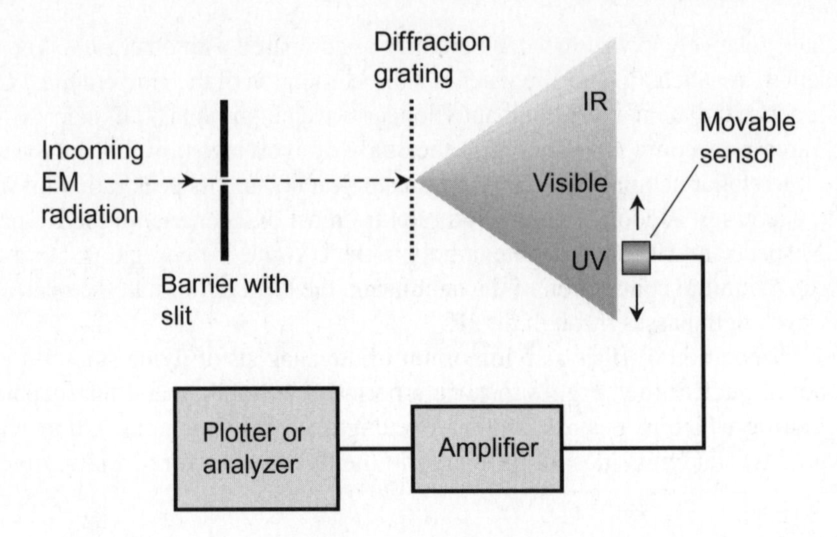

Figure 8-14 Functional diagram of a wideband spectrophotometer.

the temperature rises. If our eyes suddenly became responsive to a range of wavelengths only half as long as they actually are (about 200 to 400 nm), the disk of the sun would appear slightly larger than we see it in the visible range. The sun would also seem dimmer, because its peak emission wavelength is longer than the UV.

X RAYS

The X-ray spectrum consists of EM energy at wavelengths from approximately 1 nm down to 0.01 nm. This translates to two mathematical *orders of magnitude* (powers of 10). Proportionately, the X-ray spectrum is vast as compared with the visible range, which spans less than one order of magnitude from red to violet.

As the X-ray wavelength becomes shorter, it gets increasingly difficult to direct and focus them because of the penetrating power of the energetic photons. Short-wavelength, high-energy X rays pass straight through glass lenses and mirrors without refracting or reflecting. A piece of paper with a tiny hole can work for UV photography, but in the X-ray spectrum, the radiation would pass through the paper as if it were transparent. However, if X rays encounter the smooth surface of a sheet of metal foil at a nearly grazing angle, reflection will occur (Fig. 8-15A). As the wavelength decreases, the angle relative to the surface must also decrease if reflection is to take place.

Figure 8-15B is a rough illustration of how an *X-ray telescope* achieves its focusing. The focusing mirror is tapered in the shape of an elongated paraboloid. As parallel X rays enter the aperture of the reflector, they strike its inner surface at a

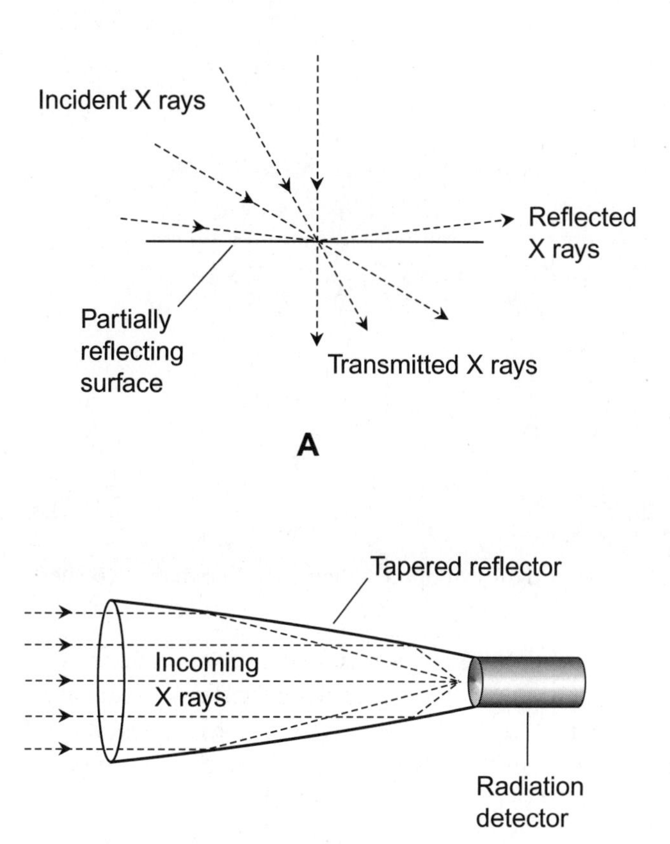

Figure 8-15 At A, incident X rays are reflected from a surface only when they strike at a grazing angle. At B, a functional diagram of an X-ray focusing and observing device.

grazing angle. The X rays are brought to a focal point, where a radiation counter or detector is placed. The resolving power of this device is not as good as that obtainable with conventional telescopes in the visible range, but it facilitates the observation of some celestial X-ray sources. As is the case with hard UV, X rays from space must be viewed from above the earth's atmosphere. X-ray telescopes aboard rockets and satellites send their information back to the earth by radio telemetry.

GAMMA RAYS

As the wavelength of EM energy becomes shorter than the hardest X rays, images become increasingly difficult to obtain. The cutoff point where the X-ray region ends and the gamma-ray region begins is approximately 0.01 nm. Gamma rays can

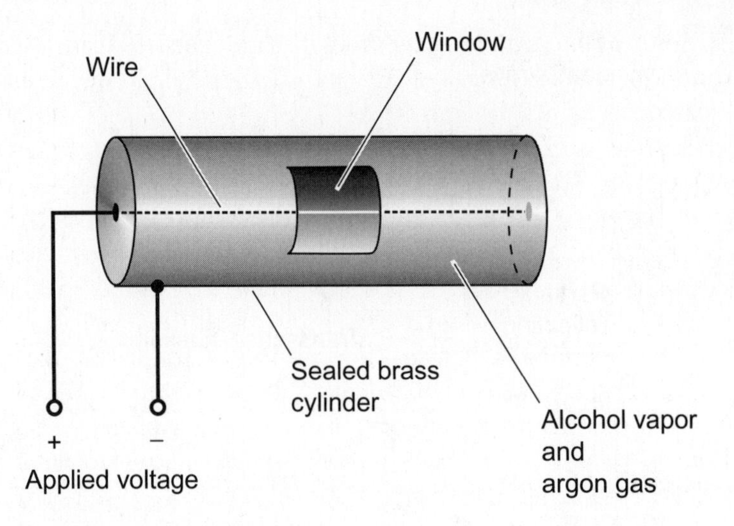

Figure 8-16 Simplified diagram of a radiation counter.

get shorter than this without limit, at least in theory. *Hard gamma rays* (those at the shortest wavelengths) can penetrate several centimeters of solid lead or more than a meter of concrete. They are damaging to living tissue. Gamma rays come from radioactive materials, both natural (such as radon) and human-made (such as plutonium). They can also come from certain catastrophic events in the cosmos.

Radiation counters are the primary means of detecting and observing sources of gamma rays. One type of radiation counter consists of a thin wire strung within a sealed, cylindrical metal tube filled with certain gases. When a gamma-ray photon enters the tube, the gas is ionized for a moment, causing it to conduct electric current. A voltage is applied between the wire and the outer cylinder, so a pulse of current occurs whenever the gas is ionized. This pulse produces a "click" in the output of an amplifier connected to the device.

Figure 8-16 is a simplified diagram of a radiation counter. A glass window with a metal sliding shutter is cut in the cylinder. The shutter can be opened to admit radiation of lower energy, and closed to allow only the most energetic radiation to get inside. Subatomic particles such as protons, neutrons, or *alpha particles* (helium nuclei), which are tiny yet massive for their size, pass easily through the window glass if they are moving fast enough. Gamma rays penetrate into the tube even when the shutter is closed.

COSMIC PARTICLES

If you sit in a room with no radioactive materials present, and switch on a radiation counter with the window of the tube closed, you'll notice an occasional click from

the device. Some of the particles come from the earth because there are radioactive elements in the ground almost everywhere (usually in small quantities). Some of the radiation comes indirectly from space; particles strike atoms in the atmosphere, and these atoms in turn eject other particles that arrive at the counter tube. The direction of arrival of high-speed subatomic particles can be determined, to a certain extent, by means of a device called a *cloud chamber*. The air in a small enclosure can be treated especially to produce condensation when a subatomic particle enters, and the path of the particle shows up as a sort of "vapor trail."

In the early 1900s, physicists noticed particle radiation that was apparently coming from space. They found that the strange background radiation increased in intensity when observations were made at high altitude; the radiation level decreased when observations were taken from underground or underwater. This space radiation has been called *secondary cosmic radiation* or *secondary cosmic particles*. The actual particles from space, called *primary cosmic particles*, usually do not penetrate far into the atmosphere before they collide with, and break up, the nuclei of atoms. To observe primary cosmic particles, it is necessary to ascend to great heights, and as with the UV and X-ray investigations, this was not possible until the advent of the space rocket.

Radio waves, IR, visible light, UV, X rays, and gamma rays consist of photons traveling at the speed of light. But cosmic particles are actually little bits of matter. They travel through space at speeds almost, but not quite, as fast as light. At such high speeds, the protons, neutrons, and other heavy particles gain mass because of relativistic effects. As they encounter the upper atmosphere, they come to us in nearly perfect straight-line paths because they have high inertia. The earth's magnetic field can exert little deflecting effect even on electrically charged particles moving at relativistic speeds. By carefully observing the trails of the particles in a cloud chamber aboard a low-orbiting space ship, it is possible to ascertain the direction from which they have come. Over time, cosmic-particle maps of the heavens can be generated and compared with maps at various EM wavelengths.

NEUTRINOS

Perhaps the strangest known form of "radiation" consists of particles similar to electrons, but lacking electrical charge. In 1931, the physicist *Wolfgang Pauli* inferred their existence on the basis of atomic theory. The particle was given the name *neutrino* by *Enrico Fermi* in 1934. The existence of neutrinos was confirmed by various experiments throughout the middle part of the 20th century. Neutrinos were originally thought to be massless, but later they were found to have tiny mass. They travel at almost the speed of light, spanning intergalactic distances and penetrating most material objects as if those objects were transparent.

Neutrinos are emitted by all stars, including the sun. Exploding stars known as *supernovae* are particularly intense sources. Neutrinos are also produced by nuclear reactors, decaying radioactive minerals in the earth's interior, and the interaction of cosmic rays with atoms in the earth's atmosphere. They are so numerous that trillions of them pass through your body every second. Yet they are so elusive that it is difficult to detect even a single one. Because of fantastic properties such as these, neutrinos seem to many lay people like mere artifacts of science-fiction writers' imaginations. But neutrinos are important to astronomers and cosmologists, because their behavior may hold clues that can help resolve mysteries concerning the origin and evolution of the universe.

The earliest neutrino detectors were tanks containing liquefied chlorine. Later detectors used purified water. One of the most well-known neutrino detectors is the *Super-Kamiokande* in Hida, Japan, which holds 50,000 tons of water. The mass and volume of the detector must be large, because neutrinos hardly ever interact with the atoms of any barrier they encounter. Neutrino detectors are placed deep underground so that all other cosmic particles are blocked. Therefore, scientists can be reasonably confident that any detected particle coming from space is indeed a neutrino.

In 2007, the U.S. National Science Foundation proposed that a new underground laboratory be built in the old Homestake gold mine in Lead, South Dakota. If and when completed, the *Sanford Underground Laboratory* will house a neutrino detector, along with hardware for conducting other scientific and engineering experiments, at 2 km below the surface—twice the depth of the Super-Kamiokande. The facility would give astronomers and cosmologists a chance to observe neutrinos of cosmic origin with less "background noise" than ever before.

When a neutrino interacts with a water molecule, a burst of visible light called *Cerenkov radiation* is produced. Cerenkov radiation occurs when certain subatomic particles pass through a substance at a speed greater than the speed of visible-wavelength EM radiation in that substance. It's the optical equivalent of a *sonic boom*, but extremely faint. The interior surface of the detector tank in the Super-Kamiokande is lined with more than 11,000 individual photodetectors whose output is amplified by electronic circuits and analyzed with computers. If a neutrino interacts with an atom in the tank, a cone-shaped burst of Cerenkov radiation strikes the photodetector array, and the computers calculate the direction from which the neutrino arrived.

PROBLEM 8-4

Is neutrino radiation harmful or dangerous? If neutrinos hardly ever interact with matter they encounter, why should scientists be so interested in them?

SOLUTION 8-4

Neutrinos pose no known threat to biological life, even though they are incredibly numerous and have more penetrating power than any other known particles. Neutrinos

are of special interest to astronomers and cosmologists because they pass through obstructions that block other forms of radiation. For example, if a "neutrino telescope" can be developed and a "neutrino map" of the heavens made in sufficient detail, astronomers will be able to scrutinize the core of our galaxy to an extent impossible with other observational devices.

Quiz

This is an open book quiz. You may refer to the text in this chapter. A good score is 8 correct. Answers are in the back of the book.

1. Observing apparatus must generally be placed above the earth's atmosphere if the emissions from cosmic objects are to be analyzed at wavelengths shorter than approximately
   (a) 400 nm.
   (b) 700 nm.
   (c) 29 nm.
   (d) 290 nm.
2. The illusion of depth can be replicated to some extent with two ordinary photographs and a special binocular image-combining scope, but such a rendition cannot portray
   (a) parallax differences between nearby and more distant objects.
   (b) color; only grayscale images can be reproduced clearly.
   (c) eclipsing that occurs when the observer moves around.
   (d) images in more than one dimension.
3. Which of the following types of radiation has the greatest penetrating power?
   (a) IR rays
   (b) Gamma rays
   (c) X rays
   (d) Neutrinos

4. When viewed from various distances, a hologram can reproduce aspects of a 3D scene on a 2D film surface, including
   (a) relative size, eclipsing, and depth in a modified way.
   (b) relative size, but not eclipsing or depth.
   (c) eclipsing, but not relative size or depth.
   (d) relative size and depth in a modified way, but not eclipsing.
5. The purpose of a beam splitter in a hologram-making arrangement is to
   (a) ensure that the photographic emulsion is not exposed to excessive illumination in any single region at the expense of other regions.
   (b) spread out the laser beam while retaining the coherence of the light.
   (c) split the laser output into various wavelengths for the faithful reproduction of color images.
   (d) allow two light beams to reach the film at different angles, thereby producing a fine interference pattern.
6. A "breathalyzer" for determining the amount of alcohol in the blood might be constructed using a
   (a) hologram-making apparatus.
   (b) laser spectrometer.
   (c) blackbody radiation analyzer.
   (d) diffraction grating.
7. A hologram does not result from a focused real image, but is instead a
   (a) set of interference patterns recorded on film.
   (b) virtual image that has passed through a tiny hole.
   (c) spectrum that has been generated by passing laser light through a diffraction grating.
   (d) graph of wavelength as a function of peak intensity.
8. The ideal way to view a hologram is to use
   (a) a microscope equipped with binocular lenses to enhance depth perception.
   (b) diffuse light to provide illumination of equal intensity from all possible angles.
   (c) a point source of light at roughly the same location as the laser used to expose the film.
   (d) a source of UV light to provide the shortest possible wavelength, thereby enhancing the image resolution.

9. An absorption spectrum typically appears as a
   (a) smooth curve on a graph of intensity versus wavelength.
   (b) set of vertical "pips" or "spikes" on a graph of intensity versus wavelength.
   (c) gradually decreasing curve on a graph of intensity versus time.
   (d) bright band with numerous dark lines.
10. "Fine-grained" film is better than "coarse-grained" film for making holograms because the finer emulsion
    (a) can reproduce interference fringes more clearly.
    (b) requires less exposure time.
    (c) can be used with incoherent-light lasers.
    (d) is less sensitive to vibrations in the hologram-making apparatus.

CHAPTER 9

Optics to Heal and Defend

Optical devices can improve the quality and length of human lives through diagnostic and surgical medicine. Optical technology can also be used in security systems and military hardware.

Medical Diagnosis and Treatment

In medicine, optical devices are used to diagnose and treat conditions as diverse as cancer, heart disease, skin problems, digestive disorders, and lung infections. Lasers have been used in general surgery, neurosurgery, and acupuncture.

ENDOSCOPE

Certain interior regions of the body can be viewed with a device called an *endoscope*, which consists of a flexible tube containing an optical fiber that carries a beam of light to illuminate tissues near the end of the tube. Other openings provide for dispersal of gas and removal of fluid. A controllable forceps at the end of the tube facilitates the removal of tissue for later analysis. A laser, guided by means of another optical fiber, is used for surgery or removal of tissue for biopsy. The entire set of end effectors can be directed and manipulated by viewing through a return optical fiber and by sending control signals through a wire or another optical fiber (Fig. 9-1).

The endoscope is useful in treating certain gastrointestinal ailments. The device is inserted into the mouth or nose, and run down the esophagus. Bleeding ulcers or ruptured blood vessels in the esophagus can be repaired. Some types of tumors can be destroyed or removed. In the large intestine, endoscopy allows physicians to detect certain chronic degenerative conditions. *Diverticulosis* involves the formation of small balloon-like outgrowths in the intestinal wall. When such outgrowths become inflamed or infected, the condition is called *diverticulitis*. A general inflammation of

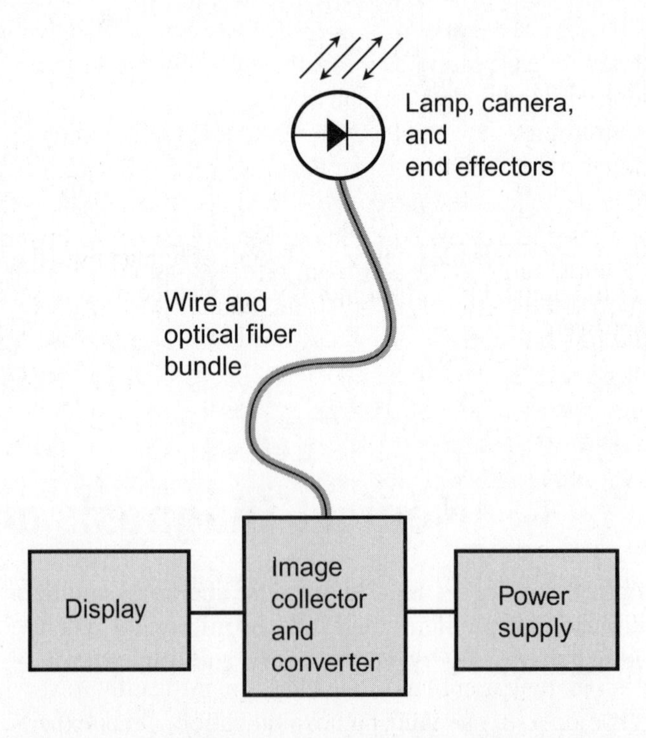

Figure 9-1 Simplified functional diagram of an endoscope.

the large intestine is known as *colitis*. To diagnose conditions such as these, the patient is given a laxative or a series of enemas followed by insertion of the endoscope into the rectum.

Some brain tumors and spinal tumors can be removed by endoscopy without causing permanent damage to essential tissue. The endoscope is inserted into the vascular system. It is possible to visually explore the interior of an artery or vein, finding and correcting conditions in which obstructions may impede circulation or in which blood clots present a risk. Damaged or diseased lungs can also be examined and treated using the endoscope. The endoscope is inserted into the bronchial passages. Fluid can be removed from the lungs, and small samples of lung tissues can be removed for biopsy.

LASERS AND CANCER

Another application of the laser in diagnostic medicine is the detection of cancerous lesions. Early diagnosis of cancer minimizes surgical risks and reduces the danger that the cancer will spread to other parts of the body. A sample is taken from the patient and a biopsy is done to evaluate the sample for signs of malignancy. The sample is stained by a dye that fluoresces when exposed to coherent light at certain wavelengths. The dye is more readily absorbed by the malignant cells than by normal, healthy cells. When a laser beam is directed at the diseased sample, the malignant cells fluoresce more than the healthy cells.

Lasers can be employed to treat cancer without actually removing tumors. Specially formulated dye, which is absorbed more readily by cancer cells than by nonmalignant cells, is injected into the body. The suspect tissue is then subjected to the light of a laser at a wavelength that is readily absorbed by the dye, heating the tissue to a temperature sufficiently high to kill the cancer cells. The normal cells, containing little or no dye, are heated much less and are therefore not affected by the laser (Fig. 9-2). It is necessary to ensure that the intensity of the irradiating laser beam is not high enough to damage normal cells, while still being sufficient to cause destruction of the malignant cells. When the dye breaks down because of the laser light, molecular fragments are produced, interfering with the functioning of the cancer cells. Otherwise inoperable tumors may shrink after laser irradiation to the extent that they can be removed by conventional surgery or laser surgery.

Some experts believe that lasers will someday supersede conventional surgical tools for dealing with malignant tumors. When lasers are used in place of scalpels, there is less bleeding and a lower risk of infection. Because some laser surgery can be performed on an outpatient basis, the cost is significantly lower than with conventional surgical operations requiring hospitalization. Other experts, however, are skeptical of laser treatment and surgery in general.

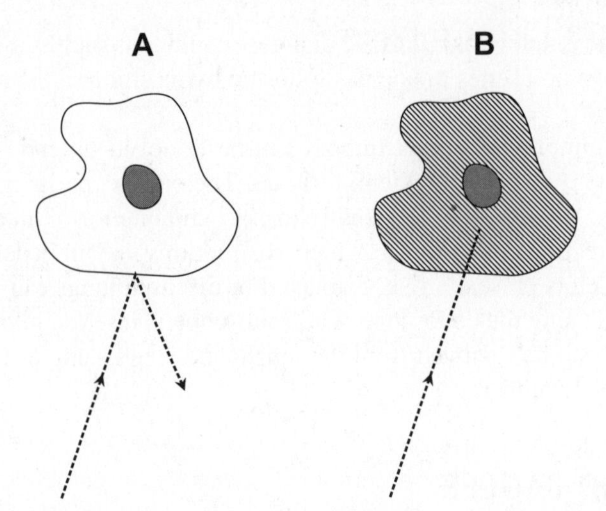

Figure 9-2 At A, a normal cell does not absorb dye and reflects a laser beam that strikes it. At B, a cancer cell, having absorbed the dye, heats up when the laser beam strikes it.

LASER ACUPUNCTURE

In the Orient, needles inserted in various points on the skin are believed to produce certain physiological effects. This form of treatment is known as *acupuncture*. In the West, acupuncture is less well understood. Some physicians doubt that acupuncture has any scientific basis. Experience has shown, however, that applying localized intense pressure at certain places on the body can cause specific and predictable things to happen, often far from the pressure point.

Acupuncture is not entirely without risk. The needles must be meticulously sterilized prior to each use to prevent hepatitis or other infection. An alternative way to perform acupuncture is to employ a small, focused or narrow-beam laser in place of each needle. This method has been used to advantage in China, and is gaining popularity in Western Europe as well. Most so-called *laser acupuncture* is done in well-equipped medical centers of large cities. Some practitioners think that lasers work better than needles. The laser certainly reduces the risk of infection, because no hardware penetrates the skin.

An example of laser acupuncture is the application of laser light to pressure points located on the toes. This practice may cause a fetus to rotate into the proper position in the last 3 months of a pregnancy. Other conditions treated with laser acupuncture include bronchitis, asthma, and chronic maladies involving pain or discomfort such as migraine and fibromyalgia. In Europe, lasers are used in some acupuncture treatments; the percentage is higher for children and nervous people. There is basically no sensation (not even warmth) from the low-powered lasers. The medical establishment in the United States recognizes that this field is susceptible to exploitation by quacks, and remains cautiously skeptical, although there is some recognition of the practice.

CARDIOVASCULAR DISEASE

Until the latter part of the 20th century, laser technology was rarely used in the treatment of cardiovascular disease, mainly because of concern that healthy tissue might be inadvertently damaged. The main fear was that a blood vessel wall might be inadvertently perforated by a misdirected laser beam. In the 1980s, physicians gained a better understanding of potential problems and how to best avoid them.

A mass of arterial plaque is often yellowish in color, so it was thought that a laser demonstrating maximum absorption by this color of material might be used to "drill" through an occlusion. Evidence indicated that coherent light of the correct wavelength would present little risk of damage to the arteries themselves, while allowing destruction of the plaque. The plaque material could be stained, so that it would readily absorb laser light of a certain wavelength at which the blood vessel wall would remain unaffected. Special fiber-bundle transmission devices were designed, along with metal-tipped fibers that result in long-wave IR emanating from the end of the fiber. The use of a metallic laser-heated probe is illustrated in Fig. 9-3. Most arterial procedures using lasers have been done in the large vessels of the legs.

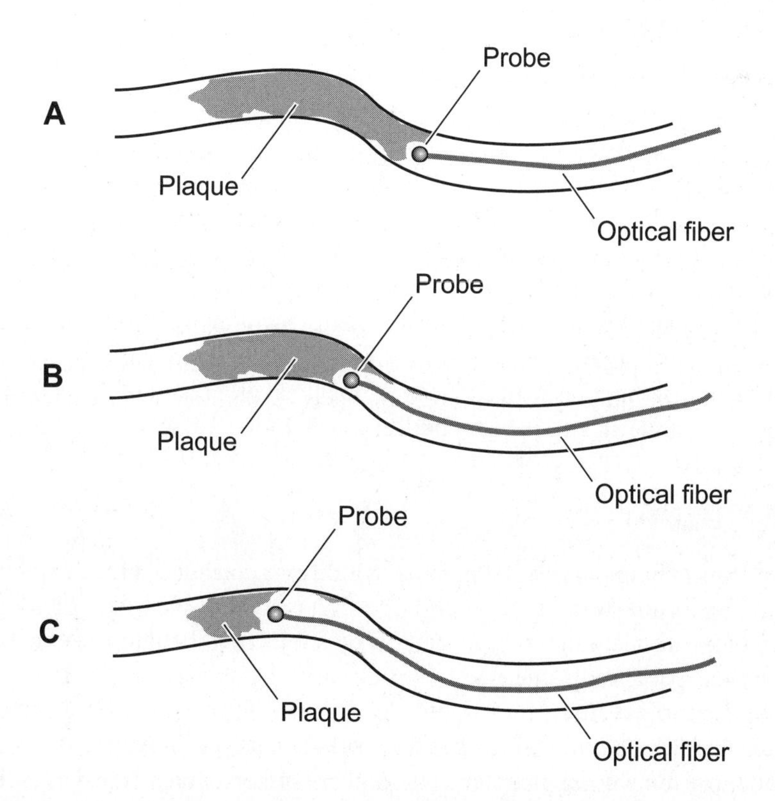

Figure 9-3 A laser-heated probe can be used to remove an arterial plaque. Progressive stages are shown at A, B, and C.

The *coronary arteries* supply blood to the heart muscle. Blockage of these arteries leads to heart attacks. Various ideas have been put forth as to the best types of devices for use in these arteries. A laser has been considered that would allow the heart muscle to get its blood directly from the heart chamber. To accomplish this, it would be necessary to make small holes or "channels" in the heart, running from the outer surface directly into the ventricles. The problem in the past, when such channels have been made using conventional surgery, has been that they tend to close in time. When the holes are made using the laser, they seem to last longer.

The laser is being considered for use in correcting chronic *heart arrhythmias* (irregular heartbeat). A focused beam of coherent light is used to interfere with the functioning of the conductive tissue bundle that controls the heart rate. Lasers are also being developed for the treatment of an abnormally enlarged heart muscle. In the past, treatment has involved invasive surgery. The laser can be employed to join blood vessels. The heat of the laser causes the tissues to bond together. The main advantage of this procedure over the use of sutures ("stitches") is that the laser method does not involve the introduction of foreign material into the body.

GASTROENTEROLOGY

The digestive tract can be reached easily by means of an endoscope, and is therefore well suited for the use of lasers to perform minor operations. Lasers can be used to destroy benign, precancerous, or malignant tumors. Flat polyps have been "burned off" using a neodymium laser and an endoscope. Advanced tumors of the colon or esophagus have also been partially removed using lasers. These operations are done to open the passageways and do not alleviate the underlying disease itself.

Vascular lesions sometimes develop in the gastrointestinal tract. If they bleed, the laser is especially useful in treating them. Swollen veins in the esophagus, known as *varices*, can be made to shrink and coagulate by means of the laser. Varices develop in conjunction with cirrhosis of the liver, as a result of the body attempting to bypass an inefficient liver by building new blood vessels around it.

DERMATOLOGY

Skin lesions were among the first medical conditions considered for treatment using the laser. This is not surprising, because the skin is easily accessible. Many kinds of lesions can be literally burned or frozen off by subjecting them to extreme temperatures for short periods of time.

One of the earliest applications of the laser in dermatology was in the removal of so-called *port wine stains*, which get their name from their characteristic dark red color. These lesions occur in places where there are too many blood vessels in the skin. They are most often found on the head, neck, hands, and arms, where they are

visible and can cause embarrassment. In some cases, much of the discoloration can be removed by means of irradiation by a laser. The procedure does not require much time, and is essentially painless. Usually, there is little or no discernible stain or scarring after successful treatment.

Other skin abnormalities that may be treated by means of the laser are ordinary *moles*, *birthmarks*, *age spots*, and *spider veins*. The *rosacea* (reddening of the nose and face) that sometimes follows plastic surgery can sometimes be alleviated using the laser. Lasers have been employed to remove tattoos. The laser can be used to treat precancerous skin lesions known as *actinic keratoses*. In some cases, the laser can remove *basal-cell carcinomas* and *squamous-cell carcinomas* (cancerous lesions) as well. The laser alone may not be sufficient to deal with *melanoma*, however, because this type of skin cancer can spread to other parts of the body.

PULMONARY MEDICINE

The laser has gained some acceptance in the treatment of disorders of the trachea and bronchi, and it has been used to remove malignant blockages in these regions. The proper choice of laser produces little or no bleeding, and can be readily transmitted through an endoscope. One method of treatment uses both surgical removal and laser radiation therapy to unblock the bronchial passageway. Then the remaining cancer is treated by conventional means such as *chemotherapy* to minimize the risk of recurrence, or at least prolong the time before recurrence. Applied or injected dye collects in malignancies, enhancing laser radiation treatment by maximizing absorption of the laser light by cancer cells while not affecting normal cells.

Benign tumors of the trachea and bronchi are sometimes treated with a carbon dioxide (CO_2) laser if they are within range of the endoscopes required for the transmission of this type of laser light. The endoscope must be inserted through the mouth, and then from there into the throat and windpipe. These types of tumors can be treated if they are relatively small and have not affected the cartilage in the trachea. Tracheal blockages have also been treated by using the CO_2 laser.

MISCELLANEOUS TREATMENTS

A benign growth known as a *papilloma* can be treated by means of a laser. These tumors sometimes grow in the nose, throat, or other regions of the head and neck, are caused by a virus, and tend to recur often. Treatment with the laser may result in a lower recurrence rate than conventional treatment methods.

Tonsillectomy (removal of the tonsils) is not performed today as routinely as it was a few decades ago, but this operation is still deemed necessary in certain instances. The laser can sometimes be used for this operation, and can also be used to treat chronic infections of the tonsils without removing them.

Ear infections can be treated using lasers. A tiny hole or cut is made in the eardrum, which allows the fluid from the middle ear to drain, relieving the pressure and helping the damaged tissue to heal. In the past, eardrum perforation was done by mechanical means, causing exquisite pain. The laser process inflicts far less pain and is essentially bloodless.

PROBLEM 9-1

What are the principal advantages of laser treatment for dealing with cancerous lesions? What is the main limitation of such treatment?

SOLUTION 9-1

Laser treatments may cost less than older methods of treatment, take less time, involve less pain, can often be done without anesthesia, produce less bleeding, and inflict less collateral tissue damage. The main limitation of laser treatment is the fact that in some cases, the tumors are so large or have spread to such an extent that the use of a laser alone will not produce the desired outcome. In such situations, conventional surgery, in conjunction with other treatments such as radiation and chemotherapy, is indicated.

Medical Surgery

A laser can function as a precision surgical cutting tool. The practical design of surgical lasers involves concentrating the beam into a nearly perfect geometric point. The focused energy can "burn" tissue apart on a microscopic scale.

EYE SURGERY

One of the earliest surgical applications of the laser was in the repair of a *detached retina* in the eyeball. Other eye problems can sometimes be corrected by means of lasers focused onto microscopic regions. The effect of a surgical laser is determined by the temperature to which the laser beam heats the tissues, and also on the amount of time that the tissue is exposed to the light.

In a process called *photoradiation*, the temperature of a given sample is raised slightly, producing a sort of localized fever. The exposure time is about 30 minutes. This process is used to destroy eye tumors. In *photocoagulation*, the temperature is raised to about 65°C. The heat causes bleeding to slow and then stop, a useful phenomenon in cases of retinal hemorrhage. A more powerful laser can raise the temperature far higher to effect *photovaporization* (at about 400°C) and *photodisruption* (at about 20,000°C). The process of photovaporization can be used to destroy malignant tumors, removing all the cancer cells so none remain to spread.

Photodisruption causes tissue to part; the action is like that of a tiny scalpel. Specifically designated cuts can be made for removing tumors or for other surgical procedures on a microscopic scale. These high-energy lasers are operated in single pulses for periods measured in microseconds or nanoseconds.

Besides the removal of tumors and the repairing of broken blood vessels or detached retinas, laser eye surgery can be used to implant artificial lenses to improve vision. This is done in cases of *cataracts*, where the eyes' natural lenses have lost much of their transparency. For relief of *glaucoma* (excessive pressure in the eyeball), a tiny hole can be burned in the *iris*, the colored part of the eye around the dark *pupil*, allowing excess fluid to escape.

The main advantages of laser eye surgery over more conventional methods are increased precision (a given area can be vaporized or cut without affecting any other area), painlessness, eliminating the need for anesthetics, and reduced risk of catastrophe in the event the patient happens to flinch during the procedure.

LARYNX SURGERY

Until the latter part of the 20th century, cancer of the *larynx* (or voice box) made it necessary to surgically remove the entire larynx. After this procedure, a patient could not speak out loud unless assisted by an artificial sound-generating device. However, small lesions can now be removed by vaporization using a laser. The noncancerous portion of the larynx is undamaged, and the patient keeps his or her voice largely unchanged. The rehabilitation time is shortened and the cost is reduced compared with older, more radical treatment methods.

The patient is first placed under general anesthesia. An endoscope is inserted, if necessary, into the airway. The laser light is beamed through the endoscope at the specific target area. Everyone in the operating area wears safety goggles because of the intensity of the laser light. The patient wears pads over the eyes. While accidents are not likely, the intensity of the laser light is comparable to that of full sunshine at midday, and precautions must be taken. The operation can be completed in about an hour, and the hospital stay is rarely more than 24 hours.

NEUROSURGERY

The laser has been used to treat some disorders of the central nervous system. In the brain, near the base of the skull, tumors may develop that interfere with many brain functions. Surgical instruments are used to get to the tumor in the conventional way. Lasers can then be used to remove lesions without affecting surrounding tissue. This level of precision is especially important in brain surgery. Side effects of the surgery, which have in the past included degradation in function of the central nervous system, are minimal.

Lasers and traditional surgery have been used together to treat and remove benign and malignant spinal tumors. As with brain surgery, the laser in spinal surgery has less effect on surrounding tissues and the rate of complications is less than with other surgical methods. For example, paralysis appears to be less likely to result from laser operations on the spinal cord as compared to conventional surgery. While complete cures for malignant tumors might not be any more likely with laser surgery than with older methods, the trauma and recuperation times may be reduced when lasers are used.

UROLOGY

The laser has been used in the treatment of certain genitourinary system problems. Infectious warts known as *condylomas* can develop on the surface tissues of the genital organs, and also at times on the insides of these organs. A laser can be used to treat the lesion areas following removal, a process called *flashing*. The treatment destroys most or all of the latent viruses that remain following removal of the warts, reducing the probability of a near-term recurrence. In the long term, however, recurrence is likely in patients predisposed to the condition.

Urinary tract tumors have been treated using lasers. Bladder tumors called *transition-cell carcinomas* are removed by directing laser light at the tumors through a flexible endoscope (Fig. 9-4). This procedure appears to reduce the rate of recurrence for this

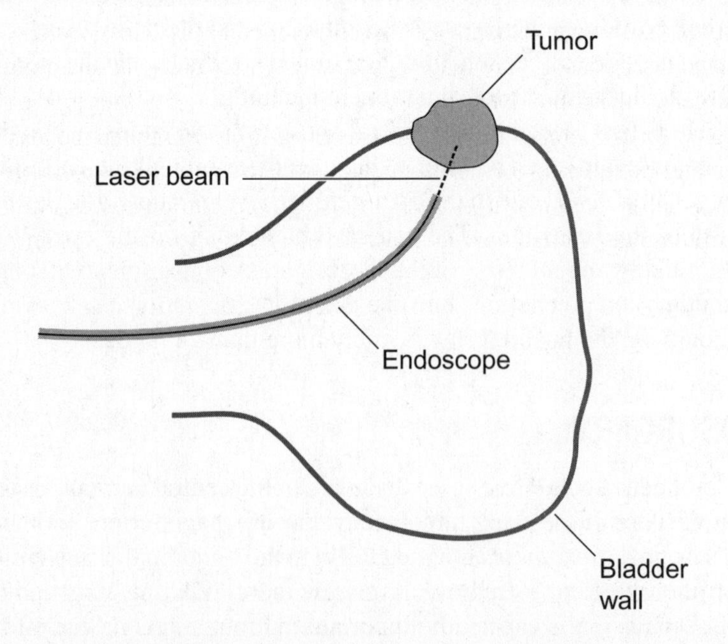

Figure 9-4 A specialized endoscope with a laser attachment can be used to remove some malignant tumors of the bladder.

type of malignancy, compared with older surgical methods. More advanced bladder tumors have been treated using the laser in conjunction with *electrosurgery*. The tumor is removed to the greatest extent possible using electrosurgery, and then the laser is directed at the region to destroy whatever cancerous tissue may remain. Specialized dye can be employed to enhance the absorption of the coherent radiation by the cancer cells while not affecting normal cells.

OTHER SURGICAL APPLICATIONS

The laser may someday be the tool of choice in most surgical operations. A good example is general removal of small tumors. Surgery with laser-equipped endoscopes can provide a clear view, self-sterilization, reduced chance of complications, shorter healing time, and minimal bleeding. A scalpel causes bleeding from severed capillaries; the blood can make it difficult for the surgeon to observe the work area. A laser cut often cauterizes itself, allowing better visibility.

The transmission of magnified images through optical fibers allows greater precision using laser surgery than is possible using conventional techniques. Tiny areas can be treated without affecting surrounding tissue. In the case of a malignant tumor, the laser method can reduce the amount of healthy tissue that must be cut out surrounding a tumor. The precision of the laser method allows a surgeon to remove much smaller tumors than is possible with conventional methods.

Lasers will likely gain acceptance in various medical applications as technology improves and laser consoles are made smaller and easier to use in operating rooms. In general surgery, the applications are almost unlimited, and the laser may even make it possible to perform operations that cannot be done by conventional means. One of the most intriguing examples involves modification of individual cells. Precision *nanoscale* (extremely small) lasers might be used to change the chromosome configuration or DNA in cellular nuclei, thereby affecting all the cells that reproduce from it.

PROBLEM 9-2

What are the most significant advantages of laser surgery over conventional surgery? What are the chief limitations of laser surgery?

SOLUTION 9-2

The laser produces less bleeding, allowing for a dry field and reducing the need for postoperative blood transfusions. The laser seals off severed blood vessels and lymphatic vessels, inflicts little or no trauma, gives rise to minimal side effects, reduces the risk of postoperative infection, causes less postoperative swelling, produces less postoperative pain, and can often be transmitted through flexible optical fibers. For the removal of malignant tumors, the laser offers precision and minimal damage to healthy tissue. The chief limitation of laser surgery is the fact that in certain situations, the region of interest cannot be reached in any other way than by mechanically cutting through the surrounding tissues. Another limitation is the fact that in some cases, tumors or diseased areas are too large or extensive to be dealt with by lasers alone.

Military Optics

When lasers became newsworthy in the 1960s, the public imagined "death rays" that could burn through walls, destroy armored tanks, shoot down aircraft, and vaporize hostile satellites. Some lasers have destructive power, but their military applicability transcends brute force. Lasers can help to guide missiles, provide communications, blind surveillance cameras, and perform some routine military operations more quickly and precisely than older methods.

A FICTIONAL LASER WEAPON

A science-fiction laser weapon might employ a huge resonant cavity surrounded by helical energizers, such as the design shown in Fig. 9-5. The power supply is massive,

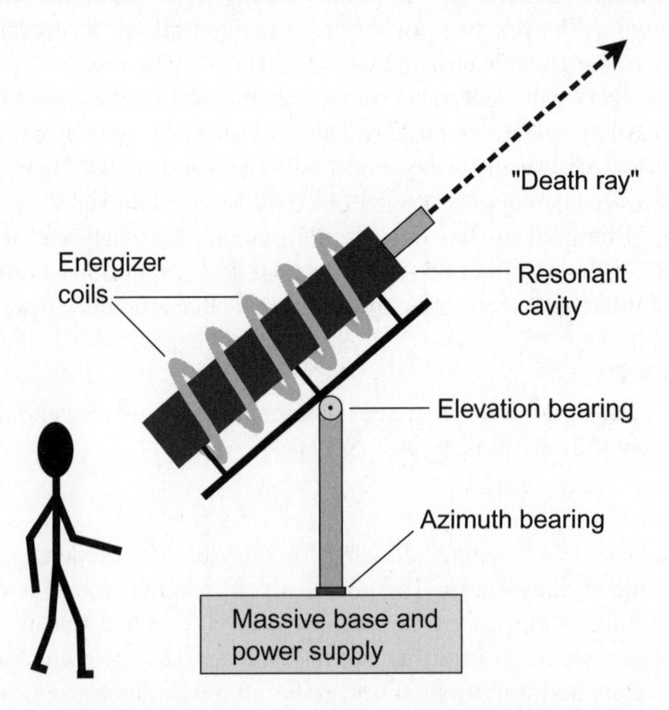

Figure 9-5 A high-powered laser "ray gun" of the sort described in science fiction. Note the humanoid figure for size comparison.

sometimes consisting of a complete power plant for a single weapon. A mobile laser of this sort can actually be developed if the funds are appropriated. Such a device might be capable of shooting down aircraft and low-earth-orbiting (LEO) satellites, but probably with no greater effectiveness than a rocket-propelled warhead.

Photons cause heating of a target when the energy is absorbed, but not when the energy is reflected. An unpainted or white-painted metal object reflects most of the visible light that strikes it, and is therefore difficult to damage by means of a laser. Protection against the laser weapons of science fiction would be simple: Coat an aircraft or spacecraft with reflective material or white paint!

Photons exert radiation pressure on a surface when they strike, as we have learned. If this force were great enough, a space ship could be "blown" off course whether its surface reflected the laser or not. However, the power required to generate a light beam of the necessary intensity is not within the reach of known technology. Even a brilliant beam of EM radiation, regardless of the wavelength, produces relatively little direct radiation pressure.

GENERAL DESTRUCTION

High-powered lasers can cause outright destruction. A good example is the scenario of a laser being used to destroy an oil refinery. The storage tanks are large, fixed in location, and make easy targets. Once started by a laser, a fire in such a facility might quickly rage out of control. The heat from the laser might be used to ignite some fuel in a part of the refinery where a fire or explosion would be especially likely to spread. Facilities could be protected against such attacks by coating all surfaces with white or reflective paint, and/or by enclosing the entire site in a gigantic, inflatable, reflective bubble.

Lasers might be used to start fires in dry, grassy, or wooded areas. The smoke from such fires would reduce visibility and make further use of lasers less effective, but any variable can be used either to advantage or suffered as a disadvantage. Smoke might make excellent cover for an invasion by foot soldiers equipped with gas masks, for example. Laser weapons could be used to cut power lines, damage roadways, runways, and railroad tracks, and blow up aircraft while they are on the ground. All of these applications would require lasers of extreme power, but the precision would exceed that of even the most accurate "smart bombs." Lasers would also inflict less collateral damage than explosive devices.

NUCLEAR ATTACK

Consider the difficulties of defending against a large-scale nuclear assault by a hostile power equipped with numerous long-range nuclear missiles. Warheads would come primarily from *intercontinental ballistic missiles* (ICBMs) launched from land-based silos, submarines, ships, and aircraft. Some aircraft might carry bombs

directly to their targets. Each missile could carry several dozen warheads and at least as many *decoys* (empty warheads designed to make effective tracking and defense difficult). In total, a full-scale attack carried out by a hostile superpower might involve thousands of individual warheads.

Antiballistic missiles (ABMs) might, with luck, destroy some enemy missiles before their multiple warheads could be launched, but in many cases the multiple warheads would deploy before the missiles came into range of defensive radar units. This vulnerability has prompted military scientists to look for early warning devices, such as *over-the-horizon radar*, to allow more time for launching accurate ABMs. However, such a radar system does not provide great accuracy. It can indicate that missiles have been launched, and may give defenders a rough idea of their whereabouts. But the radio frequencies involved, and the interaction of the ionosphere with emitted signals and echoes, limits the accuracy of this type of radar system.

Missile launches from a hostile power might be detected by video cameras and radar sets aboard earth-orbiting satellites, but only if those satellites were sufficient in number, and only if they were not destroyed prior to the main attack. A sophisticated enemy could employ laser weapons to put such satellites out of commission in advance, crippling the defense apparatus.

NUCLEAR DEFENSE

In the event of an all-out nuclear assault of the sort widely feared during much of the 20th century, the sheer number of incoming warheads, their speed, and the unpredictable nature of their movements would create an almost indefensible threat. Consider a situation in which a single long-range rocket deploys several warheads and many dummies in all directions (Fig. 9-6). Then multiply this by, say, 100. How

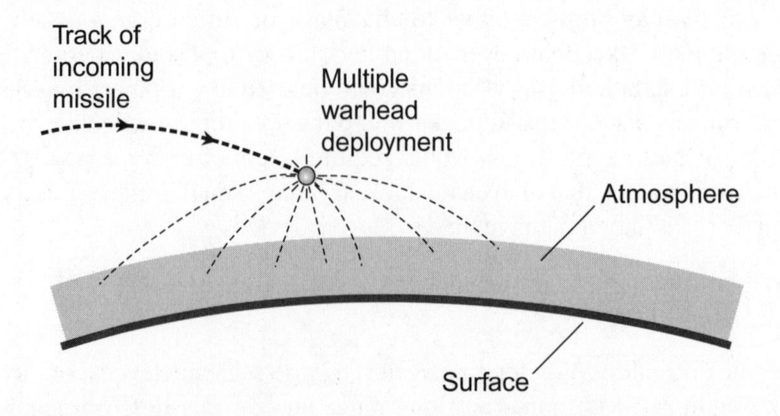

Figure 9-6 A long-range missile can be equipped with multiple warheads that are ejected as the missile approaches the target region.

would a defender ever manage to destroy every one of these bombs? If there were 700 bombs and 90 percent of them were disabled before they could explode, the victim nation would still face 70 viable warheads, each with enough nuclear energy to destroy a city the size of Pittsburgh or Dallas.

The unpredictable movements of the warheads (even the military commanders in the attacking country would not be aware of where all of them would go) are all the more difficult to track because of the enormous speeds involved, in the order of 10 km/s. At such speeds, a timing error of 0.01 s would result in a position error of 100 m. A laser with a beam only a few centimeters wide would have to be well-placed indeed to have the desired effect, and the warhead would have to be exposed to the radiation for a sufficient length of time.

Defensive laser devices could be located on the surface of the earth, either at ground stations or on ships; they could also be carried on aircraft or satellites. Ground-based lasers would be the easiest to put together for several reasons. Primary among these considerations would be the relative simplicity of the aiming apparatus; only the motion of the warhead with respect to the earth itself would have to be taken into account. Power supplies at fixed locations could be huge, because they would not have to be hauled around. However, only those regions of the sky visible on a line of sight from a given ground station could be covered with the laser at any given time. The atmosphere would inflict some beam attenuation, even in the absence of clouds. The resulting energy loss would be a particular problem at low firing angles because of the large amount of air through which the beams would have to travel (Fig. 9-7). Cloudy or foggy weather would preclude the use of lasers for shooting down warheads speeding high above the atmosphere.

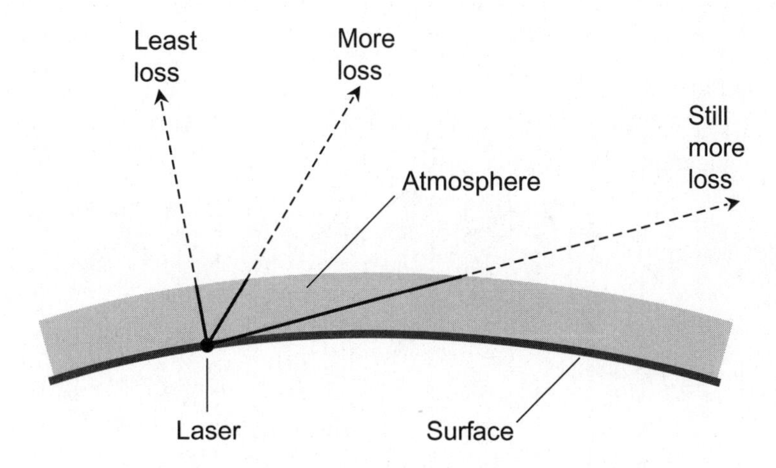

Figure 9-7 A laser beam directed into space must pass through more air at a shallow firing angle than at a steep firing angle.

A space station equipped with defensive lasers could utilize solar power, and could be placed in an orbit with predictable and calculated paths. Numerous stations, positioned strategically in diverse orbits, could disable hostile missiles as they were launched, as long as the weather in the vicinity of the launch site did not obscure visibility. It might be possible to conduct a preemptive strike and destroy enemy missile silos without notice before the commencement of hostilities. Laser weapons on spacecraft would not be hampered by atmospheric attenuation once enemy missiles had gained sufficient altitude, but they would have to be aimed with extreme precision. System redundancy and "overengineering" would be necessary to ensure that none of the missiles got past this level of defense.

A REALISTIC LASER WEAPON

The *free-electron laser* (FEL) is capable of producing high-power, coherent radiation at wavelengths ranging from the microwave radio-frequency spectrum to the X-ray band. The device operates by means of a phenomenon called *synchrotron radiation*, which occurs when electrons are accelerated at relativistic speeds. The beam emerges in a sharp, intense cone whose axis lies in the direction of the instantaneous electron velocity, as shown in Fig. 9-8.

The high-power output capability of the FEL makes it well suited for shooting down missiles. Such a device could produce intense IR beams at wavelengths that propagate efficiently through clear air. Orbiting reflectors could be used to direct the beams from surface-based FEL stations toward enemy missile silos located thousands of kilometers from the laser sites (Fig. 9-9). The computers that aim the laser beams would also be capable of orienting the mirrors quickly by remote control. Turning the mirrors would have to be done in such a way so as not to upset the stability of the satellites to which they were attached. Several reflecting mirrors

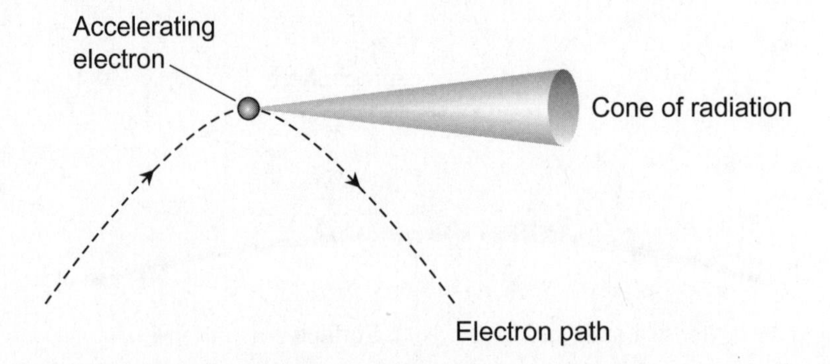

Figure 9-8 The free-electron laser exploits the energy produced by accelerating electrons to generate coherent radiation over a wide range of wavelengths.

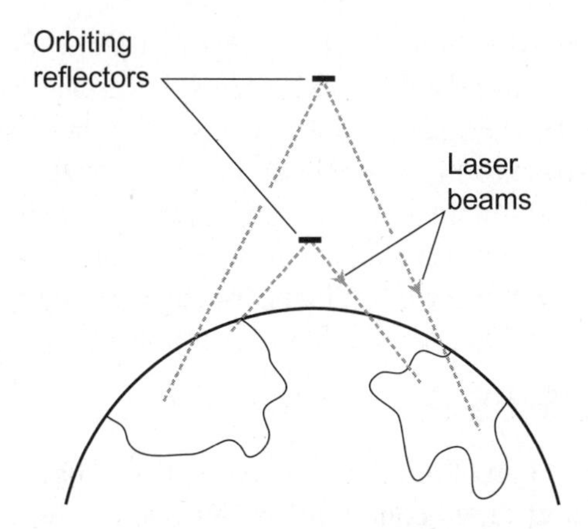

Figure 9-9 Reflectors in earth orbit could be used to guide laser beams to distant surface targets.

might be used with a single high-powered laser to direct multiple beams in independent directions. The laser at the surface would have to provide sufficient power to make all of these beams intense enough to carry out the requisite operations.

LASERS TO GUIDE MISSILES

A laser beam, especially in the near IR where the waves propagate through the atmosphere with minimal attenuation, can be used to "illuminate" a target so that thermal sensors can be employed for homing. The intensity of such a laser beam can be far lower than that required to cause direct destruction. A laser of moderate intensity is aimed at the object designated the target. A heat-seeking device on the attack missile guides that missile to the target, whereupon the target is destroyed (Fig. 9-10).

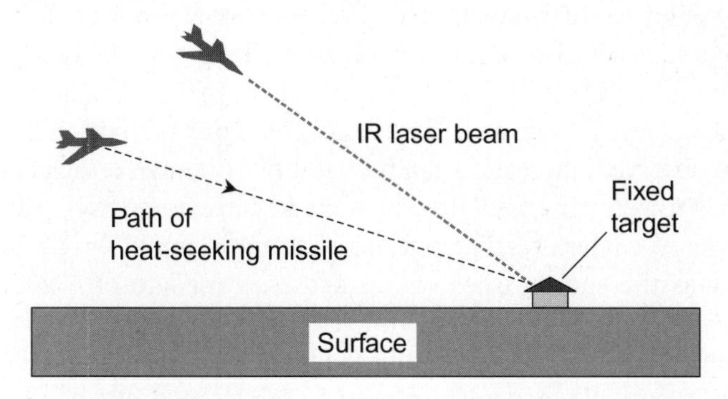

Figure 9-10 An IR laser beam from an aircraft "illuminates" a target for a heat-seeking missile launched from a second aircraft.

Laser missile-guidance systems have been perfected for air-to-surface missiles, as well as for air-to-air, surface-to-air, and surface-to-surface missiles. The problem is, as with any laser, that a line of sight is needed for the system to be effective. Clouds and other obstructions interfere with operation. Reflectors at the target may reduce or eliminate the "illumination" caused by the laser. A principal advantage of using lasers to guide weapons is the fact that such systems minimize collateral damage, and also reduce the number of humans directly exposed to peril.

SUBMARINE LASERS

Water is relatively transparent to certain wavelengths of light. This fact may make lasers practical for short-range communication between ships and submarines, as well as for target location and guidance systems. *Extremely low frequency* (ELF) radio signals, with transmitters on land designed to resonate inside the earth, can send signals only at slow data transfer rates, because of the narrow bandwidths of their modulators. With visible light there is no such limitation.

Underwater, the principle of total internal reflection might be employed to provide communication around obstructions such as hills on the bottom of the sea. The detector would be aimed at the image point on the surface. A sensitive receiver with multiple sensors would be necessary because the light from such a reflected source would appear to scintillate. Messages arriving at a single sensor would likely be garbled by severe amplitude "fading" similar to the *libration* effects that occur with microwave radio signals reflected from the moon.

LOW-POWER MILITARY LASERS

Lasers having modest power, not capable of starting fires or explosions or of melting asphalt roads, are still useful in warfare. We have already seen one example, in the guiding of missiles to their targets. However, even lower-power devices might be used, for example, to disable or distract visual tracking gear. Only a few milliwatts of laser power are necessary to create a beam that looks as bright as the sun from several kilometers away. Such a brilliant beam could divert a person's attention even if not directly observed, increasing the probability of human error in critical operations. If the beam were observed directly, it could cause temporary blindness.

Reconnaissance cameras can be put out of commission by bright light, in much the same way as the sun "washes out" a camera exposure or a television image taken from a bad angle. Such light can impair the operation of visual tracking gear. A good example is the use of a ground-based laser or lasers to blind a satellite camera, making it impossible to resolve images. Large lasers might be set up in or near missile bases and other strategic places, as well as in decoy locations where no weapons are kept.

An especially interesting application of low-power and medium-power lasers in the military is for communications purposes. A laser beam could be aimed at clouds, mountains, or simply into the air, and the scattered light picked up and demodulated to retrieve signals. The equipment, if properly disguised, would look nothing like communications apparatus, so the enemy would think it was intended for some other, perhaps unknown, purpose. Such equipment would have limited range but would be easy to deploy and move around, making it ideal for use by ground-based armies. The laser could serve the dual purpose of effecting communications and blinding aircraft pilots or cameras.

Low-powered lasers are commonly used to determine the distance to a target. This is accomplished by sending a narrow beam to a target, where the beam is reflected and the returning beam detected by sensors. The device is pulsed at a high frequency, and connected to a timer that automatically displays the range by measuring the time required for the beam to complete the circuit. This time interval can be measured by a digital clock, or by counting the number of pulses that occur between the laser emission and the returned beam.

A specialized laser device can be employed in training exercises for ground-based personnel. Soldiers wear devices on their uniforms that detect laser emissions at specific wavelengths. If such emissions are detected, the soldier is considered "hit." Each individual is designated by a certain code transmitted to a central computer if a "hit" occurs (Fig. 9-11), so the computer knows which soldiers are "casualties" and which are not, and which weapons are to be thereafter disabled. Such a system can help officers to conceive and optimize tactics to deal with specific scenarios (urban warfare in hot weather, for example), without risking life or property.

Laser-equipped rifles can be used for developing marksmanship. The laser is mounted on the barrel of a rifle-like device. The pulses from the laser simulate bullets being fired. The rifle makes a popping sound to simulate being fired. Devices of this type can also be used to simulate larger weapons on ships, aircraft, tanks, and other motorized vehicles.

PROBLEM 9-3

A manned LEO space station equipped with laser weapons could serve as an excellent defensive or offensive military base. What significant precautions would have to be taken to ensure the viability of such a base?

SOLUTION 9-3

An orbiting space station is a large satellite. Any LEO satellite is highly vulnerable to attack from ground-based weapons. If a LEO military space station passes over a well-equipped antisatellite base on the surface in clear weather, the space station could be easily destroyed unless equipped with detection apparatus, defensive hardware, and offensive weaponry capable of destroying the ground-based station instantly.

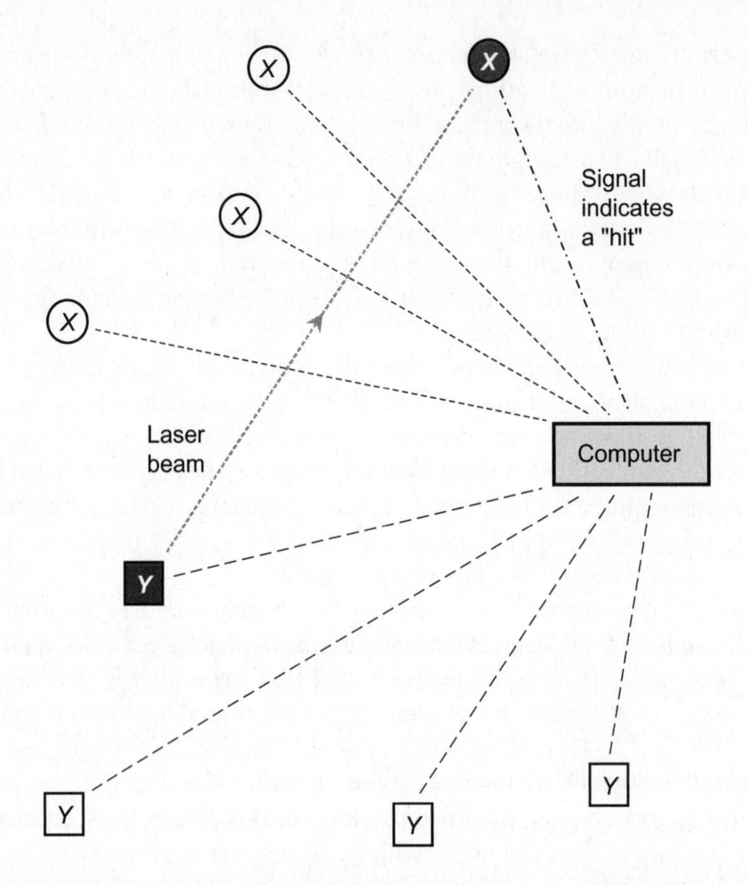

Figure 9-11 Soldiers (*X* and *Y*) are equipped with laser guns and sensors on different frequencies. Here, a solider from "*Y* squad" has hit a solder from "*X* squad." A signal is sent to the computer to indicate and record the hit.

Quiz

This is an open book quiz. You may refer to the text in this chapter. A good score is 8 correct. Answers are in the back of the book.

1. A moderate-power IR laser on board an aircraft can "illuminate" a ground target to facilitate destruction of that target by
   (a) photocombustion.
   (b) photodisruption.
   (c) photovaporization.
   (d) None of the above.

2. Lasers can sometimes be used to
   (a) promote hair growth on the scalp.
   (b) remove basal cell carcinomas.
   (c) encourage normal cuticle growth.
   (d) produce a "suntan" without causing skin damage.
3. A visible laser beam of moderate power, aimed at a surveillance satellite, can
   (a) degrade the sensitivity of the satellite's video cameras.
   (b) exert radiation pressure, disrupting the satellite's orbit.
   (c) improve the field of view of the satellite's cameras.
   (d) guide the satellite into the most favorable observing position.
4. A satellite can be designed for minimal vulnerability to laser "ray guns" by
   (a) coating the satellite's exterior surface with black paint.
   (b) covering the satellite's exterior surface with heat-absorbing material.
   (c) ensuring that the satellite's exterior surface reflects visible and IR rays.
   (d) surrounding the satellite with an ELF field.
5. One of the chief advantages of the use of lasers for acupuncture is
   (a) low probability that cancer will spread.
   (b) increased effectiveness when used with an endoscope.
   (c) the fact that it can remove large tumors.
   (d) minimal risk of infection.
6. A "cloudbounce" laser communications system
   (a) is easy to deploy in a fixed location, but cannot be easily moved.
   (b) requires the use of high-power X-ray devices to penetrate fog.
   (c) can temporarily blind an aircraft pilot who looks directly into the beam.
   (d) must be camouflaged, because its intended use is obvious to casual observers.
7. For dealing with large tumors, laser treatment
   (a) inflicts more pain than conventional surgery.
   (b) causes excessive bleeding unless a blood thinner is used.
   (c) causes significant damage to surrounding tissue.
   (d) may not work as well as conventional treatment.

8. The FEL produces coherent radiation when electrons
    (a) change orbital shells within atoms.
    (b) are heated to form a glowing plasma.
    (c) accelerate while traveling at relativistic speed.
    (d) encounter the earth's magnetic field.
9. An endoscope can be used to
    (a) observe the bronchial passages.
    (b) track the maneuvers of military troops.
    (c) detect incoming long-range missiles.
    (d) replace needles in acupuncture.
10. A ground-based laser "ray gun" would work best for shooting down satellites
    (a) below the horizon.
    (b) at or near the horizon.
    (c) midway between the horizon and the zenith.
    (d) at or near the zenith.

CHAPTER 10

Optical Illusions

Optical illusions have existed ever since people began to scribble images on paper. In this chapter, we'll examine, and attempt to explain, some of the simplest and most well-known "tricks of the pen."

Lines and Curves

Straight lines can look curved, line lengths can be distorted, collinear lines can appear displaced, and lines can distort the apparent dimensions of objects.

DOMINANCE OF BACKGROUND

When we think of the background in a scene, we imagine something less important than the main subject. But the background can divert our attention away from the main subject to such an extent that the subject's shape appears distorted. Figure 10-1 is an example. The square has straight edges, a fact that we can verify using a ruler

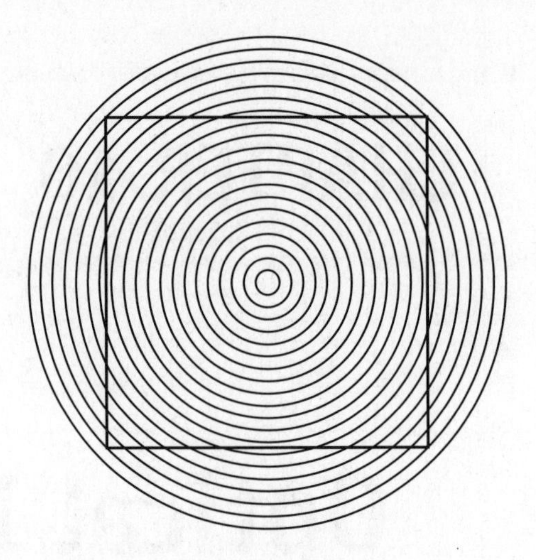

Figure 10-1 The sides of the square are straight, but most observers see them as bent slightly inward.

or a rectangular sheet of paper. Nevertheless, most people see the square's edges as slightly concave. The outward-curving circles distract the "eye/brain team," altering its notion of what constitutes "straightness." Try moving the page closer to your eyes and then farther away. Try looking at it with only one eye. Try rotating the page 45°. Does the vividness of the illusion vary?

LENGTH DISTORTION

Figure 10-2A is an example of a deceptive drawing called the *Mueller-Lyer illusion*. The upper and lower halves of the line segment are of equal length, but most observers will insist that the lower half is shorter until they measure the two halves to verify that the lengths are identical. Even the knowledge that the two halves are of equal length does not destroy the illusion.

The Mueller-Lyer illusion can be demonstrated in the reverse sense by shortening the upper half of the line segment to the extent that it attains the same apparent length as the lower half (Fig. 10-2B). Although the two halves of the line segment seem equally long, the upper portion is only 70 percent as long as the lower portion. You can measure the lengths in millimeters, and then divide the smaller figure by the larger figure to verify this fact.

The human "eye/brain team" can distort perpendicular displacements. Figure 10-3 is an example. At A, the vertical and horizontal line segments have equal lengths. If we shorten the vertical segment by 25 percent we get drawing B, where the two segments appear equally long, or at least more nearly the same. Try rotating the entire drawing 90° clockwise or counterclockwise. What do you see? Some people observe

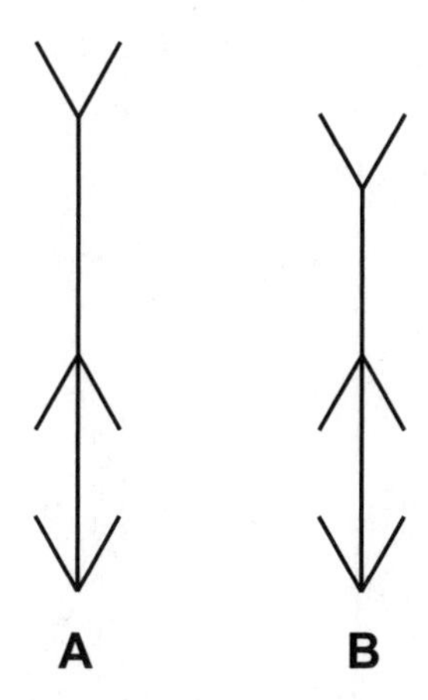

Figure 10-2 At A, the upper and lower halves of the line segment have equal length. At B, the upper portion is 70 percent as long as the lower portion.

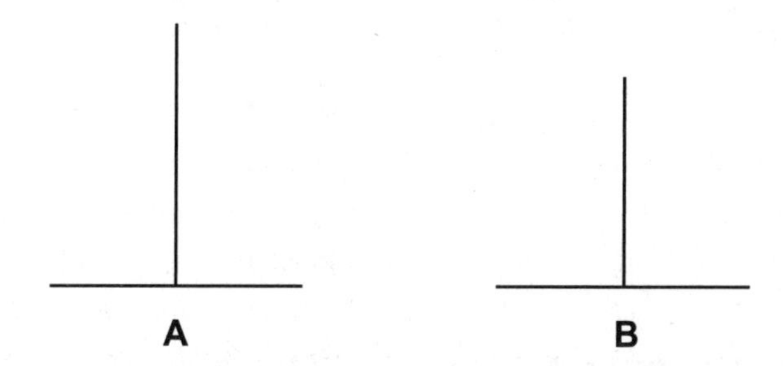

Figure 10-3 At A, the vertical and horizontal line segments have the same length. At B, the vertical line segment is 25 percent shorter than the horizontal line segment.

the same distortion to the same extent. Some people see the same distortion, but to a lesser extent. A few people lose the illusion altogether.

ALIGNMENT

An interesting optical illusion was shown to exist more than a century ago by the German physicist *Johann Christian Poggendorff*. The so-called *Poggendorff illusion* involves the apparent displacement of a line eclipsed by a barrier at an angle. The effectiveness of this illusion depends on the angle that the lines subtend with respect to the barrier. As the transversal lines become more nearly parallel to the edges of the barrier, the vividness of the illusion increases.

The Poggendorff illusion can be rendered in several ways. In one popular version (Fig. 10-4), slanting line segments are placed on either side of the barrier. The

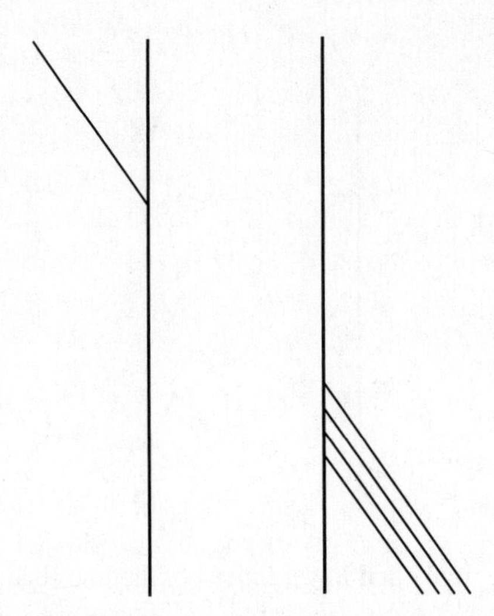

Figure 10-4 Which slanted line segment on the right is collinear with the slanted line segment on the left?

observer tries to guess which pair of line segments is collinear. When the drawing is properly done, this is impossible without the help of a straightedge. In the example of Fig. 10-4, the slanting line segment on the left is collinear with the lowermost slanting line segment on the right.

ANGLES AND CIRCLES

Basic geometric figures such as line segments and disks are easy to draw, offer geometric simplicity, and are ideal for creating optical illusions. In Fig. 10-5, two congruent circular disks *X* and *Y* are drawn within an acute angle formed by two

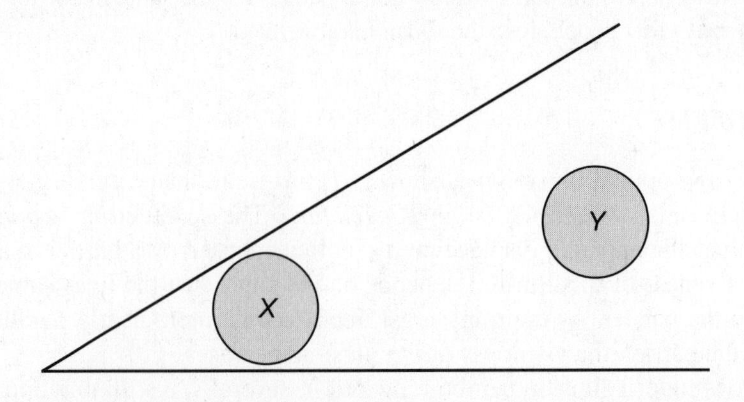

Figure 10-5 The circular disks *X* and *Y* have the same diameter.

straight line segments. Disk X is closer to the vertex of the angle than disk Y. As a result, some people think at first glance that X is larger than Y.

This illusion is a primitive manifestation of perspective, where an object near a horizon point seems larger than an object of equal size that is farther from the horizon point. Another way to explain this illusion is that the proximity of the line segments to disk X exaggerates the size of that disk, making our "eye/brain teams" expand X in an attempt to make it tangent to the line segments. With disk Y, no such subconscious maneuvering occurs, because Y is too far away from the line segments.

PROBLEM 10-1

Figure 10-6 shows a parallelogram divided into two smaller parallelograms by a line segment XW chosen so that the distances XY and XZ are the same. Most observers think that XY is shorter than XZ, even after verifying with a graduated straightedge that the two lengths are the same. How can we resolve this illusion?

SOLUTION 10-1

We can remove the line segments that distract our perception, leaving only segments XY and XZ as shown in Fig. 10-7. Then it's easy to see that line segments XY and XZ have the same measure.

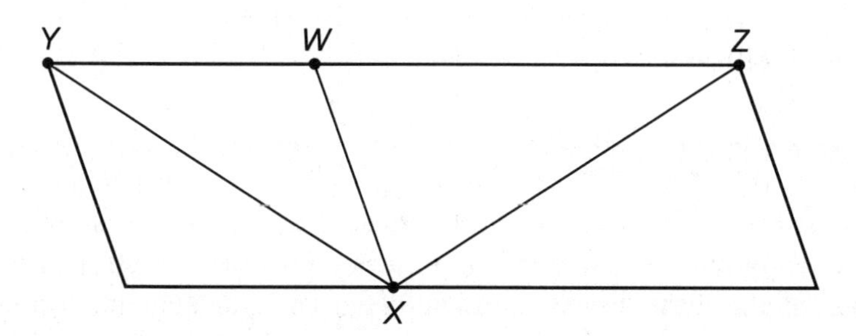

Figure 10-6 Illustration for Problem 10-1.

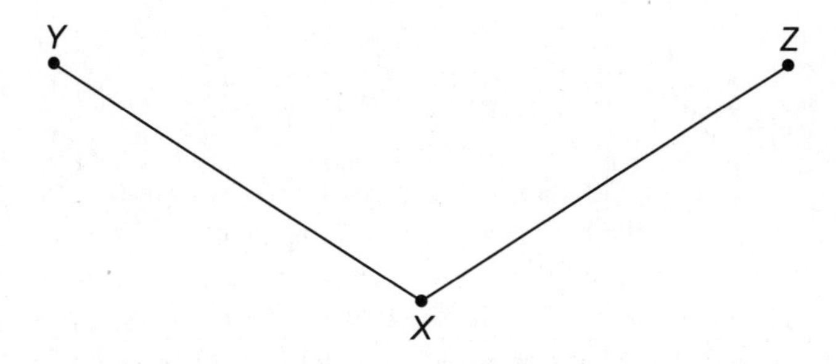

Figure 10-7 Illustration for solution of Problem 10-1.

Depth and Displacement

Everyday situations and simple line drawings can create illusions that distort dimension and orientation. The incorrect interpretation may persist even when we know better.

THE BLOATED MOON

The distance between the earth and the moon varies by a few percent, but this difference cannot account for the apparent increase in the moon's diameter when it is near the horizon, as compared with the way it looks when it is high in the sky.

According to noted physicist *S. Tolansky*, the moon's apparent diameter is exaggerated immediately after it rises and before it sets, because we expect an object to shrink in our field of view when it approaches the horizon. This effect occurs with objects near the earth such as balloons, aircraft, and clouds, because such objects are farther from us when they're near the horizon than when they're overhead. However, the size vs. distance variation does not take place with celestial objects such as the moon. The moon's disk is always pretty much the same size as seen from the earth: an angular diameter of approximately 0.53°. Tolansky recommends the use of a long tube, such as we get when finished using a roll of paper towels, to look at the moon with one eye when it is near the horizon. The "bloated moon" illusion disappears when we observe it through the tube. When we take the tube away and keep looking, the illusion returns.

Sometimes the disk of the moon appears *oblate* (flattened slightly) when it grazes the horizon. This phenomenon is not an illusion. It is genuine optical distortion, caused by refraction in the atmosphere. The geometric size of the moon's image is actually reduced, the vertical dimension shortened by refraction so that the angular height of the disk is less than 0.53°. The atmosphere, when it has any noticeable effect, acts to make the moon appear smaller, not larger, than when it is elevated well above the horizon. But even so, most people think that the moon—especially the full moon—looks unnaturally large just after it rises or before it sets.

WARPED SOLIDS

Consider the rectangular solid shown in Fig. 10-8. When we direct our gaze toward the top of the object, it seems as if we are viewing it from below and to the right. However, if we shift our gaze to the lower part of the figure, it seems as if we are looking at it from above and to the left. The whole object is mysteriously and impossibly twisted. Somehow, the "eye/brain team" concludes that this rectangular figure is warped, even though we know that if it were, its edges couldn't be straight and its faces couldn't be flat.

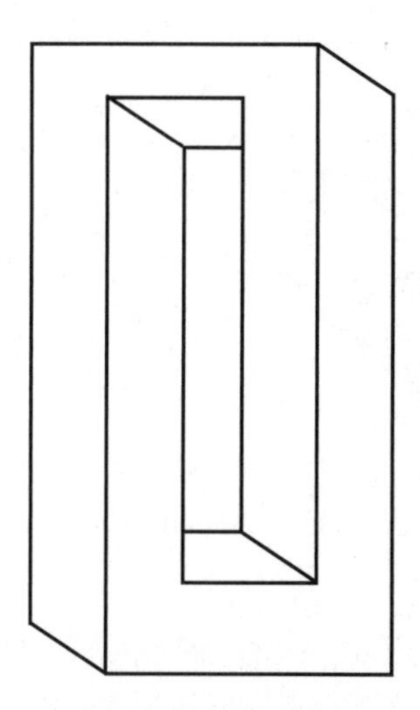

Figure 10-8 A three-dimensional object with ambiguous faces.

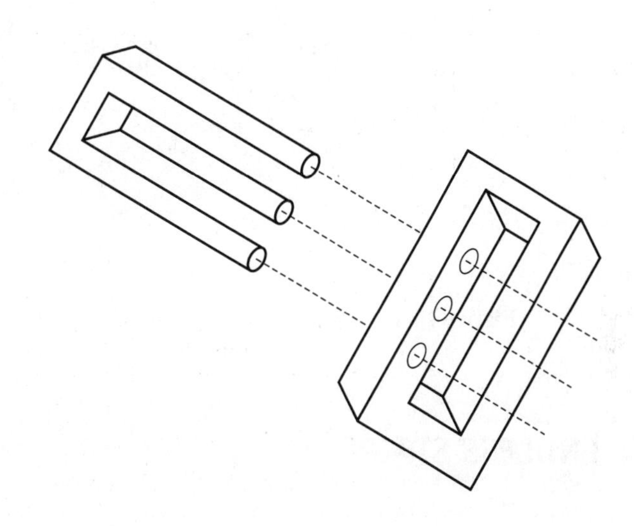

Figure 10-9 An ambiguous assembly diagram.

Figure 10-9 illustrates another impossible scenario. Imagine reading the instruction manual, which says, "Insert the ends of the fork through the holes in the rectangular ring as shown." Then ask, "Will the resulting assembly stay together in three dimensions?"

Another impossible object is rendered in Fig. 10-10. As we move our gaze clockwise around the triangular solid, we can try to follow one face even when it is hidden from view, but we're likely to become frustrated. This object resembles the well-known *Möbius band*, which consists of a flat strip given a half-twist and then pasted to itself end-to-end. The Möbius band has one face in its totality, but two faces in any localized region. The object in Fig. 10-10 has one face in its totality, but four faces in any localized region.

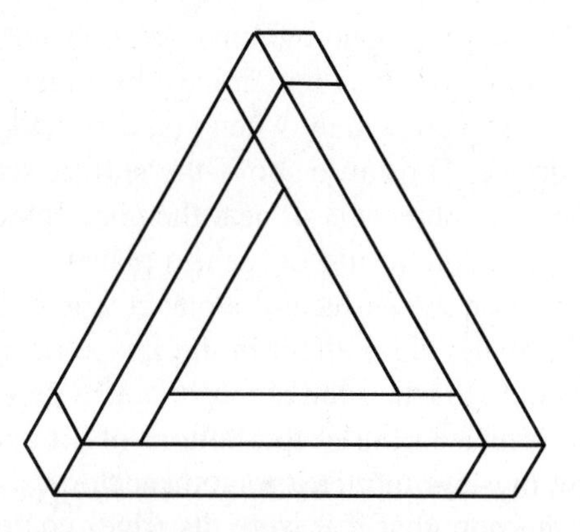

Figure 10-10 Another ambiguous three-dimensional solid.

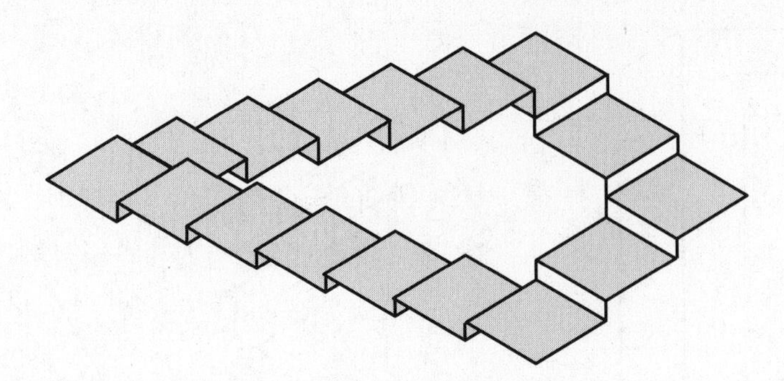

Figure 10-11 An endless staircase. Circle counterclockwise to "ascend," and circle clockwise to "descend."

ENDLESS STAIRS

Figure 10-11 shows the steps in a staircase that seems to constantly "ascend" if we move our eyes around in a counterclockwise direction, and to constantly "descend" if we move our gaze around clockwise. This concept of visual ambiguity was extensively used by the artist *M. C. Escher* to create mesmerizing scenarios, such as a power generator with a waterwheel that worked forever because the water would go down the falls and then be allowed to flow back up again to the top of the falls. The trick was that the flowing stream actually looked, at every point in the drawing, as if it were flowing downhill. Imagine replacing the three-step part of the staircase at the upper right by a waterfall with a water wheel. Then imagine replacing the rest of the stairs by a trough in which the water can always flow "downhill" in the clockwise direction from the "bottom" of the falls around to the "top" again!

EXTREME ATMOSPHERES

Most of us have seen wide-angle photographs and noticed how the scene is distorted. Straight lines appear curved. A human face becomes a bulging caricature at close range. This illusion is also evident when you dive underwater in a calm pool and look at the water surface. The entire above-the-surface scene is "squashed" into a large circular aperture, and objects at or near the horizon are greatly distorted.

An illusion of this same sort would occur if a planet's atmosphere were dense enough. Imagine standing on the surface of a planet where the density of the air is so great that it causes a wide-angle effect in the image of the sky. The horizon is farther away than it would appear if the air were not so dense. If the effect is pronounced enough, you think you are at the bottom of a huge bowl-shaped valley, because incoming light rays are refracted more near the surface than at higher altitudes (Fig. 10-12).

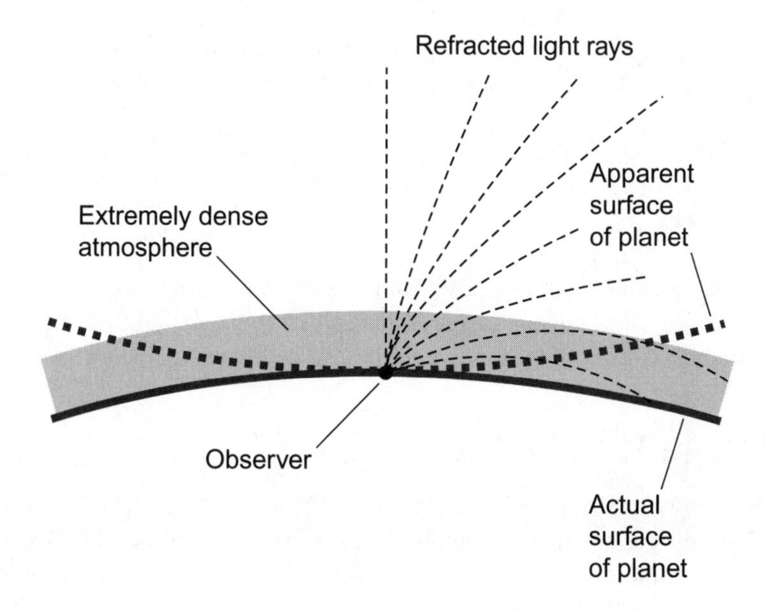

Figure 10-12 Refraction in the atmosphere of a hypothetical planet, causing the surface to appear bowl-shaped.

On a planet like this, the length of the day would be artificially increased. The image of the sun's disk would be greatly flattened just after sunrise and before sunset. There would be a limit to the extent that this illusion could manifest itself, but the apparent horizon might be hundreds of kilometers away.

PROBLEM 10-2

Are there any known "real-world" examples of the phenomenon described above at wavelengths other than those of visible light?

SOLUTION 10-2

Yes. This effect can occur when radio waves at certain frequencies are refracted in the earth's lower atmosphere near weather fronts, where warm, relatively light air hangs above cool, dense air. The cool air has a higher index of refraction than the warm air, causing radio waves to be bent downward at a considerable distance from the transmitter. This phenomenon is called *tropospheric bending*.

Radio-wave bending is often responsible for anomalies in reception of signals on the standard frequency-modulation (FM) broadcast band. Normally, reception is limited to distances of about 100 km. When tropospheric bending occurs, stations hundreds of kilometers away suddenly appear on channels that are normally vacant. Sometimes two or more distant stations come in on a single channel, interfering with each other. The effect may appear, continue for a few minutes or hours, and then vanish.

Size and Shape

Our impressions of size, area, volume, and shape are often inaccurate. The way we envision an object or scene can be prejudiced by contrast with figures nearby, or by comparison with similar scenes we've witnessed in the past.

DISTORTING THE SQUARE

When people see the square in Fig. 10-13A alone, some think that the object's width is slightly greater than its height. If these same observers then look at Fig. 10-13B alone, some think it is taller than it is wide. A few observers suspect that one of the squares is larger overall than the other. Nevertheless, the two figures have identical outer dimensions. This illusion has been explained by the theory that the "eye/brain team" imagines the squares as stacks of thin rectangles. Stacked objects apparently attain exaggerated displacement in the direction of stacking.

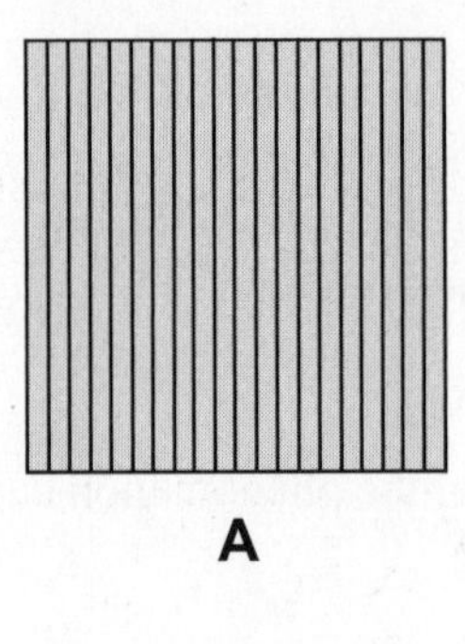

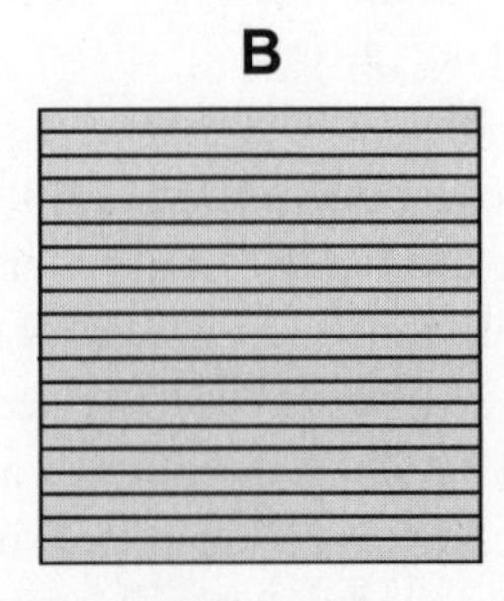

Figure 10-13 Two squares assembled from stacked rectangles. At A, horizontal stack; at B, vertical stack. Which square appears larger or taller?

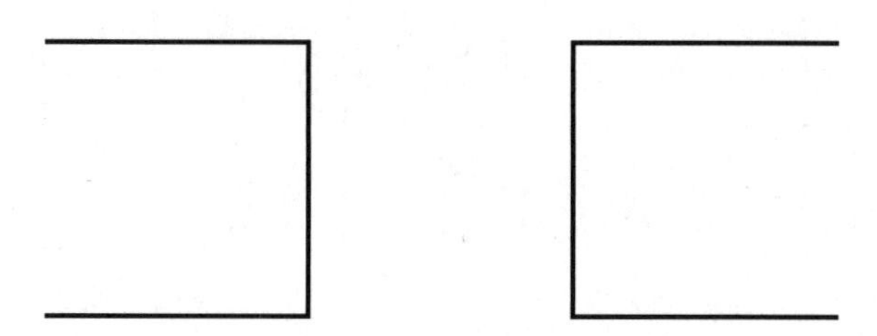

Figure 10-14 The leftmost and rightmost squares show three sides. The center square shows only two sides.

Here's an experiment you can try at home to create this illusion in three dimensions. Stack some pennies on top of each other to make a vertical cylinder. Then, next to the stack, stand another penny on its edge for height comparison. When you've built up a stack whose height is equal to the diameter of a penny, remove the height-comparison penny and look at the stack all by itself. If you're like most observers, you'll think that the stack of pennies is taller than the diameter of a single penny.

Figure 10-14 illustrates another example of square distortion. The leftmost and rightmost objects are three edges of perfect squares. The space between them would be a third square, identical to the other two, if its two missing edges were filled in. Nevertheless, many observers insist that the distance between the leftmost and rightmost figures is slightly less than the length of a square's edge.

CENTRAL CIRCLE

We perceive the size of an object by comparing it with its surroundings. Consider the three sets of concentric circles in Fig. 10-15. The inner circles are exactly the same size in all three drawings. But to most observers, the inner circle at A looks

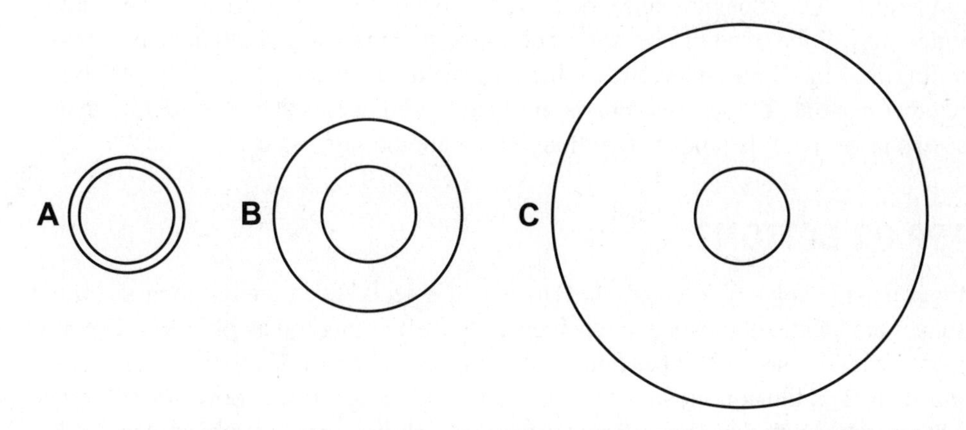

Figure 10-15 At A, B, and C, the inner circles have equal diameters.

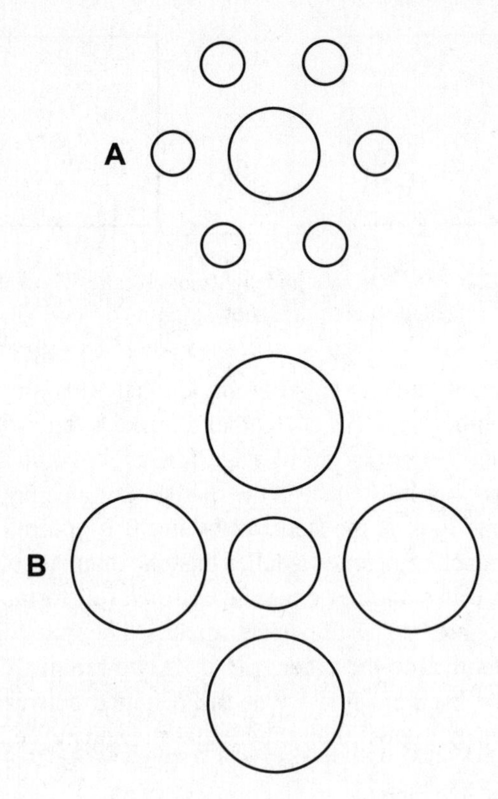

Figure 10-16 At A and B, the central circles have equal diameters.

larger than the inner circle at B, and the inner circle at B looks larger than the inner circle at C. The illusion occurs because the outer circles distort how we see the inner ones. If the outer circles were not in the drawings, we would have no trouble telling that the three inner circles have identical diameters. Figure 10-16 shows another example of a size-comparison illusion with circles. The central circle at A seems larger than the one at B, although they are the same size.

TOP OR BOTTOM?

Figure10-17 shows two congruent fan-shaped objects. Even though their sizes and shapes are identical, most people perceive object B as larger than object A. The illusion occurs because the upper figure's bottom span contrasts with the lower figure's top span. The illusion persists even if we turn the page upside-down or sideways. Most people think that object B always looks slightly larger than object A.

A related illusion is shown in Fig. 10-18. Here, the two objects are obviously not congruent, but most observers initially think that the top edge of trapezoid A is

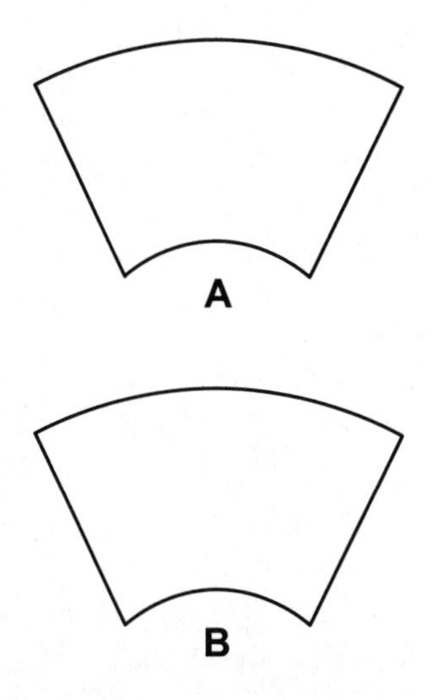

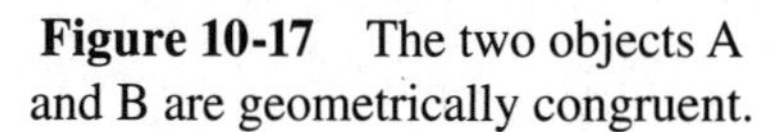

Figure 10-17 The two objects A and B are geometrically congruent.

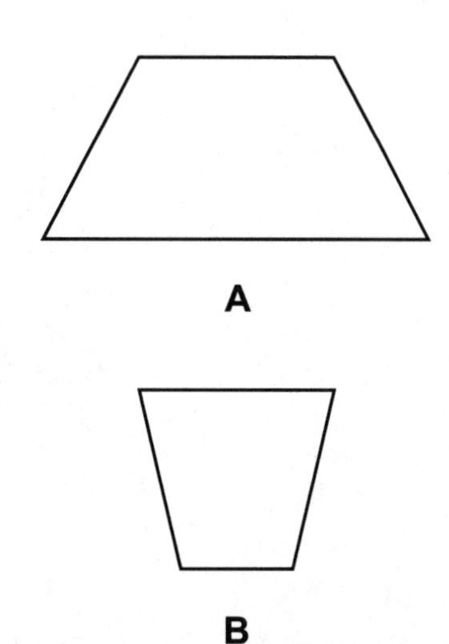

Figure 10-18 The top edges of trapezoids A and B are equally long.

longer than the top edge of trapezoid B, even though their lengths are in fact the same. The "eye/brain team" instinctively gravitates to the center of the entire scene, comparing the top edge of B with the bottom edge of A, where the contrast between the two figures is the greatest.

RADIUS OF AN ARC

Examine the curves in Fig. 10-19. The two curves are arcs of circles having equal radii, but most observers imagine that B has a larger radius than A. We see this illusion for two reasons. First, we can get a good idea of the arc radius at A because the

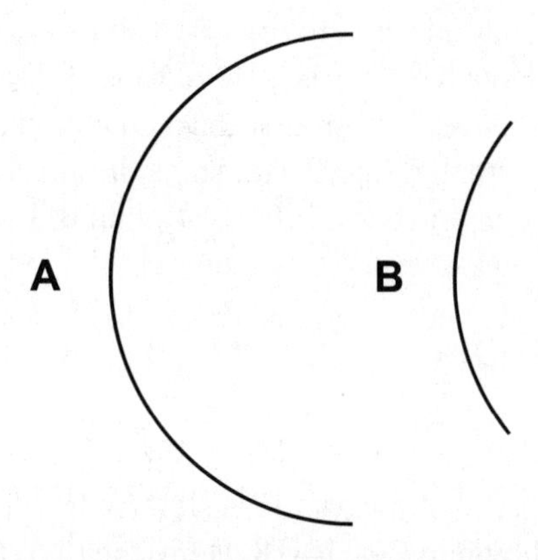

Figure 10-19 The arcs at A and B have equal radii.

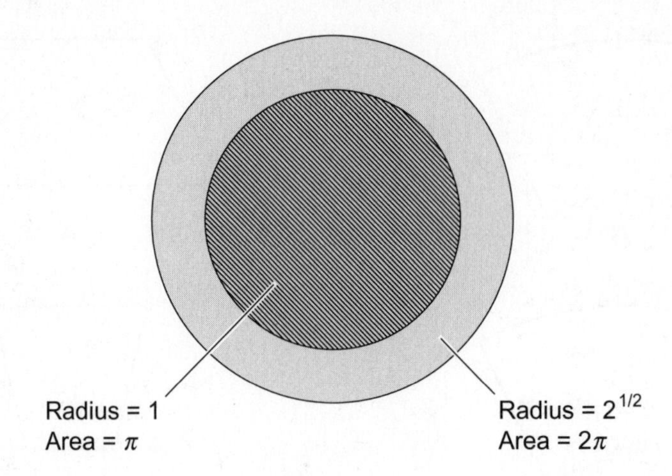

Figure 10-20 The area of the shaded disk is twice the area of the hatched disk.

location of the center is easy to intuit, while the arc at B is too small a part of a circle for us to readily perceive the center. Second, the arcs have been strategically positioned so as to exaggerate the apparent radius of B.

AREAS OF CIRCLES

Even when we perceive linear units in their true proportions, geometric areas can appear exaggerated or distorted. For example, the interior area of a circle is directly proportional to the square of its radius. The formula is

$$A = \pi r^2$$

where r is the radius in units of a specific linear size, A is the area in square units of the same linear size, and π is the ratio of a circle's circumference to diameter, a mathematical constant whose value is approximately 3.14159. If we double the radius of a circle, we quadruple its area. If we want to double the area, we must increase the radius by a factor of $2^{1/2}$ (the square root of 2, or approximately 1.414). Figure 10-20 shows two concentric circular disks, the larger disk having a radius of $2^{1/2}$ times the radius of the smaller one. That means the area of the larger disk is twice the area of the smaller one. Upon casual observation of this illustration, most observers don't think the ratio of disk areas is that large.

VOLUMES OF SPHERES

With concentric spheres, the volume ratio is exaggerated so that the illusion is even more vivid than with concentric disks. The volume of a sphere is given by the formula

$$V = 4\pi r^3/3$$

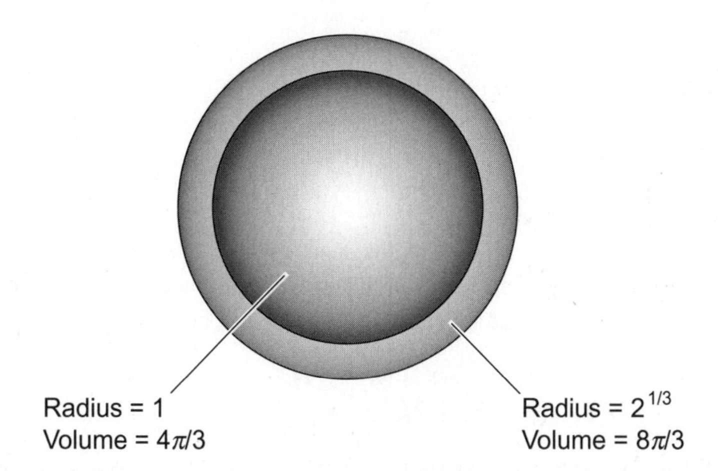

Figure 10-21 The volume of the outer sphere is twice the volume of the inner sphere.

where r is the radius in specific units, and V is the volume in cubic units of the same linear size. If we have a sphere of a given radius and we double its radius, we increase the volume by a factor of 2^3, or 8. If we increase the diameter by a factor of $2^{1/3}$ (the cube root of 2, or approximately 1.260), we double the volume. Figure 10-21 is a cross-sectional view of two concentric spheres, the larger sphere having a radius of $2^{1/3}$ times that of the smaller sphere. Most people find it difficult to imagine that the larger sphere's volume is fully *twice* the smaller sphere's volume.

PROBLEM 10-3

Glance at Fig. 10-22 for 3 seconds (but not any longer!). Then guess: How many edges does polygon A have? How many edges does polygon B have?

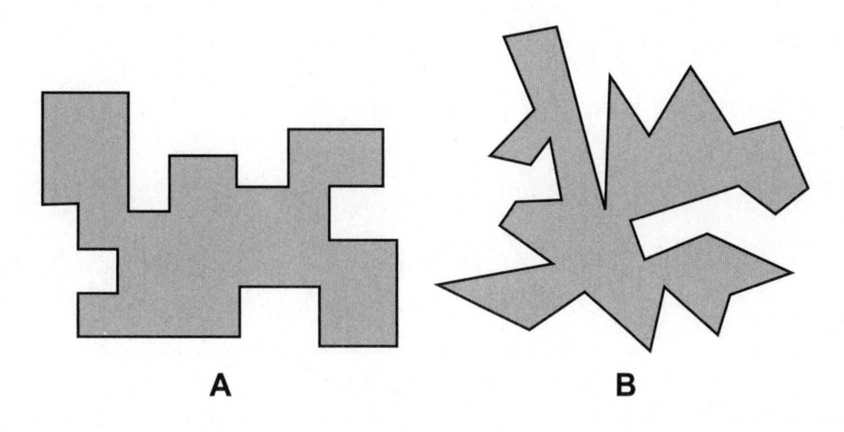

Figure 10-22 Illustration for Problem 10-3.

SOLUTION 10-3

Count the edges of each figure.

Ambiguity and "Ghosts"

Simple geometric drawings can baffle us when we lack a well-defined point of view. Such figures may seem to shimmer, rotate, pulsate, or be "haunted" by optical "ghosts."

EQUIVOCAL TETRAHEDRON

Figure 10-23 is a perspective drawing of a transparent *regular tetrahedron*. If we could actually handle the real 3D object, we would see that each of the four faces is an equilateral triangle. Most observers imagine that triangle *XYZ* is at the back of the object, farthest away, as if gazing down at the tetrahedron. But if we think of the object as floating in space above us rather than below us, triangle *XYZ* appears to be at the front, closest to us.

EQUIVOCAL CUBE

Figure 10-24A portrays an ambiguous transparent cube. When we think of the lower right-hand square as being closest to us, we get the impression of looking down at the cube. If we imagine the lower right-hand square as being farthest from us, we imagine ourselves as gazing up at the cube. If perspective and shading are added as shown in Fig. 10-24B, the ambiguity vanishes, and we are certain that the cube is below us with the light-gray face at the front.

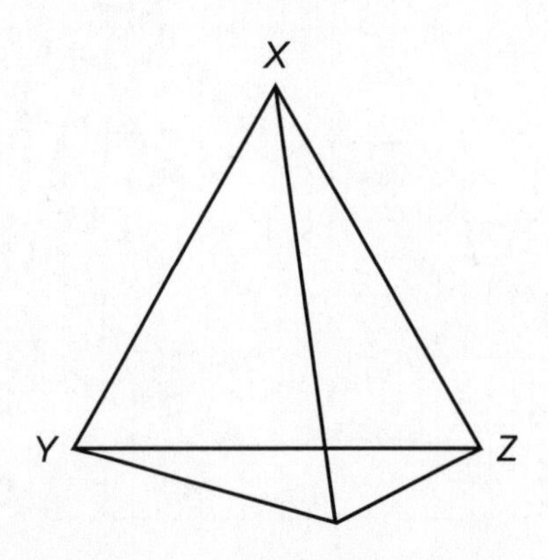

Figure 10-23 An equivocal tetrahedron.

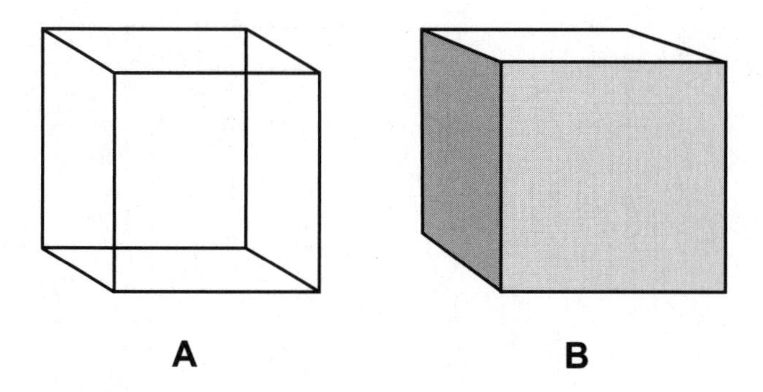

Figure 10-24 At A, an equivocal cube. At B, perspective and shading eliminate the ambiguity.

EQUIVOCAL STAIRWELL

Figure 10-25 illustrates another equivocal scene. Most observers imagine that they are looking down at a stairwell that has a sheer drop-off on the near side and a wall on the far side. But the "eye/brain team" can flip this drawing inside-out and upside-down, so that we see the stairwell as it would look from underneath, with walls on both the near and far sides. Whichever vantage point we assume, the stairs seem to descend as we move toward the right, and ascend as we move toward the left.

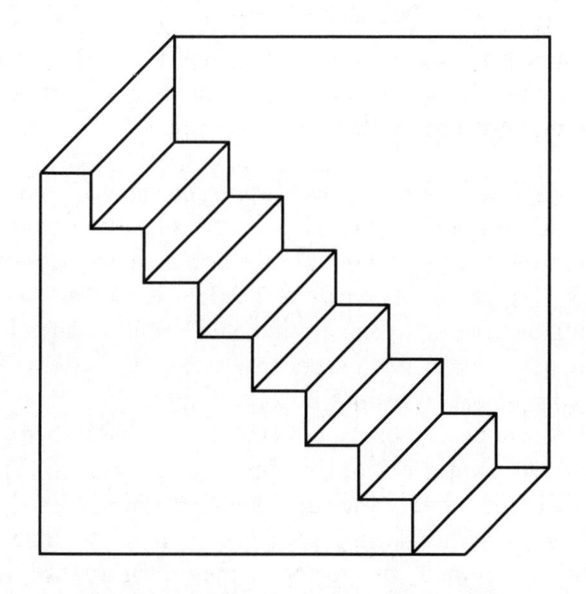

Figure 10-25 An equivocal stairwell.

Figure 10-26 Do you see two human facial profiles or a birdbath?

FACES OR BIRDBATH?

Examine Fig. 10-26. Some observers see a birdbath the first time they see this drawing. Then, given the suggestion, they see a pair of faces in profile. Other people see the faces first, and get the birdbath idea only after being told about that alternative. Which image do you see first? Do you suppose that your initial sense of this scene reveals anything about your temperament (whether you're a "people person" or an "ideas person," for example)?

AMBIGUITY OF REPETITION

Groups of closely spaced, identical objects can produce illusions of ambiguity. Figure 10-27A is an example. Which circle is the closest to us, the extreme left-hand one or the extreme right-hand one? In the first case, we imagine a transparent, corrugated length of pipe as viewed from the right front; in the second case, we seem to be looking into the same pipe from the left front. Once you have settled on one point of view or the other, it may suddenly switch. Can you make the transformation occur at will, or does it happen uncontrollably?

Figure 10-27B illustrates another set of ambiguous circles. Most people see this object from above, so point *X* is closer and point *Y* is farther away. In that case, a vector from *X* to *Y* seems to run upward, to the right, and generally away from us. With some effort, the "eye/brain team" can switch this set of circles around so that, going from point *X* to point *Y*, the vector seems to run upward, to the right, and generally toward us. In the first interpretation, the circle more nearly around point

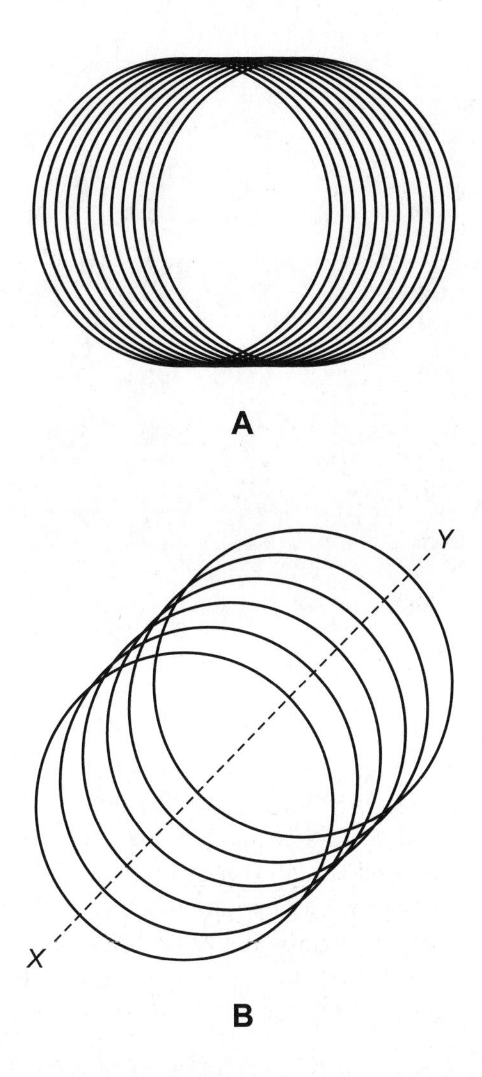

Figure 10-27 At A, a set of circles where the point of view can shift. Some observers see this figure as a spring. At B, another set of circles for which the vantage point is ambiguous.

X seems closest. In the second interpretation, the circle more nearly around point *Y* seems closest.

Some people initially think that Fig. 10-27A is a coil or spring, rather than a set of circles. This illusion is probably based on past experience with real objects. The circles are so close together that their true nature is difficult to resolve. In contrast, most observers can tell immediately that Fig. 10-27B is a set of circles, because the individual objects are far enough apart to prevent the illusion from taking place.

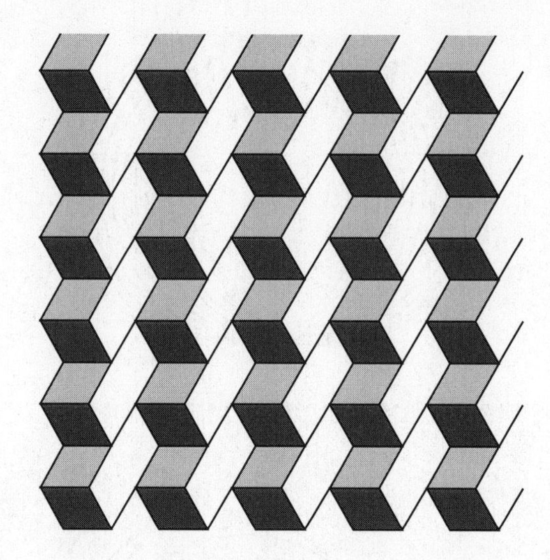

Figure 10-28 Stacked-beam illusion.

STACKED BEAMS

Figure 10-28 looks like a view of neatly stacked rods or beams. The individual beams have exposed ends, all tinted dark gray. Most observers think at first that the beams are stacked vertically, and that the dark ends are the bottoms. In the "real world," if the beams aren't packed tightly together, some of them might slide down and to the left! Other observers see the dark ends as if they are the tops of the beams; again, one or more of them could easily slide down and to the left. Still other observers imagine looking down on the scene as if the beams are stacked horizontally. In that case, the beams are not likely to slide.

SHIMMERING STAR

Some 2D figures give rise to uncontrollable visual artifacts. Figure 10-29 is a classic example, known as the *shimmering-star illusion*. The drawing is constructed from thin, black isosceles triangles with their apexes intersecting at a central point. The light and dark regions have identical angular measure. Some people see false color here. Nearly every observer sees transient "flux loops."

The vividness of the shimmering-star illusion varies with the distance from which you look at the drawing. It also depends on how much astigmatism (variation in focal length that occurs with changes in the polarization of incoming light waves) exists in the lenses of your eyes. Try holding the page approximately 15 cm (6 in) from the tip of your nose, and then gradually move it away until it is about 75 cm

Figure 10-29 Shimmering-star illusion.

(30 in) away. Try gazing at the page through both eyes, then through the left eye only, and then through the right eye only.

CONCENTRIC CIRCLES

Figure 10-30 is another "shimmering" illusion. Most people find it difficult to fix their gaze on a single point. If we look midway between the center and the periphery, we see shifting "flux zones," possibly associated with vague impressions of color. The vividness of this illusion, like that of the shimmering-star illusion, depends on the extent of astigmatism in the lenses of your eyes. If you have considerable astigmatism, you'll find these illusions more vivid than you will if your lenses are nearly perfect. For this reason, older people are more likely to see these illusions than younger people are. Some observers find that the use of cheap, nonprescription reading glasses enhances the effect by bringing the image into sharper focus. For other observers, the use of glasses or contact lenses interferes with or destroys the illusion.

Figure 10-30 Shimmering-circle illusion.

GRID OF DOTS

When we look at Fig. 10-31, our minds want to group the individual dots into rows, columns, diagonals, or squares, and other figures. Our eyes roam, and it's difficult to stop them. If we try to focus on a single dot near the center, the surrounding dots seem to move. Some observers have difficulty judging the distance to the grid. Hold the page about 15 cm (6 in) from the tip of your nose so the dots are aligned in perfectly horizontal rows and vertical columns. When you cross or uncross your eyes slightly to make pairs of adjacent columns line up, the distance illusion will suddenly spring into view. Don't continue to try this if you are unsuccessful after a few seconds. Don't stare at the dots uninterrupted for a long time either, unless you want to "see spots" for awhile and perhaps even get a headache.

SOLIDS AND SPACES

Figure 10-32A shows seven dark dots, all of equal diameter. The dark dots contrast against the white background. The distance between the outer edge of the center dot

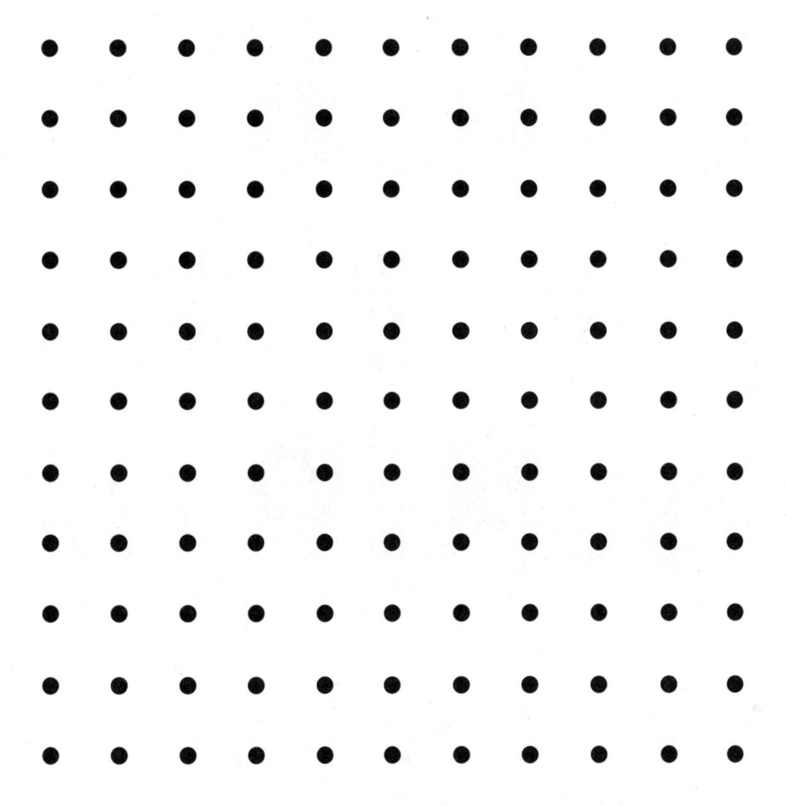

Figure 10-31 Grid-of-dots illusion.

and the outer edge of any surrounding dot (as measured along a straight line connecting the centers of the dots) is equal to the diameter of a single dot. Nevertheless, even when they know better, most observers get the sustained impression that the separation between the edges of adjacent dots is greater than the diameter of a single dot. The illusion operates to a lesser extent, if at all, when the dots are white and the background is dark, as shown in the "negative" image of Fig. 10-32B. The "eye/brain team" tends to minimize the sizes of dark, solid objects surrounded by light, open spaces.

PROBLEM 10-4

Figure 10-33 shows a grid of dark squares separated by narrow spaces. Alternatively, we can think of it as a single large, dark square divided by horizontal and vertical white lines into a set of small, dark squares. Either way, if you are like most observers, you see "gray ghosts" at the points where the corners of the small squares intersect—*except* for the corner you look directly at—when you observe the page in bright light. What might be responsible for this illusion?

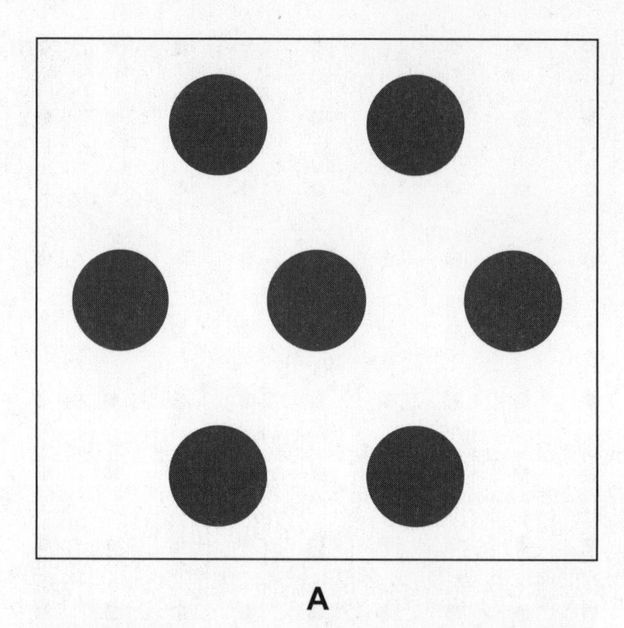

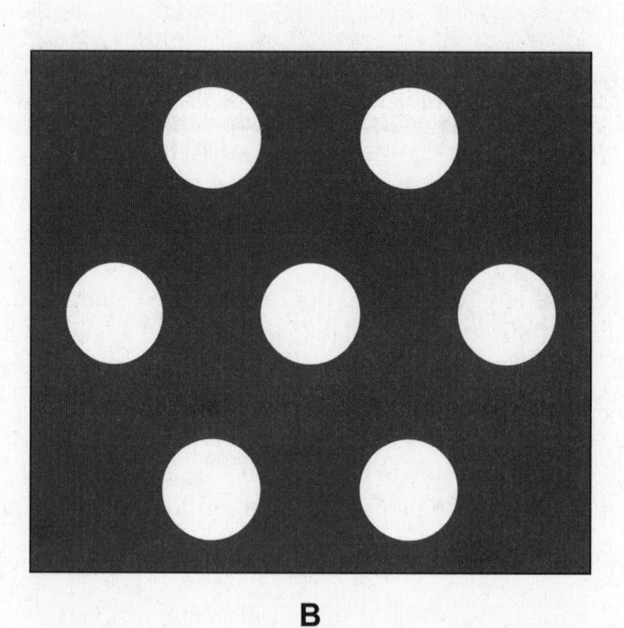

Figure 10-32 The spaces appear larger than the dot diameters, even though they are equal. Most people see the illusion at A more vividly than the one at B.

Figure 10-33 Illustration for Problem 10-4.

SOLUTION 10-4

We see "gray ghosts" in the regions away from the *fovea* (the spot on the retina representing the point at which our gaze is directed), but not at the fovea itself. The fovea has better image resolution than the other parts of the retina, but this localized advantage exists at the expense of sensitivity to dim light and changes in brightness. The sensitivity difference explains why faint stars, visible outside the fovea, disappear when we focus right on them. The sensitivity difference might also explain why we see "gray ghosts" in Fig. 10-33. The more light-sensitive parts of the retina, away from the fovea, apparently produce contrast-related illusions more easily than the less sensitive fovea does.

Quiz

Treat this quiz as a game. Select your answers based on what you initially see or believe, not on the basis of physical measurements, calculations, or geometric principles. Answers are in the back of the book. Don't worry about your score.

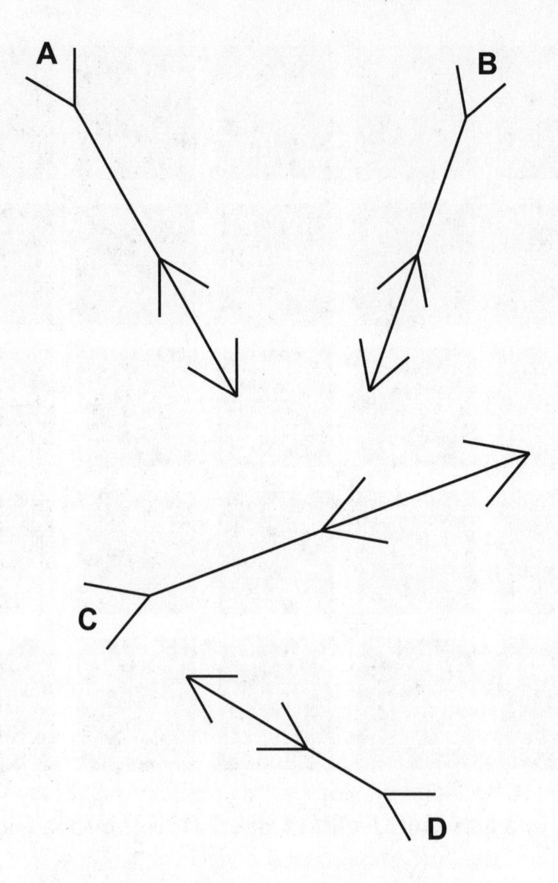

Figure 10-34 Illustration for Quiz Question 1.

1. The tip of the middle arrow is at the center of the long line segment in one of the objects in Fig. 10-34. Which one?
   (a) Object A
   (b) Object B
   (c) Object C
   (d) Object D
2. Which, if any, of the line segments above the shaded box in Fig. 10-35 lines up with the line segment below the shaded box?
   (a) None of them.
   (b) The leftmost upper line segment.
   (c) The middle upper line segment.
   (d) The rightmost upper line segment.

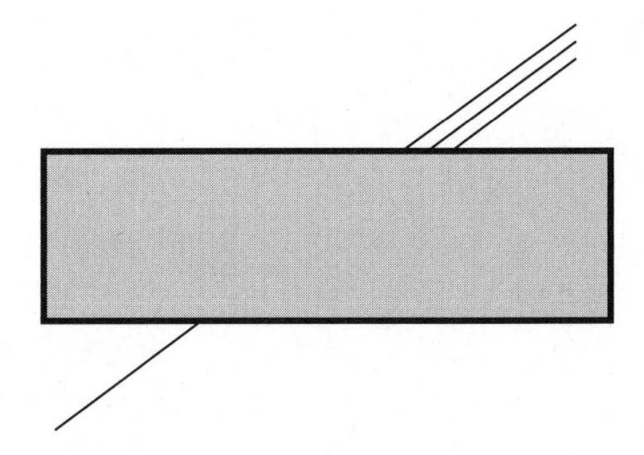

Figure 10-35 Illustration for Quiz Question 2.

3. Imagine the moon's image on the horizon as viewed from a flat plain or a calm sea. Approximately how many lunar disk images, placed side-by-side so that their outer edges just touch each other, would it take to go "full-circle" around the horizon?
    (a) 240
    (b) 360
    (c) 480
    (d) 680
4. Imagine that a rectangular strip of paper measuring 1 cm wide by 10 cm long is given a half-twist and then taped to itself end-to-end, forming a Möbius band. The surface area of the entire band is
    (a) 20 cm^2.
    (b) 10 cm^2.
    (c) 5 cm^2.
    (d) impossible to define.
5. Figure 10-36 shows a right circular cylinder. All of the coordinate axis increments are the same size. On that basis, we can conclude that the circumference of the cylinder is
    (a) twice its height.
    (b) greater than its height, but less than twice its height.
    (c) equal to its height.
    (d) less than its height.
6. Figure 10-37 shows two spheres. All of the coordinate axis increments are the same size. On that basis, we can conclude that the volume of the larger sphere is
    (a) less than twice the volume of the smaller one.
    (b) twice the volume of the smaller one.

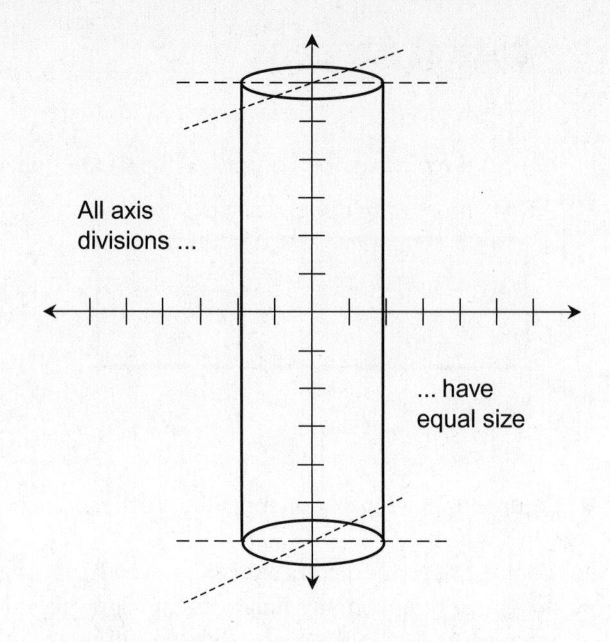

Figure 10-36 Illustration for Quiz Question 5.

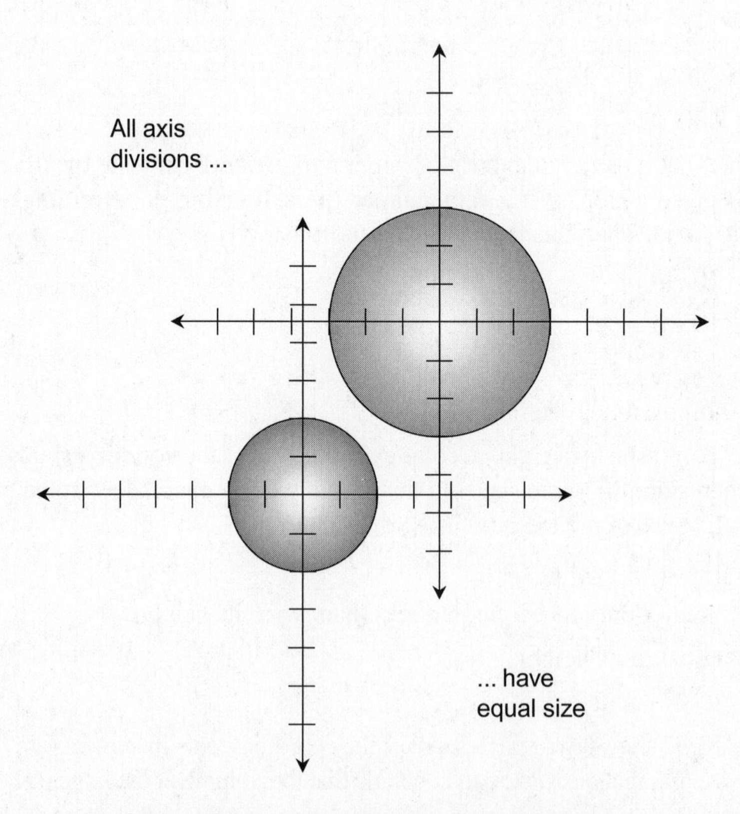

Figure 10-37 Illustration for Quiz Question 6.

(c) three times the volume of the smaller one.

(d) more than three times the volume of the smaller one.

7. Imagine the object in Fig. 10-38 as a transparent cube. The point marked *T* can represent either of two vertices: the one closest to us, or the one farthest away. (Our vantage point is such that these two vertices precisely line up.) Suppose that *T* is the vertex closest to us. In this situation, the vertices of the two vertical square faces oriented generally toward us are

   (a) *TVWX* and *TXYZ*.

   (b) *TVUZ* and *TWXY*.

   (c) *TUZY* and *TUVW*.

   (d) More than one of the above.

8. Keep looking at Fig. 10-38. Imagine that point *T* represents the vertex of the cube farthest away from us. In this case, the vertices of the top horizontal square face of the cube are

   (a) *TWXY*.

   (b) *TXYZ*.

   (c) *TVUZ*.

   (d) None of the above.

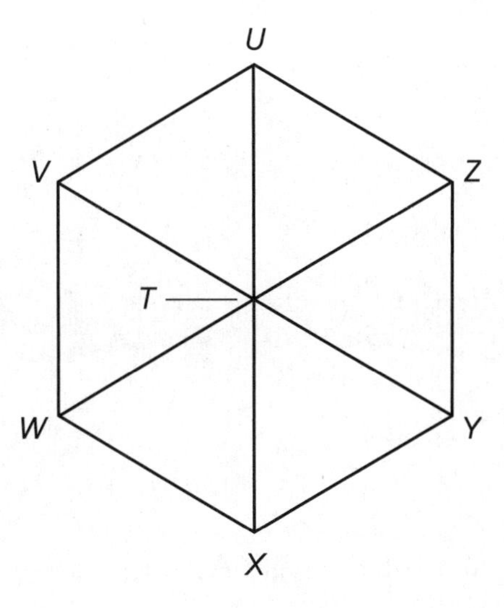

Figure 10-38 Illustration for Quiz Questions 7 and 8.

9. Based on the refractive phenomena implied by Fig. 10-39, assuming that the atmosphere is transparent and cloud-free, an observer on the surface of the planet would be able to see
   (a) less of outer space than she would see if there were no atmosphere.
   (b) the same amount of outer space as she would see if there were no atmosphere.
   (c) more of outer space than she would see if there were no atmosphere.
   (d) none of outer space.
10. In the scenario of Fig. 10-39, suppose that the index of refraction abruptly drops at the top of the atmosphere. What phenomenon might an observer on the surface see under such conditions?
   (a) Total internal reflection, producing an upside-down "false surface" near the horizon.
   (b) Total external reflection, causing the sky to appear pitch-black even in the daytime.
   (c) Negative refraction, reducing the amount of outer space that could be seen.
   (d) A red sky, caused by scattering of short-wavelength visible light.

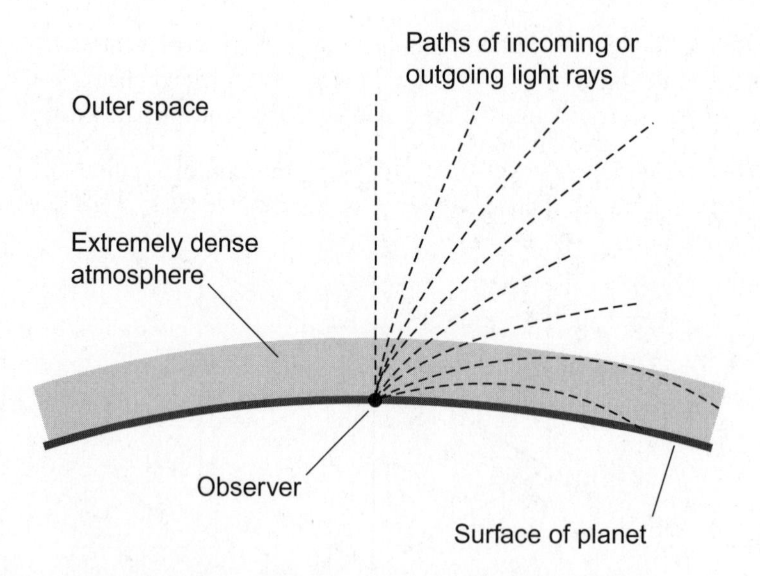

Figure 10-39 Illustration for Quiz Questions 9 and 10.

Final Exam

Do not refer to the text when taking this exam. A good score is at least 75 correct. Answers are in the back of the book. It's best to have a friend check your score the first time, so you won't memorize the answers if you want to take the exam again.

1. When we observe the sky on a cloudless day, the polarization of the light reaching our eyes depends on where in the sky we look. What causes this polarization?
   (a) Rayleigh scattering of sunlight as it passes through the atmosphere.
   (b) Ionization of atoms in the upper atmosphere by solar UV radiation.
   (c) Refraction of sunlight as it passes through the atmosphere.
   (d) Partial absorption of sunlight by the lower atmosphere.
   (e) The premise is wrong. The light from the sky on a clear day is not polarized at all.

2. A liquid-crystal display (LCD) contains a fluid whose light-transmitting and light-reflecting properties depend on
    (a) the intensity of applied visible light.
    (b) the intensity of applied UV.
    (c) the contrast ratio of the real image focused within it.
    (d) the intensity and orientation of an applied electric field.
    (e) the intensity of the output of an electron gun.
3. Which of the following is *not* an advantage of fiberoptic communications systems over conventional current-carrying cable systems?
    (a) Fiberoptic systems do not require repeaters, while conventional cable systems do.
    (b) Fiberoptic systems are less vulnerable than conventional cable systems to the destructive effects of an EM pulse.
    (c) Fiberoptic systems are less vulnerable than conventional cable systems to EM interference.
    (d) Fiberoptic systems do not suffer from crosstalk among the signals in the cable, while conventional systems do unless each wire has EM shielding.
    (e) Fiberoptic cables do not corrode as rapidly as conventional cables do.
4. Figure Exam-1 illustrates the Fizeau wheel apparatus for measuring the speed of light. In order to maximize the accuracy of the measurement, we should
    (a) make the wheel rotate as fast as possible.
    (b) ensure that the light source is monochromatic.
    (c) conduct the experiment in a dark environment, preferably at night.
    (d) make the distance d between the assembly and the mirror as large as possible.
    (e) use a laser diode as the light source.

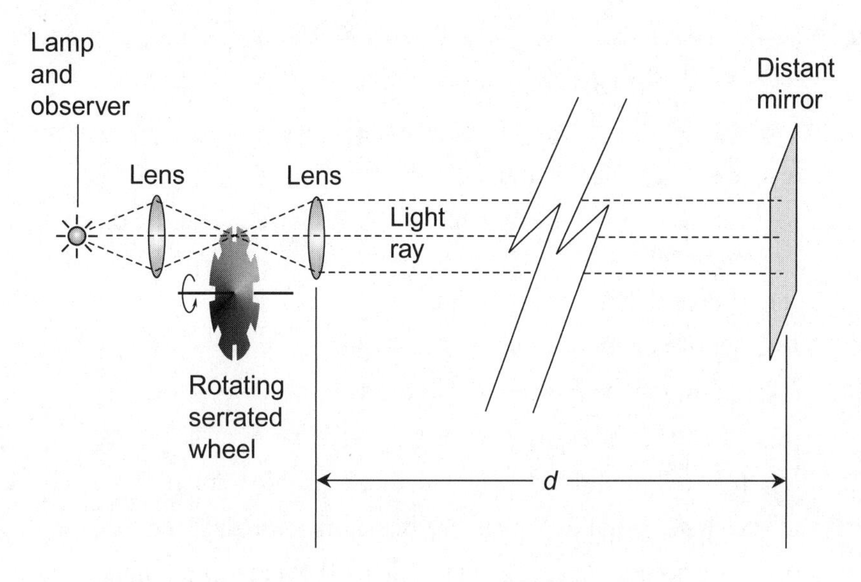

Figure Exam-1 Illustration for Final Exam Question 4.

5. Which of the following is a significant advantage of a laser device over a blowtorch for small-scale "outdoor" welding in space?
   - (a) Laser welders are less likely than blowtorches to malfunction in a weightless environment.
   - (b) It is impractical to carry the power supply for a blowtorch into outer space.
   - (c) Blowtorches are dangerous when operated in a low-pressure environment or in a vacuum.
   - (d) Laser welders do not require a supply of oxygen, which blowtorches do need, and which must be transported and stored in heavy, bulky tanks.
   - (e) The premise is wrong. A blowtorch works better than a laser for small-scale "outdoor" welding in space.
6. Three-dimensional optical range plotting is typically done in
   - (a) rectangular plane coordinates.
   - (b) polar plane coordinates.
   - (c) spherical spatial coordinates.
   - (d) rectangular spatial coordinates.
   - (e) curvilinear spatial coordinates.

7. Pigment is defined as a property that causes a surface or object to
   (a) absorb light at certain wavelengths.
   (b) radiate light at certain wavelengths.
   (c) transmit light at certain wavelengths.
   (d) appear brighter than it really is when viewed against a dark background.
   (e) appear tinted when it is actually white.
8. A laser interferometer is commonly used to
   (a) determine the temperature of a distant blackbody.
   (b) map the surface of the earth from orbiting satellites.
   (c) transmit modulated-light data through optical fibers.
   (d) convey digital data over vast distances through space.
   (e) measure tiny changes in the distance between two points.
9. Stimulated photon emission occurs in a gas laser because
   (a) the gas absorbs more energy than it receives.
   (b) a population inversion exists in the gas.
   (c) the atoms contain an excess of electrons in low-energy shells.
   (d) all of the electrons have been stripped from the atomic nuclei.
   (e) the gas fluoresces when an electrical current passes through it.
10. Imagine that you stand in front of a vertical, flat, wall-mounted mirror that extends all the way from the floor to the ceiling as shown in Fig. Exam-2. Suppose that you are 1.800 m tall. The distance x
    (a) is 0.4500 m.
    (b) is 0.9000 m.
    (c) is 1.273 m.
    (d) depends on how far you stand from the mirror.
    (e) depends on how far your eyes are from the top of your head.

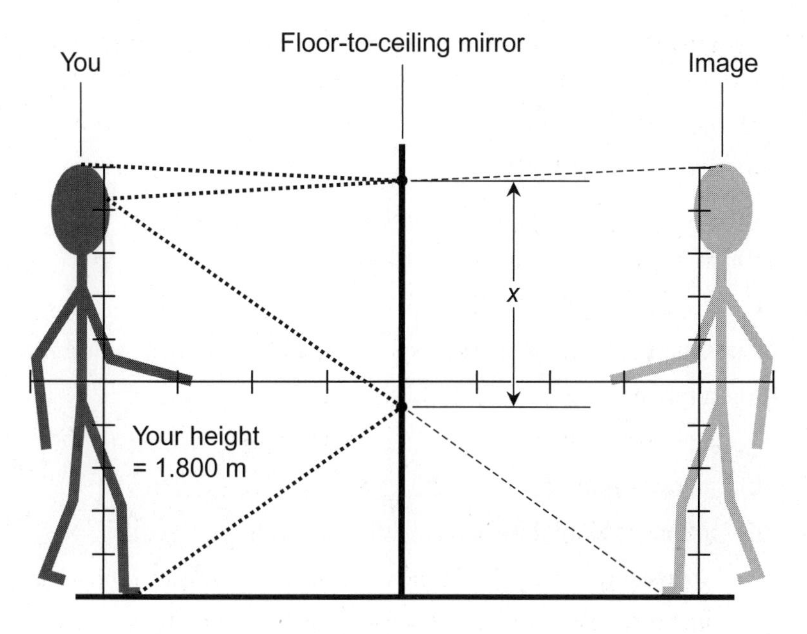

Figure Exam-2 Illustration for Final Exam Question 10.

11. Which of the following technologies can be useful for detecting the presence or measuring the concentrations of certain toxic chemicals in the air?
    (a) Laser interferometry
    (b) Laser range plotting
    (c) Laser gyroscopy
    (d) Laser holography
    (e) Laser spectroscopy

12. When a laser beam strikes a compact disc (CD) in a digital playback device or data reader,
    (a) all of the photons reflect from the surface as if from a mirror.
    (b) all of the photons scatter from the surface as if from a white sheet of paper.
    (c) some of the photons reflect from the surface as if from a mirror, while other photons scatter from the surface as if from a white sheet of paper.
    (d) some of the photons scatter from the surface as if from a white sheet of paper, while other photons are completely absorbed by the medium.
    (e) some of the photons reflect from the surface as if from a mirror, while other photons pass through the medium as if it were a pane of glass.

13. The intensity of the light emission from an LED depends on the forward current. As the current rises, the output brightness
    (a) increases indefinitely.
    (b) increases until the saturation point is reached.
    (c) increases until forward breakover occurs.
    (d) increases until avalanche breakdown occurs.
    (e) does not change at all.
14. The image produced by the paraboloidal mirror as shown in Fig. Exam-3 is
    (a) a real image.
    (b) a rendered image.
    (c) an optical image.
    (d) a transcendental image.
    (e) a virtual image.
15. In the situation of Fig. Exam-3, suppose that you initially stand at a distance of 7 m from the mirror. The mirror has a focal length of 1 m. You back away from the mirror a little, so you're a few centimeters farther from it, say 7.2 m. This movement causes the distance between the image and the mirror to
    (a) increase greatly.
    (b) increase slightly.
    (c) remain unchanged.
    (d) decrease slightly.
    (e) decrease greatly.

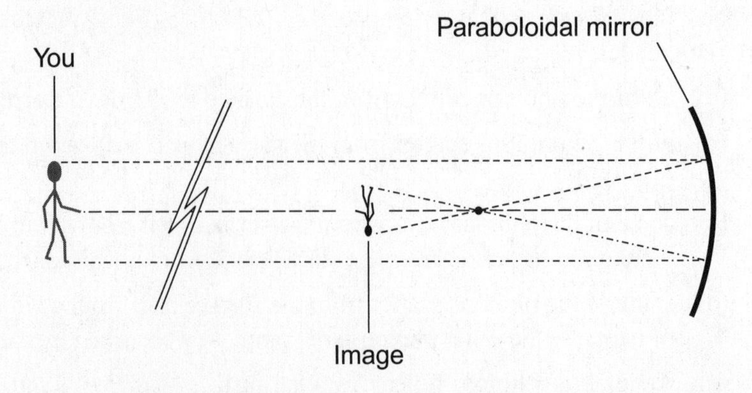

Figure Exam-3 Illustration for Final Exam Questions 14 and 15.

16. The speed of light in a vacuum, as seen from a nonaccelerating reference frame,
    (a) is approximately 3.00×10^8 km/s.
    (b) is approximately 3.00×10^5 km/s.
    (c) is approximately 1.86×10^8 km/s.
    (d) is approximately 1.86×10^5 km/s.
    (e) depends on the speed of the observer relative to the source.
17. Suppose that the peak output power from a neodymium-glass laser is 800 GW. If the average output power is 48.0 W and we assume that the output pulses are rectangular, what is the duty cycle expressed as a percentage?
    (a) 3.84×10^{-11} percent
    (b) 6.00×10^{-9} percent
    (c) 1.67×10^{-9} percent
    (d) 1.67×10^{-7} percent
    (e) We need more information to answer this question.
18. If the average input power to the device described in Question 18 is 960 watts, what is the efficiency expressed as a percentage?
    (a) 2.20 percent
    (b) 3.33 percent
    (c) 5.00 percent
    (d) 12.0 percent
    (e) We need more information to answer this question.
19. Consider the scenario shown by Fig. Exam-4, in which a ray of light passes from amber into air. The formula relating the angles and refractive indices is

$$\sin \phi / \sin \theta = r/s$$

Amber has refractive index $r = 1.55$, and air has refractive index $s = 1.00$. Suppose that the angle of incidence θ at the boundary is equal to 20°. On the basis of this information, the angle of refraction ϕ is
    (a) 24°.
    (b) 32°.
    (c) 48°.
    (d) 64°.
    (e) nonexistent, because total internal reflection occurs.

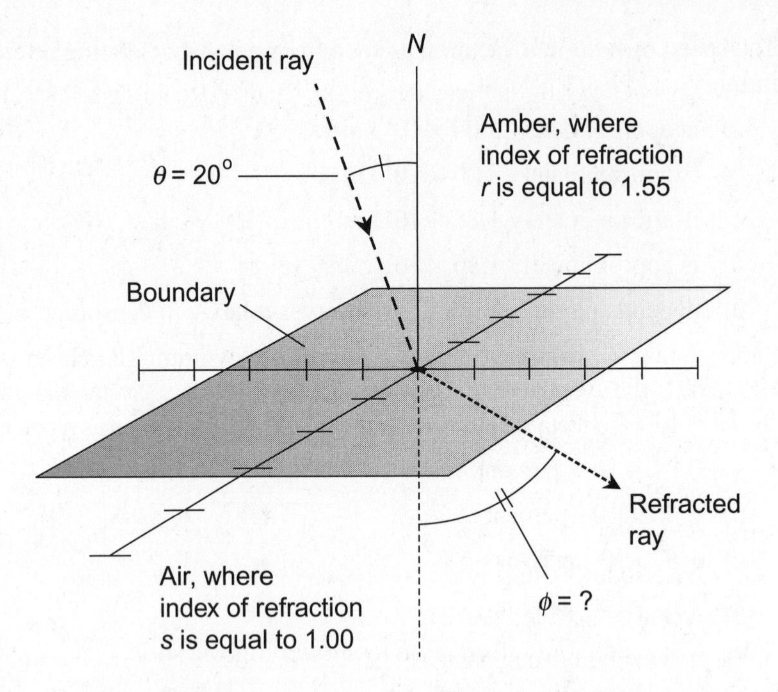

Figure Exam-4 Illustration for Final Exam Question 19.

20. In industry, lasers are commonly used for all of the following applications, *except*
    (a) examining solder joints.
    (b) examining textile samples.
    (c) drilling holes in rubber.
    (d) cutting through plastics.
    (e) curing meat, fish, and poultry.
21. The blue appearance of the sky on a clear day is the result of
    (a) Rayleigh scattering of visible light by air molecules.
    (b) electromagnetic refraction caused by water vapor in the air.
    (c) color dispersion produced by the variation of air density with altitude.
    (d) wave interference produced as light reflects from air molecules.
    (e) the fluorescing of ionized atoms in the upper atmosphere.

22. Relativistic time dilation can be explained by the fact that
    (a) the speed of light always appears the same from any nonaccelerating point of view.
    (b) the speed of light appears to decrease as the speed of the observer increases.
    (c) the speed of light appears to increase as the speed of the observer increases.
    (d) time flows smoothly and continuously, and the speed of light depends on the rate of its flow.
    (e) it is an optical illusion produced when objects move at extreme speeds.
23. Figure Exam-5 is a functional diagram of
    (a) an image orthicon.
    (b) a vidicon.
    (c) a cathode-ray tube.
    (d) a charge-coupled device.
    (e) a holographic camera tube.

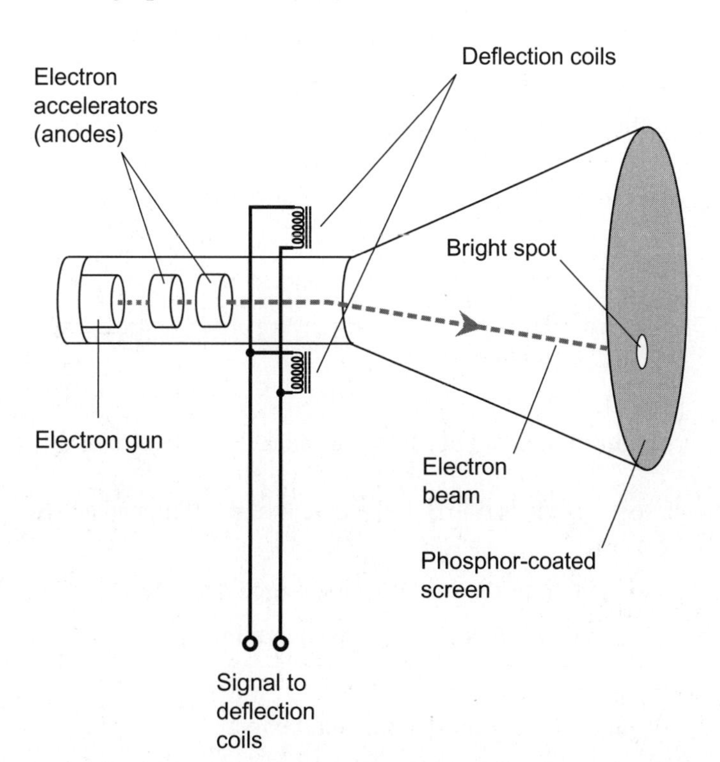

Figure Exam-5 Illustration for Final Exam Question 23.

24. What's the fundamental difference between stereoscopic photography and holography?
    - (a) There is no difference. The two terms refer to the same technology.
    - (b) A hologram requires the use of transparencies, while a stereoscopic photograph can be rendered on photographic paper.
    - (c) A hologram can be made with an ordinary source of visible light, but a stereoscopic photograph requires a laser.
    - (d) Holography requires two separate photographic exposures, but stereoscopic photography requires only one.
    - (e) The appearance of a hologram changes as the observer moves, but the appearance of a stereoscopic photograph does not.
25. Figure Exam-6 is a simplified diagram of a Galilean refracting telescope, also known as a "spy glass." Suppose that the telescope is aimed at a distant object, and that object appears in good focus. If the distance between the objective and the eyepiece increases, then
    - (a) the distant object will get out of focus.
    - (b) the magnification factor will decrease.
    - (c) the magnification factor will increase.
    - (d) the distant object will appear brighter.
    - (e) nothing will change at all.

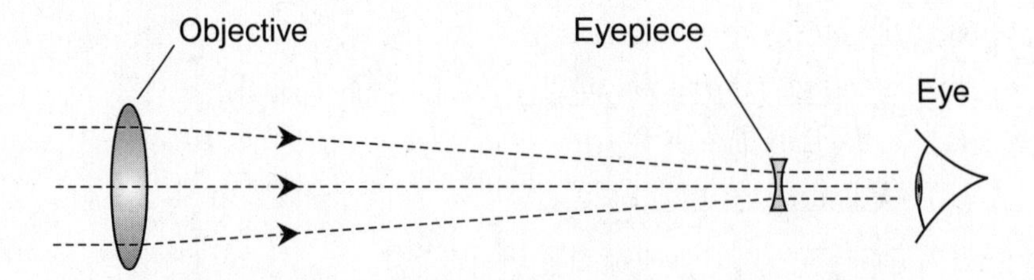

Figure Exam-6 Illustration for Final Exam Question 25.

26. When a scene is viewed in 3D, the extent of parallax and eclipsing effects increase as the
    - (a) wavelength of the illumination source decreases.
    - (b) brightness of the illumination source increases.
    - (c) distance between the left and right viewpoints increases.
    - (d) distance to the illumination source increases.
    - (e) number of objects in the scene decreases.

27. When all other factors are equal, the kinetic energy of a photon is *directly proportional* to its
    (a) period.
    (b) frequency.
    (c) wavelength.
    (d) speed.
    (e) radius.
28. Suppose you observe a source of nonpolarized light through a pair of polarizing lenses, placing one lens in front of the other and then rotating one of them. The maximum amount of light will pass through the pair of lenses when the angle between their polarization axes is
    (a) 90°.
    (b) 60°.
    (c) 45°.
    (d) 30°.
    (e) 0°.
29. Suppose a Keplerian refracting telescope has an objective with a focal length of 1000 mm and a diameter of 60 mm. An eyepiece with a focal length of 20 mm is used initially. If that eyepiece is taken out and replaced with one having a focal length of 10 mm, the telescope's *magnification*
    (a) doubles.
    (b) increases by a factor of $2^{1/2}$.
    (c) stays the same.
    (d) decreases by a factor of $2^{1/2}$.
    (e) becomes half as great.

30. Figure Exam-7 illustrates a Crookes radiometer. Why do the vanes rotate with the white sides leading when this device is exposed to visible light?
    (a) Incident photons exert greater radiation pressure on the dark faces than on the white faces, because the collisions of photons with the dark faces are inelastic while the collisions with the white faces are elastic.
    (b) Electromagnetic waves vibrate at a higher frequency against the dark faces than the white faces, so the air molecules near the dark faces impart more kinetic energy to the vanes than the air molecules near the white faces.
    (c) More of the incident photon momentum is transferred to the dark faces than is transferred to the white faces, because the dark faces absorb most of the photons while the white faces reflect most of them.
    (d) The dark faces become warmer than the white faces, so the rarefied air molecules near the dark faces impart more kinetic energy to the vanes than the molecules near the white faces.
    (e) The premise stated in the question is wrong. The vanes actually rotate with the dark sides leading when a Crookes radiometer is exposed to visible light.

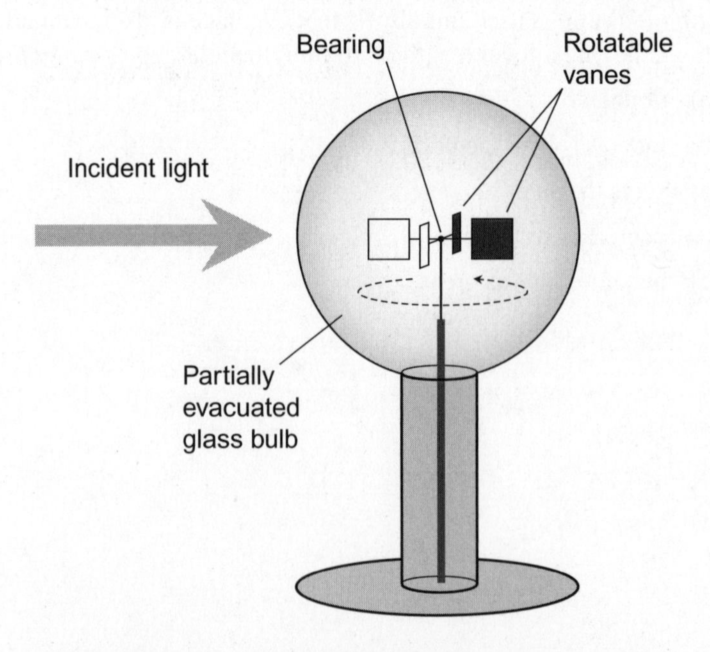

Figure Exam-7 Illustration for Final Exam Question 30.

31. Suppose a compound microscope employs an objective lens with a magnification factor of 36× and an eyepiece with a magnification factor of 4×. The magnification factor of the entire instrument, assuming that it is operated in an air medium, is
    (a) 9×.
    (b) 12×.
    (c) 40×.
    (d) 144×.
    (e) 576×.
32. The image information contained in a hologram becomes more complete and detailed as
    (a) the number of rendered points of view increases.
    (b) the number of illumination sources increases.
    (c) the wavelengths of the illuminating light sources increase.
    (d) the distance between the left and right viewpoints increases.
    (e) All of the above
33. The spectral temperature of a distant star can be estimated by examining the intensity of its radiant EM energy as a function of wavelength, assuming that
    (a) most of its radiant energy occurs in the visible region.
    (b) it does not produce emission lines.
    (c) cosmic background radiation is absent.
    (d) it behaves as a blackbody.
    (e) it has not begun to run out of nuclear fuel.
34. Figure Exam-8 shows an example of
    (a) dispersion.
    (b) Rayleigh scattering.
    (c) diffraction.
    (d) Mie scattering.
    (e) heterodyning.

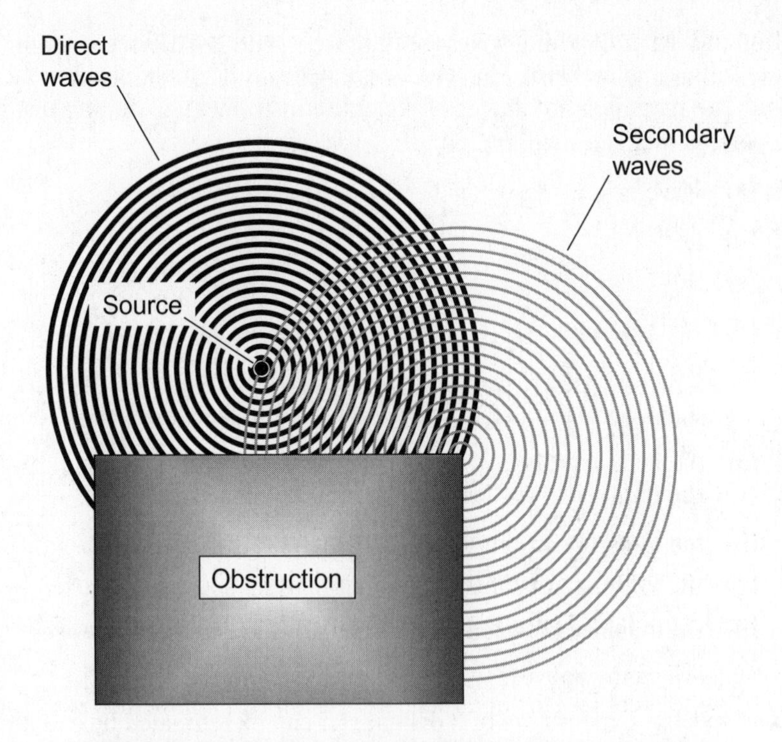

Figure Exam-8 Illustration for Final Exam Question 34.

35. High-energy X rays cannot be focused with conventional glass lenses. Why?
    (a) Glass is essentially opaque to high-energy X rays.
    (b) High-energy X rays pass straight through glass lenses without refracting.
    (c) Glass lenses tend to spread high-energy X rays rather than bringing them to a focus.
    (d) Excessive wavelength dispersion occurs, blurring the image beyond recognition.
    (e) The premise is incorrect. High-energy X rays can be focused with conventional glass lenses.
36. Both of the mirrors in a helium-neon laser are always
    (a) concave.
    (b) convex.
    (c) flat.
    (d) first-surface types.
    (e) 100 percent reflective.

37. Complete the following sentence to make it true: "When a hologram is made, parallax information is recorded because of differences in the ________ among individual light rays as they strike the film at various distances from a given point on the scene."
    (a) wavelength
    (b) phase
    (c) brightness
    (d) frequency
    (e) polarization
38. In a convex lens, astigmatism is a condition in which
    (a) the focal length varies with the wavelength of the light passing through.
    (b) the transparency varies with the wavelength of the light passing through.
    (c) the polarization varies with the wavelength of the light passing through.
    (d) the focal length varies with the polarization of the light passing through.
    (e) the transparency varies with the polarization of the light passing through.
39. As a ray of monochromatic light passes from air into glass or vice versa, which of the following parameters are *both* affected?
    (a) The frequency and the period
    (b) The wavelength and the period
    (c) The frequency and the speed of propagation
    (d) The wavelength and the speed of propagation
    (e) The wavelength and the frequency
40. Figure Exam-9 shows an example of
    (a) dispersion.
    (b) Rayleigh scattering.
    (c) diffraction.
    (d) Mie scattering.
    (e) heterodyning.

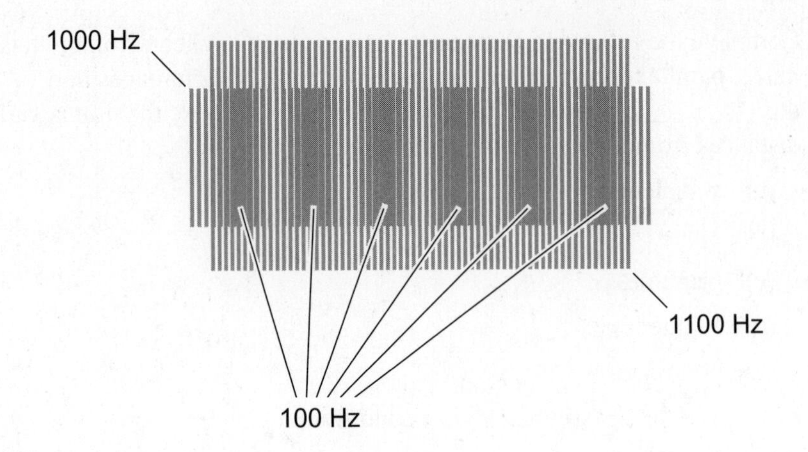

Figure Exam-9 Illustration for Final Exam Question 40.

41. When a ray of white visible light passes through a diffraction grating having dark lines spaced 750 nm apart, the effect is similar to that produced by
    (a) a tiny circular aperture.
    (b) a concave lens.
    (c) a convex lens.
    (d) a glass prism.
    (e) the surface of a calm pool of water.
42. The interior walls of the large intestine can be examined using
    (a) an endoscope.
    (b) a spectrophotometer.
    (c) a free-electron laser.
    (d) a Super-Kamiokande.
    (e) a Cerenkov detector.
43. Figure Exam-10 illustrates a bound electron whose principal quantum number changes from $n = 2$ to $n = 3$. At the moment the "jump" occurs, the electron *might*
    (a) emit a photon.
    (b) become more negatively charged.
    (c) absorb a photon.
    (d) become less negatively charged.
    (e) become more massive.

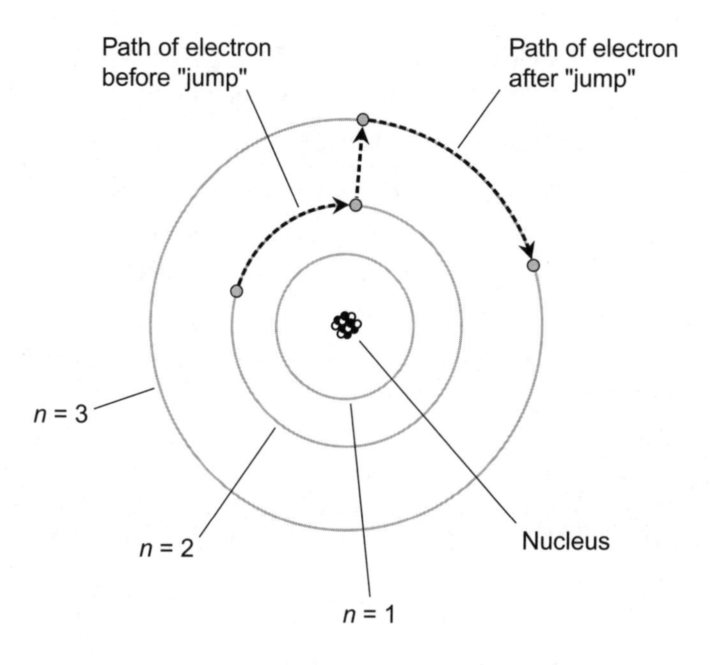

Figure Exam-10 Illustration for Final Exam Question 43.

44. Why must equipment intended for the detection of celestial neutrinos be placed in the earth's orbit or in outer space?
    (a) The atmosphere emits neutrinos that would otherwise interfere with the observation of those that come from space.
    (b) The atmosphere is opaque to neutrinos.
    (c) Detectors must be removed from the planet to prevent neutrinos of terrestrial origin from interfering with the observation of those that come from space.
    (d) Neutrino detectors function properly only when operated in a weightless environment.
    (e) The premise is wrong. Equipment intended for the detection of celestial neutrinos is usually placed deep underground.
45. In simple glass lenses, chromatic aberration is caused by
    (a) excessively long focal lengths.
    (b) imperfect surface curvature.
    (c) total internal reflection.
    (d) color dispersion.
    (e) excessive incident light intensity.

46. Why are some physicians in the United States wary of laser acupuncture?
    (a) The treatment carries a high risk of infection.
    (b) The lasers cause intense pain and scarring.
    (c) The field is susceptible to exploitation by quacks.
    (d) The equipment is expensive, bulky, and massive.
    (e) The premise is wrong. Most physicians in the United States have fully accepted laser acupuncture.
47. The momentum of a photon traveling through a vacuum is *inversely proportional* to its
    (a) frequency.
    (b) kinetic energy.
    (c) wavelength.
    (d) speed.
    (e) radius.
48. If an object is placed closer to a converging lens than the focal length, a real image
    (a) forms at a distance less than the focal length, but on the opposite side of the lens.
    (b) forms at a distance less than the focal length, but on the same side of the lens.
    (c) forms at a distance greater than the focal length, but on the opposite side of the lens.
    (d) forms at a distance greater than the focal length, but on the same side of the lens.
    (e) does not form.
49. In dermatology, lasers have been used to treat
    (a) port wine stains.
    (b) diverticulosis.
    (c) arterial plaques.
    (d) coronary occlusions.
    (e) papilloma.

50. A cavity laser consists of a resonant lasing medium
    (a) with a totally silvered mirror at one end and a partially silvered mirror at the other end.
    (b) with all of the gas evacuated so the electrons can propagate freely.
    (c) surrounded by a coil of current-carrying wire that produces a powerful magnetic field inside the cavity.
    (d) surrounded by an array of permanent magnets that produce a powerful magnetic field inside the cavity.
    (e) with a positively charged electrode at one end and a negatively charged electrode at the other end.
51. In a modulated-light receiver designed for free-space operation, the photodiode should not be exposed to a bright external source of illumination, because the stray light might
    (a) reverse-bias the photodiode.
    (b) render the photodiode insensitive to the desired signal.
    (c) cause a destructive current to flow through the P-N junction.
    (d) cause the photodiode to undergo forward breakover.
    (e) distort the wavelength response.
52. Solar radiation can cause ionization of the atoms in the earth's upper atmosphere. This effect is largely responsible for the fact that
    (a) the sky appears blue on a clear day but nearly black on a clear night.
    (b) worldwide communication is possible at certain radio frequencies.
    (c) the sun and moon sometimes appear red when rising or setting.
    (d) the sun and moon sometimes appear oblate ("flattened") when rising or setting.
    (e) most of the deadly UV radiation from the sun fails to reach the earth's surface.
53. Figure Exam-11 is a functional diagram of a specialized type of
    (a) injection laser.
    (b) cavity laser.
    (c) gas laser.
    (d) neodymium laser.
    (e) quartz laser.

54. "Component X" in Fig. Exam-11 ensures that
    (a) the voltage across the P-N junction exceeds the avalanche level.
    (b) the current through the P-N junction is sufficient to produce lasing.
    (c) the battery can be easily recharged by an external source.
    (d) the lasing medium can be optically pumped without overheating the quantum well.
    (e) the device is not forced to carry too much current.

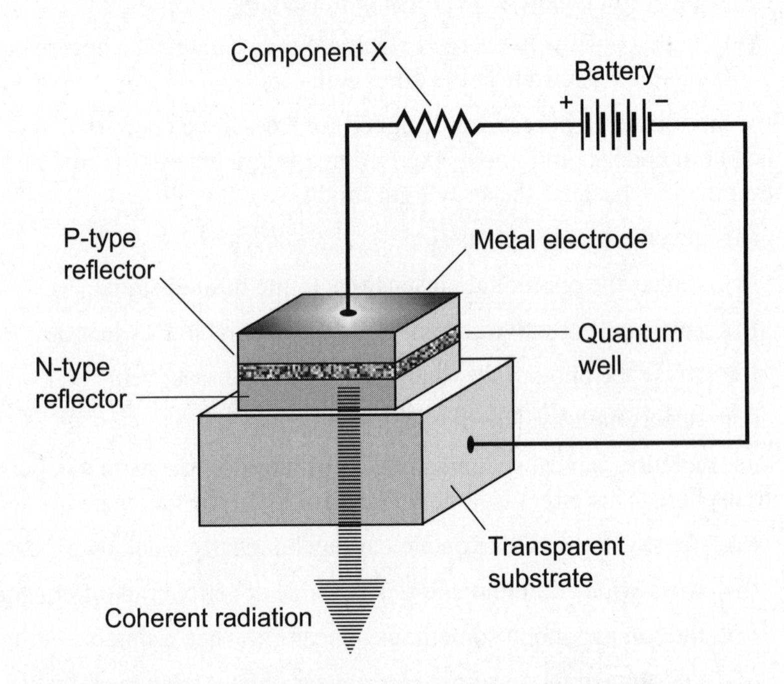

Figure Exam-11 Illustration for Final Exam Questions 53 and 54.

55. Some ear infections can be treated using a laser by
    (a) irradiating the inner ear with coherent light, killing the bacteria.
    (b) injecting special dye into the middle ear, and then using an endoscope to direct a small laser device into the region where the dyed tissue can be heated.
    (c) destroying the cochlea, which typically harbors most of the bacteria.
    (d) making a tiny hole in the eardrum, allowing fluid from the middle ear to drain.
    (e) sterilizing the ear canal and the outer surface of the eardrum.

56. Which of the following optical processes, phenomena, or technologies has *not* been used to treat disorders of the eye?
    (a) Photodisruption
    (b) Photocoagulation
    (c) Photoradiation
    (d) Photovaporization
    (e) Photoconduction
57. Which of the following could be an advantage of a laser weapon over an ordinary bomb for destroying targets with precision?
    (a) Lasers would likely cause less collateral damage than conventional explosives.
    (b) Even a relatively dim laser can exert significant radiation pressure.
    (c) Lasers cost less than conventional bombs.
    (d) Lasers can work even when fog or smoke makes conventional bombing impractical.
    (e) Lasers can inflict more widespread destruction than ordinary bombs.
58. Which of the curves in Fig. Exam-12 is typical of a photodiode that is reverse biased at a constant voltage?
    (a) Curve R
    (b) Curve S
    (c) Curve T
    (d) Curve U
    (e) Curve V

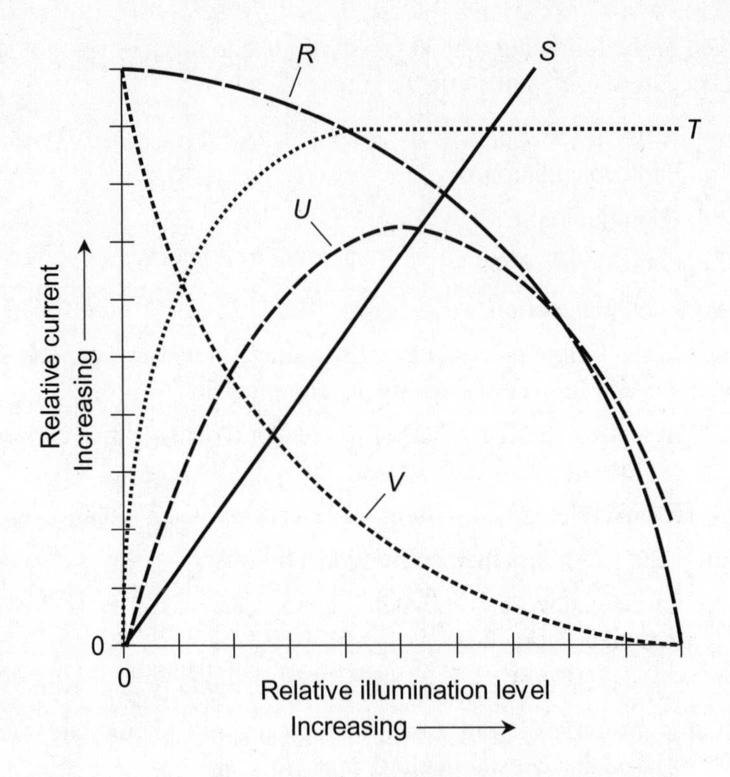

Figure Exam-12 Illustration for Final Exam Question 58.

59. Suppose that you mount a flat sheet of photographic film directly on a wall, with no lenses or mirrors to focus light on it. The rendition of the surrounding environment on the film is
    (a) a real image.
    (b) a rendered image.
    (c) an optical image.
    (d) a transcendental image.
    (e) a virtual image.

60. How does the force produced by the solar wind differ from the force produced by true solar radiation pressure?
    - (a) It doesn't. The terms "solar wind" and "true solar radiation pressure" refer to exactly the same phenomenon.
    - (b) The solar wind consists of particles with definable rest mass moving at less than the speed of light, while true solar radiation pressure is caused by transfer of momentum from particles having undefinable rest mass and moving at the speed of light.
    - (c) The solar wind consists of particles with undefinable rest mass moving at the speed of light, while true solar radiation pressure is caused by transfer of momentum from particles having definable rest mass and moving at less than the speed of light.
    - (d) The solar wind consists of particles with undefinable rest mass moving at less than the speed of light, while true solar radiation pressure is caused by transfer of momentum from particles having definable rest mass and moving at the speed of light.
    - (e) The solar wind consists of particles with definable rest mass moving at the speed of light, while true solar radiation pressure is caused by transfer of momentum from particles having undefinable rest mass and moving at less than the speed of light.
61. Figure Exam-13 is a functional diagram of
    - (a) an optical shaft encoder.
    - (b) an optoisolator.
    - (c) an electric eye.
    - (d) a modulated-light data link.
    - (e) an epipolar ranging system.

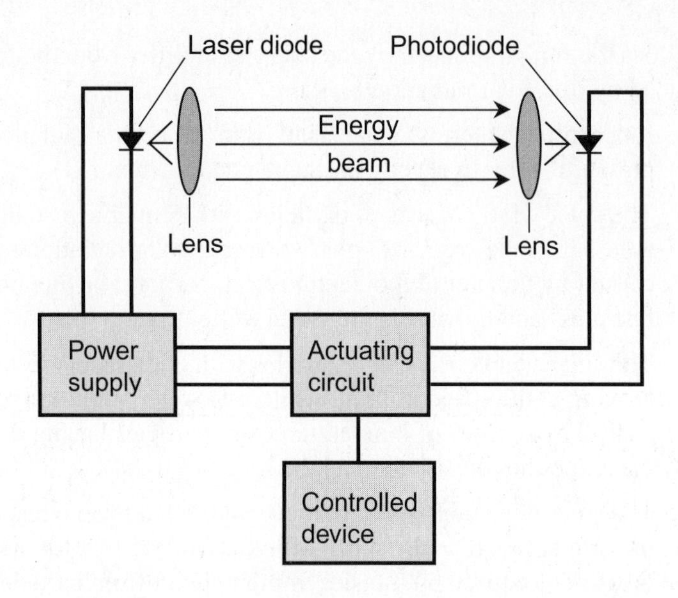

Figure Exam-13 Illustration for Final Exam Question 61.

62. A real image can be produced without a lens or mirror to bring the light to a focus by
    (a) combining red, blue, and green colors in equal intensities and allowing it to shine on a screen or film.
    (b) combining cyan, magenta, and yellow pigments in equal concentrations and allowing light to reflect from the mixture.
    (c) allowing two light beams having different wavelengths to strike a single spot on a screen or film.
    (d) punching a tiny hole in an opaque barrier and allowing the light from a scene to shine through it and strike a screen or film.
    (e) no known means.
63. Figure Exam-14 is a simplified functional diagram of a 2D optical triangulation system for a mobile robot. The boxes marked *X* represent
    (a) prisms.
    (b) spherical convex mirrors.
    (c) spherical concave mirrors.
    (d) flat mirrors.
    (e) tricorner reflectors.

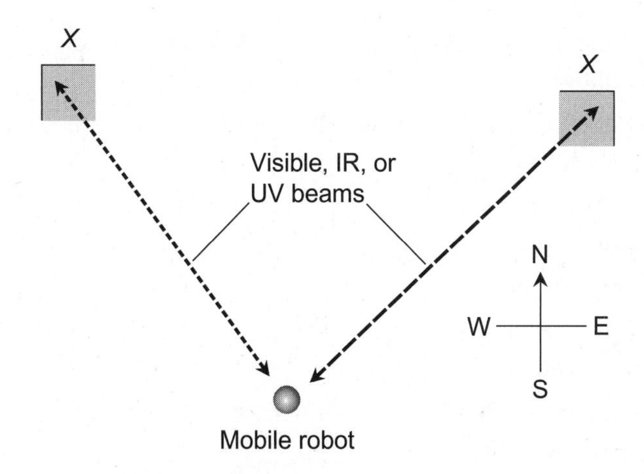

Figure Exam-14 Illustration for Final Exam Question 63.

64. Suppose a Keplerian refracting telescope has an objective with a focal length of 1000 mm and a diameter of 60 mm. An eyepiece with a focal length of 20 mm is used initially. If that eyepiece is taken out and replaced with one having a focal length of 10 mm, the telescope's *light-gathering area*
    (a) increases by a factor of 4.
    (b) doubles.
    (c) stays the same.
    (d) becomes half as great.
    (e) becomes 1/4 as great.
65. The path of a light ray can be curved or bent by
    (a) a change in the index of refraction of the medium through which the ray travels.
    (b) extreme acceleration.
    (c) extreme gravitation.
    (d) the use of flexible optical fibers to carry the light.
    (e) Any of the above.
66. Glass is difficult to drill through with a laser primarily because glass
    (a) does not melt or vaporize.
    (b) transmits most of the laser energy.
    (c) is extremely hard.
    (d) is extremely brittle.
    (e) crystallizes when heated.

67. Which of the following is a significant difference between a photoFET and a bipolar phototransistor?
    (a) A photoFET has a translucent package, while a phototransistor has an opaque package.
    (b) A photoFET is a PNP device, while a phototransistor is an NPN device.
    (c) A photoFET responds to shorter wavelengths than a bipolar phototransistor does.
    (d) A photoFET has higher input impedance than a bipolar phototransistor does.
    (e) A photoFET has lower input resistance than a bipolar phototransistor does.
68. Suppose that a source of 1.0×10^4 V DC is connected to a simple, nonlasing gas-discharge tube, ionizing the gas and causing a current of 0.0050 A to flow. How much power is dissipated in the gas?
    (a) 2.0×10^6 watts
    (b) 5.0×10^4 watts
    (c) 2.0×10^3 watts
    (d) 50 watts
    (e) We need more information to answer this question.
69. Figure Exam-15 illustrates an arrangement that can be used to measure the distance to a tricorner reflector. Assuming that the pulsed laser, the sensors, and the oscilloscope are located in close proximity so their separation does not affect the apparent pulse delay of 6.00 ms, what is the distance *d* between the measuring apparatus and the tricorner reflector?
    (a) 20.0 km
    (b) 90 km
    (c) 200 km
    (d) 600 km
    (e) 900 km

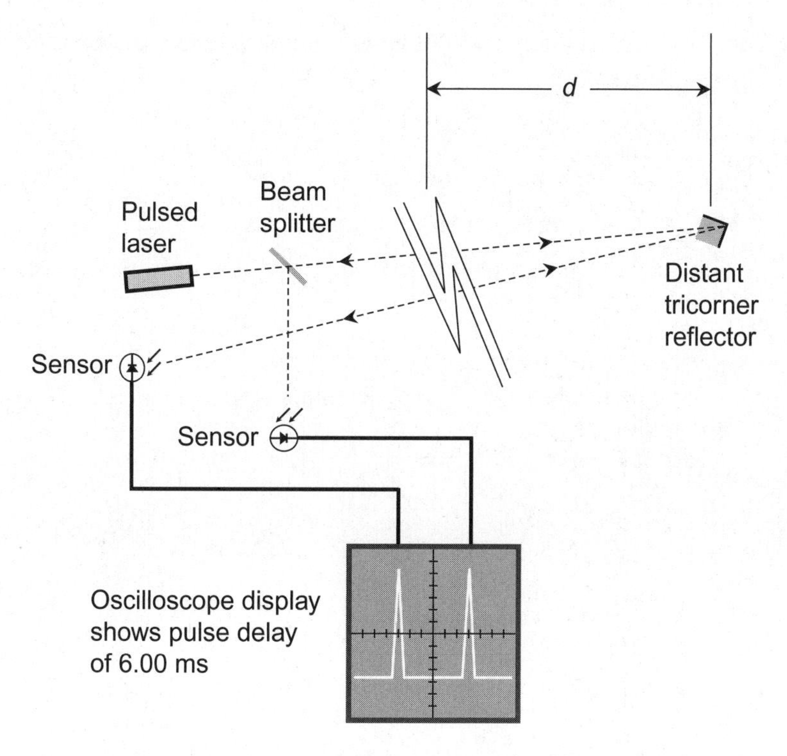

Figure Exam-15 Illustration for Final Exam Question 69.

70. A He-Ne laser produces output in the visible red range, giving it a decided advantage over shorter-wavelength sources for use in
    (a) line-of-sight modulated-light communications in the earth's lower atmosphere.
    (b) optical shaft encoders designed for control of digital systems.
    (c) epipolar navigation systems intended for use in outer space.
    (d) long-distance optical ranging systems and radar devices.
    (e) electric eyes intended for remote control of robotic space vessels.
71. The duty cycle of a pulsed laser is
    (a) the proportion or percentage of the total output energy that exists as coherent waves.
    (b) the ratio of the input power to the output power.
    (c) the proportion or percentage of the input power that contributes to lasing.
    (d) the ratio of the applied voltage to the current drawn.
    (e) the proportion or percentage of the time that the laser is active.

72. Figure Exam-16 is a functional diagram of a device that can be used to
    (a) make holograms.
    (b) detect rotational motion.
    (c) calibrate Newtonian reflecting telescopes.
    (d) measure the distances to celestial objects.
    (e) evaluate the spectral characteristics of the light from interstellar nebulae.

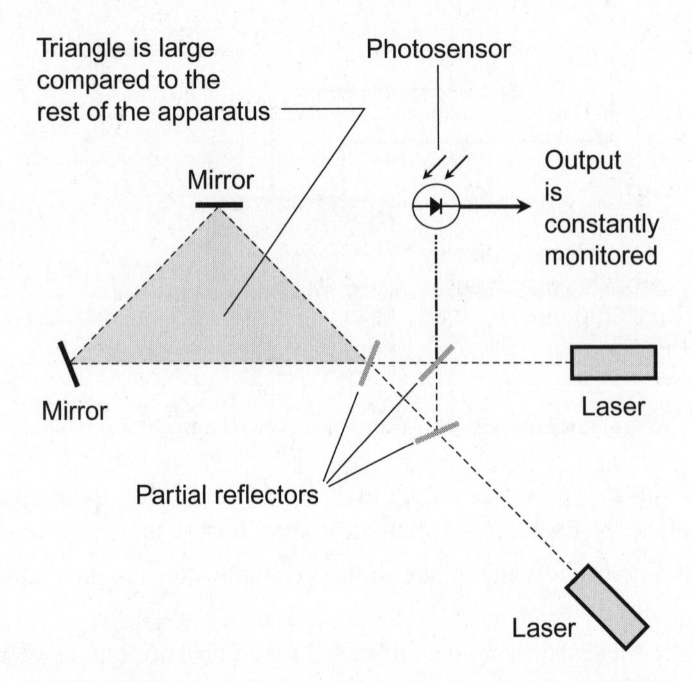

Figure Exam-16 Illustration for Final Exam Question 72.

73. When sound waves travel through certain crystalline substances along with visible light rays, the acoustic vibrations can cause
    (a) an increase in the photon frequency.
    (b) an increase in the photon wavelength.
    (c) modulation of the light intensity.
    (d) blockage of the light rays.
    (e) reflection of the light rays.

74. Figure Exam-17 is a functional diagram of
    (a) a Cassegrain camera.
    (b) a laser spectrometer.
    (c) a laser interferometer.
    (d) an arrangement for making a hologram.
    (e) an arrangement for detecting rotational motion.
75. Suppose that the concave lens in the system of Fig. Exam-17 is replaced by a convex lens that spreads the transmitted light so the rays, although inverted, land in similar locations on the object and the film. How will this affect the performance of the apparatus?
    (a) The system will work in the same way.
    (b) The image resolution will be degraded.
    (c) The wavelength resolution will be improved.
    (d) The color dispersion will improve.
    (e) The system will not work at all.

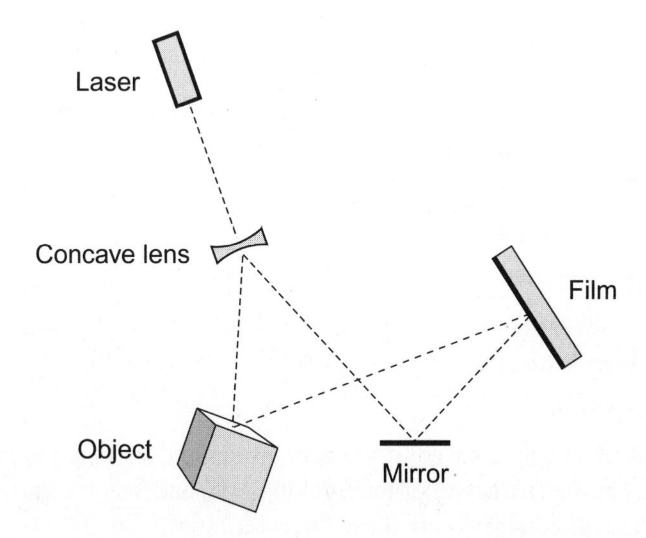

Figure Exam-17 Illustration for Final Exam Questions 74 and 75.

76. Based on the data shown in Table Exam-1 and our general knowledge of refractive phenomena, total internal reflection *cannot* occur when a ray of light passes
    (a) from fresh liquid water into air.
    (b) from fresh liquid water into diamond.
    (c) from crown glass into fresh liquid water.
    (d) from diamond into crown glass.
    (e) from amber into air.

Table Exam-1 Table for Final Exam Question 76.

Substance	Index of Refraction
Air	1.00
Amber	1.55
Diamond	2.42
Crown Glass	1.52
Fresh Liquid Water	1.33

77. An astronomer would most likely measure the brightness of a celestial object using
    (a) a wavelength analyzer.
    (b) a holograph.
    (c) an interferometer.
    (d) a spectrometer.
    (e) a photometer.
78. Suppose that c is the speed of a monochromatic light beam in meters per second, f is the frequency of the light in hertz, and λ is the wavelength of the light in meters. How are these three variables related?
    (a) $c = f\lambda$
    (b) $c^2 = f\lambda$
    (c) $c = f\lambda^{-1}$
    (d) $c^2 = f\lambda^{-1}$
    (e) $c^2 = f^{-1}\lambda$

79. Suppose that a pulsed laser generates rectangular pulses with a period of 8.00 ms, a duration of 200 μs, and a peak power output power of 1.00 kW. If the pulse period is increased to 32.0 ms while the other two parameters remain unchanged, what happens to the amount of energy produced by the laser in 1.00 s?
    (a) It becomes 1/4 as great.
    (b) It is cut in half.
    (c) It does not change.
    (d) It doubles.
    (e) It becomes 4 times as great.
80. In glass, color dispersion occurs because the index of refraction
    (a) increases as the wavelength of transmitted visible light increases.
    (b) increases as the wavelength of transmitted visible light decreases.
    (c) varies depending on the angle of incidence.
    (d) varies depending on the intensity of the transmitted light.
    (e) varies depending on the index of refraction of the external environment.
81. The arrangement shown in Fig. Exam-18 would most likely be used to create
    (a) a digital photograph.
    (b) a hologram.
    (c) a spectrograph.
    (d) a laser light show.
    (e) an interferogram.

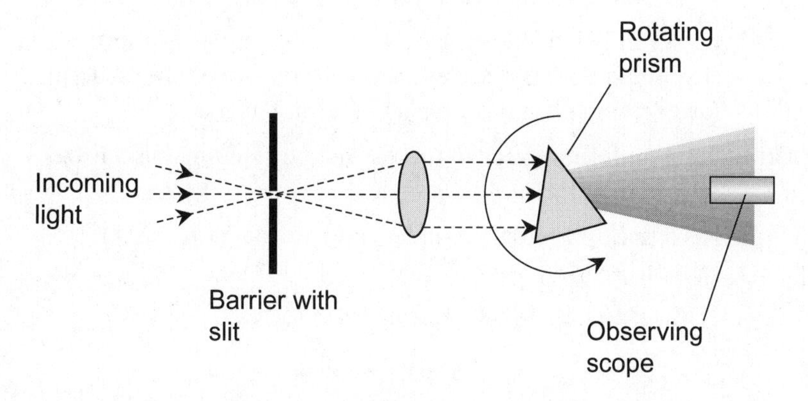

Figure Exam-18 Illustration for Final Exam Question 81.

82. Suppose we generate a "rainbow" spectrum using a prism made of diamond rather than glass. Diamond has a higher refractive index than glass. Therefore, we can expect that the colors in the "rainbow" spectrum emerging from the diamond prism will be
    (a) more spread out than the colors in the spectrum emerging from a glass prism.
    (b) less spread out than the colors in the spectrum emerging from a glass prism.
    (c) more vivid than the colors in the spectrum emerging from a glass prism.
    (d) less vivid than the colors in the spectrum emerging from a glass prism.
    (e) no different, in any respect, than the colors in the spectrum emerging from a glass prism.
83. Suppose a Keplerian refracting telescope has an objective with a focal length of 1000 mm and a diameter of 60 mm. An eyepiece with a focal length of 20 mm is used initially. If that eyepiece is taken out and replaced with one having a focal length of 10 mm, the telescope's *absolute field of view*
    (a) doubles.
    (b) increases by a factor of $2^{1/2}$.
    (c) stays the same.
    (d) decreases by a factor of $2^{1/2}$.
    (e) becomes half as great.
84. In a graded-index optical fiber, the core
    (a) has a refractive index that is maximum along the central axis and steadily decreases outward from the center, and the cladding has a lower refractive index than any part of the core.
    (b) has a refractive index that is minimum along the central axis and steadily increases outward from the center, and the cladding has a lower refractive index than any part of the core.
    (c) has a uniform refractive index, and the cladding has a lower refractive index.
    (d) has a uniform refractive index, and the cladding has a higher refractive index.
    (e) is transparent, and the cladding is opaque.

85. Suppose we have a point light source of constant intensity in free space. The power density of the radiation from the source
    (a) varies directly in proportion to the distance from the source.
    (b) varies directly in proportion to the square of the distance from the source.
    (c) varies inversely in proportion to the distance from the source.
    (d) varies inversely in proportion to the square of the distance from the source.
    (e) is independent of the distance from the source.
86. The rest mass of a photon is
    (a) mathematically zero.
    (b) the same as the rest mass of an electron.
    (c) the same as the rest mass of a proton.
    (d) the same as the rest mass of a neutron.
    (e) not definable in the conventional sense.
87. The apparatus shown in Fig. Exam-19 would most likely be used in
    (a) radio astronomy.
    (b) IR astronomy.
    (c) long-wave UV astronomy.
    (d) short-wave UV astronomy.
    (e) X-ray astronomy.

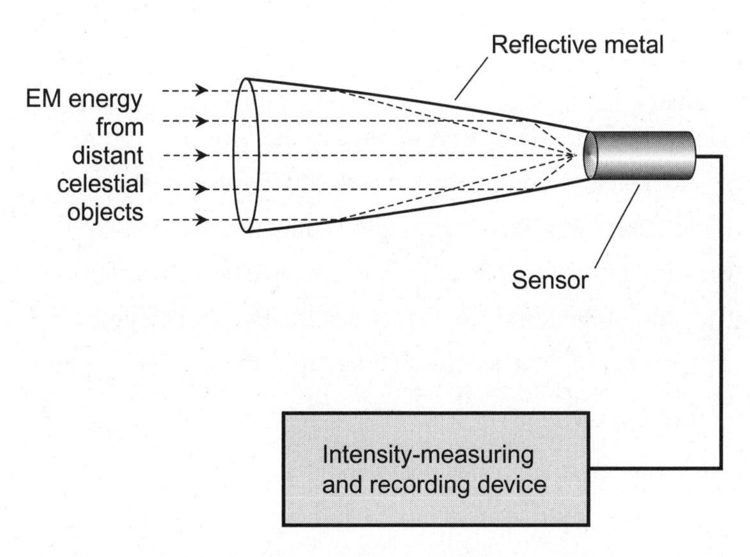

Figure Exam-19 Illustration for Final Exam Question 87.

88. Suppose you place a tiny object along the axis of a convex lens at a distance r_o from the center of the lens having a focal length f. The distance r_i of the object's real image is related to r_o and f by the formula

$$1/f = 1/r_i + 1/r_o$$

as long as f, r_i, and r_o are expressed in the same units. Given this information, how far from the lens should the object be placed so that the image forms at the same distance from the lens as the object?

    (a) A distance of 2 focal lengths.
    (b) A distance of $2^{1/2}$ (the square root of 2) focal lengths.
    (c) A distance of 1 focal length.
    (d) A distance of $1/2^{1/2}$ focal lengths.
    (e) A distance of 1/2 focal length.

89. Suppose it takes a laser with 18 W of average power output 10 s to drill a 12-mm hole through a piece of solid material. How much average output power would a laser of the same wavelength need to drill a 4-mm hole through the same piece of material in 10 s?

    (a) 2 W
    (b) 6 W
    (c) 18 W
    (d) 54 W
    (e) 162 W

90. In a camera that uses a convex lens along with a photosensitive internal screen or film, the *f*-number is defined as the

    (a) focal length for external objects at infinite distances.
    (b) effective lens diameter as allowed by the shutter.
    (c) ratio of the focal length to the effective lens diameter.
    (d) ratio of the lens diameter to the effective focal length.
    (e) product of the lens diameter and the effective focal length.

91. Epipolar navigation is typically used to determine all of the following variables as an aircraft travels through the earth's stratosphere, *except*
    (a) the speed of the aircraft relative to the surface.
    (b) the direction in which the aircraft is traveling relative to the surface.
    (c) the geographic latitude and longitude of the aircraft.
    (d) the transparency of the atmosphere between the aircraft and the surface.
    (e) the altitude of the aircraft above the surface.
92. Suppose that d is the distance in meters that a ray of light travels, t is the time in seconds that the ray takes to travel that distance, and c is the speed of light in meters per second. Assuming that the point of view is not accelerating, the relationship between these three variables can be expressed as
    (a) $c = dt$
    (b) $c = dt^2$
    (c) $d = ct$
    (d) $t = cd$
    (e) $t = c^2d$
93. Synchrotron radiation is the basis for the operation of
    (a) a helium-neon laser.
    (b) a diode laser.
    (c) a neodymium laser.
    (d) a free-electron laser.
    (e) All of the above
94. Which of the following (a), (b), (c), or (d), if any, is *never* used as the objective in a telescope?
    (a) A concave lens.
    (b) A convex lens.
    (c) A concave spherical mirror.
    (d) A concave paraboloidal mirror.
    (e) Any of the above can be used as the objective in a telescope.

95. Which of the following factors (a), (b), (c), or (d), if any, is *never* a significant problem with large Keplerian refracting telescopes?
    (a) Spherical aberration, in which the focus of a spherical objective is not a perfect point.
    (b) Chromatic aberration, in which the index of refraction varies with the wavelength.
    (c) Lens sag, in which the shape of a large objective lens is distorted by its own weight.
    (d) Awkwardness of viewing through an eyepiece located in the side of the telescope tube.
    (e) All of the above can be significant problems with large Keplerian refractors.
96. According to the electromagnetic theory, UV waves are similar to visible light waves, but UV waves
    (a) have longer wavelengths than visible light waves in a vacuum.
    (b) have lower frequencies than visible light waves.
    (c) travel faster than visible light waves in a vacuum.
    (d) have shorter periods than visible light waves.
    (e) can propagate for longer distances than visible light waves in a vacuum.
97. Consider two waves that combine by heterodyning. Their frequencies are 8.00 MHz and 12.0 MHz. At what frequency or frequencies do mixing products occur?
    (a) 4.00 MHz, the difference frequency.
    (b) 9.80 MHz, the geometric mean frequency.
    (c) 10.0 MHz, the arithmetic mean frequency.
    (d) 20.0 MHz, the sum frequency.
    (e) 4.00 MHz and 20.0 MHz, the difference and sum frequencies.
98. Which of the following *never* alters the polarization of incident sunlight?
    (a) A dichroic plastic, as the light passes through it
    (b) The surface of a calm pond, as the light is reflected from it
    (c) The earth's atmosphere, as the light is scattered by it
    (d) A first-surface mirror, as the light is reflected from it
    (e) The surface of a glass window, as the light is reflected from it

99. Figure Exam-20 is a simplified illustration that portrays
    (a) Mueller-Lyer radiation.
    (b) synchrotron radiation.
    (c) Cerenkov radiation.
    (d) stimulated emmision in a gas laser.
    (e) EM radiation from an ionized gas.

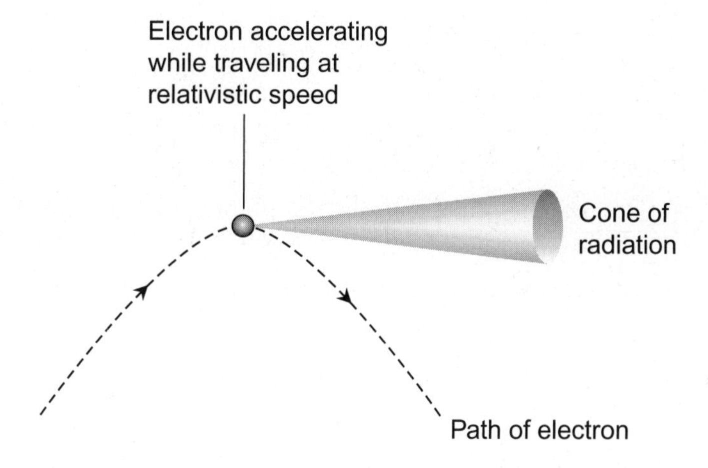

Figure Exam-20 Illustration for Final Exam Question 99.

100. Fill in the blank in the following statement to make it true: "Regardless of the reference frame, a ray of light always follows ________ between two points in space."
    (a) a straight line
    (b) an accelerating path
    (c) the shortest possible path
    (d) a constant-wavelength path
    (e) an accelerating or decelerating path

Answers to Quiz and Exam Questions

Chapter 1

1. b	2. d	3. d	4. a	5. c
6. a	7. c	8. b	9. b	10. d

Chapter 2

1. c	2. a	3. c	4. d	5. b
6. a	7. a	8. b	9. d	10. a

Chapter 3

1. c	2. b	3. a	4. c	5. c
6. c	7. b	8. a	9. d	10. d

Chapter 4

1. c	2. c	3. a	4. b	5. b
6. b	7. b	8. a	9. d	10. c

Chapter 5

1. d	2. c	3. d	4. a	5. a
6. b	7. d	8. b	9. c	10. a

Chapter 6

1. d	2. d	3. c	4. a	5. c
6. a	7. d	8. c	9. b	10. c

Chapter 7

1. b	2. a	3. b	4. c	5. d
6. b	7. a	8. d	9. d	10. d

Chapter 8

1. d	2. c	3. d	4. a	5. d
6. b	7. a	8. c	9. d	10. a

Chapter 9

1. d	2. b	3. a	4. c	5. d
6. c	7. d	8. c	9. a	10. d

Chapter 10

1. b	2. c	3. d	4. a	5. c
6. d	7. d	8. c	9. c	10. a

Final Exam

1. a	2. d	3. a	4. d	5. d
6. c	7. a	8. e	9. b	10. b
11. e	12. c	13. b	14. a	15. d
16. b	17. b	18. c	19. b	20. e
21. a	22. a	23. c	24. e	25. c
26. c	27. b	28. e	29. a	30. d
31. d	32. a	33. d	34. c	35. b
36. d	37. b	38. d	39. d	40. e
41. d	42. a	43. c	44. e	45. d
46. c	47. c	48. e	49. a	50. a
51. b	52. b	53. a	54. e	55. d
56. e	57. a	58. c	59. e	60. b
61. c	62. d	63. e	64. c	65. e
66. b	67. d	68. d	69. e	70. a
71. e	72. b	73. c	74. d	75. a
76. b	77. e	78. a	79. a	80. b
81. c	82. a	83. e	84. a	85. d
86. e	87. e	88. a	89. a	90. c
91. d	92. c	93. d	94. a	95. d
96. d	97. e	98. d	99. b	100. c

Suggested Additional Reading

Boyd, Robert. *Nonlinear Optics,* 2nd ed. Burlington, Mass.: Elsevier/Academic Press, 2002.

Cloud, Gary. *Optical Methods of Engineering Analysis*. Cambridge, England: Cambridge University Press, 1998.

Fisher, Robert and Tadic, Bijana. *Optical System Design*. New York, N.Y.: McGraw-Hill, 2000.

Gerry, Christopher and Knight, Peter. *Introductory Quantum Optics*. Cambridge, England: Cambridge University Press, 2004.

Gibilisco, Stan. *Advanced Physics Demystified.* New York, N.Y.: McGraw-Hill, 2007.

Gibilisco, Stan. *Astronomy Demystified.* New York, N.Y.: McGraw-Hill, 2003.

Gibilisco, Stan. *Physics Demystified.* New York, N.Y.: McGraw-Hill, 2002.

Halpern, Alvin. *Beginning Physics II: Waves, Electromagnetism, Optics and Modern Physics*. New York, N.Y.: McGraw-Hill, 1998.

Hecht, Eugene. *Optics,* 4th ed. Boston, Mass.: Addison Wesley, 2001.

Hunsperger, Robert. *Integrated Optics,* 5th ed. Berlin, Germany: Springer, 2002.

Keating, Michael. *Geometric, Physical, and Visual Optics,* 2nd ed. Burlington, Mass: Elsevier/Butterworth-Heinemann, 2002.

Parker, Michael. *Physics of Optoelectronics*. Boca Raton, Fla.: CRC Press, 2005.

Rogers, Alan. *Essentials of Optoelectronics,* 2nd ed. Boca Raton, Fla.: CRC Press, 2007.

Shannon, Robert. *The Art and Science of Optical Design*. Cambridge, England: Cambridge University Press, 1997.

Smith, Warren. *Modern Optical Engineering,* 3rd ed. New York, N.Y.: McGraw-Hill, 2000.

Winston, Roland, et al. *Nonimaging Optics*. Burlington, Mass.: Elsevier/Academic Press, 2004.

INDEX

C

D

E

F

G

Q

R

S

T

U

V

W

XYZ